SCOPE 38: Ecotoxicology and Climate with Specia
1989, 432 pp
SCOPE 39: Evolution of the Global Biogeochemica
SCOPE 40: Methods for Assessing and Reducing I
320 pp (SGOMSEC 6)
SCOPE 41: Short-Term Toxicity Tests for Nor
(SGOMSEC 4)
SCOPE 42: Biogeochemistry of Major World Rivers, 1991, [illegible] pp
SCOPE 43: Stable Isotopes: Natural and Anthropogenic Sulphur in the Environment, 1991, 472 pp
SCOPE 44: Introduction of Genetically Modified Organisms into the Environment, 1990, 224 pp
SCOPE 45: Ecosystem Experiments, 1991, 296 pp
SCOPE 46: Methods for Assessing Exposure of Human and Non-Human Biota, 1991, 448 pp (SGOMSEC 5)
SCOPE 47: Long-Term Ecological Research. An International Perspective, 1991, 312 pp
SCOPE 48: Sulphur Cycling on the Continents: Wetlands, Terrestrial Ecosystems and Associated Water Bodies, 1992, 345 pp
SCOPE 49: Methods to Assess Adverse Effects of Pesticides on Non-Target Organisms, 1992, 264 pp (SGOMSEC 7)
SCOPE 50: Radioecology After Chernobyl, 1993, 367 pp
SCOPE 51: Biogeochemistry of Small Catchments: A Tool for Environmental Research, 1993, 432 pp
SCOPE 52: Methods to Assess DNA Damage and Repair: Interspecies Comparisons, 1994, 304 pp (SGOMSEC 8)
SCOPE 53: Methods to Assess the Effects of Chemicals on Ecosystems, 1995, 436 pp (SGOMSEC 10)

Funds to meet SCOPE expenses are provided by contributions from SCOPE Committees, an annual subvention from ICSU (and through ICSU, from UNESCO), an annual subvention from the French Ministère de l'Environnement, contracts with UN Bodies, particularly UNEP, and grants from Foundations and industrial enterprises.

SCOPE 53
IPCS Joint Activity 23
SGOMSEC 10

Methods to Assess the Effects of Chemicals on Ecosystems

SCOPE 53
IPCS JOINT ACTIVITY 23
SGOMSEC 10

Methods to Assess the Effects of Chemicals on Ecosystems

Edited by

RICK A. LINTHURST
US Environmental Protection Agency, Research Triangle Park, North Carolina, USA

PHILIPPE BOURDEAU
Commission of the European Communities, and Université Libre de Bruxelles, Brussels, Belgium

ROBERT G. TARDIFF
EA Engineering, Science, and Technology, Inc., Silver Spring, Maryland, USA

Prepared by

Scientific Group on Methodologies for the Safety Evaluation of Chemicals (SGOMSEC)

Published on behalf of the Scientific Committee on Problems of the Environment (SCOPE) of the International Council of Scientific Unions (ICSU), and the International Programme on Chemical Safety (IPCS) of the World Health Organization (WHO), the United Nations Environment Programme (UNEP), and the International Labour Organisation (ILO)

UNEP

by
JOHN WILEY & SONS
Chichester • New York • Brisbane • Toronto • Singapore

Published 1995 by John Wiley & Sons Ltd,
Baffins Lane, Chichester,
West Sussex PO19 1UD, England
Telephone National Chichester (01243) 779777
International (+44) (1243) 779777

Other Wiley Editorial Offices

John Wiley & Sons, Inc., 605 Third Avenue,
New York, NY 10158-0012, USA

Jacaranda Wiley Ltd, 33 Park Road, Milton,
Queensland 4064, Australia

John Wiley & Sons (Canada) Ltd, 22 Worcester Road,
Rexdale, Ontario M9W 1L1, Canada

John Wiley & Sons (SEA) Pte Ltd, 37 Jalan Pemimpin #05-04,
Block B, Union Industrial Building, Singapore 2057

British Library Cataloguing in Publication Data

A catalogue record for this book is available from the British Library

ISBN 0-471-95911-1

Produced from camera-ready copy supplied by the editor.
Printed and bound in Great Britain by Biddles Ltd, Guildford and King's Lynn.
This book is printed on acid-free paper responsibly manufactured from sustainable forestation,
for which at least two trees are planted for each one used for paper production.

International Council of Scientific Unions (ICSU)
Scientific Committee on Problems of the Environment (SCOPE)

SCOPE is one of a number of committees established by a non-governmental group of scientific organizations, the International Council of Scientific Unions (ICSU). The membership of the ICSU includes representatives from 92 national academies of science, 23 scientific unions and 28 other associates. To cover multi-disciplinary activities which include the interests of several unions, ISCU has established more than 20 interdisciplinary bodies, of which SCOPE, founded in 1969, is one. Currently representatives of 37 member countries and 22 unions and scientific committees participate in the work of SCOPE, which directs particular attention to the needs of developing countries.

The mandate of SCOPE is to assemble, review, and assess the information available on man-made environmental changes and the effects of these changes on man; to assess and evaluate the methodologies of measurement of environmental parameters; to provide an intelligence service on current research; and by the recruitment of the best available scientific information and constructive thinking to establish itself as a corpus of informed advice for the benefit of centres of fundamental research and of organizations and agencies operationally engaged in studies of the environment.

SCOPE is governed by a General Assembly, which meets every three years. Between such meetings its activities are directed by the Executive Committee.

R. E. Munn
Editor-in-Chief
SCOPE Publications

Secretariat: 51 boulevard de Montmorency
75016 Paris, France

Contents

PART 2 CONTRIBUTED PAPERS

Foreword

Evaluation of the effects of chemicals on the environment as well as human health is an important part of the mandate of the International Programme on Chemical Safety (IPCS), a collaborative undertaking of the United Nations Environmental Programme, the International Labour Organisation, and the World Health Organization. The development of scientifically sound methodology for the assessment of human health and environmental risk continues to be an important component of the overall IPCS activities. Without appropriate methodology, adequate protection of human health and the environment will be impossible.

Many chemicals, both man-made and natural, are capable of causing damage to ecosystems, by altering the structure or function of important components such as the trees, avian species, terrestrial animals, or soil micro-organisms. The complexity of the field and the rapid development of knowledge related to anatomic and biochemical alterations makes it extremely difficult to integrate this information into ecological management programmes. However, this can be accomplished when research scientists are given an opportunity to discuss their findings with scientists having the responsibility for the assessment and management of ecological risks from exposure to man-made and naturally occurring chemicals. By convening the workshop on "Methods to Assess the Effects of Chemicals on Ecosystems," the Scientific Group on Methodologies for the Safety Evaluation of Chemicals (SGOMSEC) has provided such an opportunity.

The in-depth scientific review made by world leaders in the field will provide much needed guidance to those asked to use the results from experimental and field studies to assess ecosystem risks from chemicals. Hopefully, the results of this SGOMSEC activity will assist in developing harmonized approaches to the assessment of ecosystem risks from chemicals and to a better integration of both human health and environmental data into a unified risk assessment framework to improve public health and environmental protection worldwide.

Michel J. Mercier
Director, International
Programme on Chemical Safety

Preface

The Scientific Group on Methodologies for the Safety Evaluation of Chemicals (SGOMSEC) was established in 1979 at the initiative of Professor Norton Nelson from New York University. SGOMSEC is a non-governmental organization sponsored by IPCS (the International Programme on Chemical Safety, established within WHO with the cooperation of UNEP and ILO), and the Scientific Committee on Problems of the Environment (SCOPE), itself a body of the International Council of Scientific Unions (ICSU).

The broad objective of SGOMSEC is to contribute to the reduction and prevention of risks caused to humans and non-human targets (ecosystems) by the introduction in the environment in increasing quantities of many natural and man-made chemicals. The specific contribution of SGOMSEC is to assess the methodologies in use for the evaluation of these hazards and risks with a view to determine their values, to identify gaps and emerging needs, and to make recommendations for future research. Previous SGOMSEC projects have dealt either with general problems of chemical exposure and effects or with specific issues such as the non-intentional effects of pesticides or the consequences of large chemical accidents.

Volume 10 of SGOMSEC is concerned with a subject basic to ecological risk assessment: the estimation of injurious effects of chemical and physical agents on large and complex ecosystems and their component populations of micro-organisms, plants, and animals. This report evaluates current knowledge of ecotoxicology. This evaluation is presented in 15 contributed papers and a joint report prepared at a workshop held at the Centre d'Ecologie Fonctionnelle et Evolutive (CNRS) located at Montpellier, France, on 7–12 March 1993. The joint report includes general recommendations aimed at improving research in this field and developing methodologies for ecotoxicology testing, taking account of the potentials and limitations of available models and approaches.

The project was co-chaired by Professors Rick Linthurst of the US Environmental Protection Agency and Marie-Madeleine Coûteaux of the Centre d'Ecologie Fonctionnelle et Evolutive. They and the authors of the contributed papers, all experts in their respective fields, are thanked for their participation. We are also most grateful to Robert G. Tardiff (EA Engineering, Science, and Technology, Inc., Silver Spring, Maryland) for his part in editing the report. Special thanks are due to Dr. F. Warembourg, Director of the Centre d'Ecologie Fonctionnelle et Evolutive, for hosting the workshop at the Centre.

SGOMSEC gratefully acknowledges the support and financial assistance of IPCS, SCOPE, NIEHS, the US Environmental Protection Agency, and the Commission of the European Communities, without which this project would not have been possible.

Philippe Bourdeau, Chairman
Bernard D. Goldstein, Vice Chairman
Scientific Group
on Methodologies for the
Safety Evaluation of Chemicals

Acknowledgements

The editors are pleased to acknowledge the assistance received from:

L. Dreves
M. Gray
D. Ludwig
T. Piccin
C. Stapleton
B. Tardiff
L. Wakefield

Scientific Group on Methodologies for the Safety Evaluation of Chemicals

***George C. Becking**
Interregional Research Unit, International Programme on Chemical Safety, World Health Organization, MD-A206, P.O. Box 12233, Research Triangle Park, NC 27709, USA (Secretary)

N. P. Bockov
Director, Institute of Medical Genetics, Kasirskoesosse 6A Moscow 115478, USSR

***Philippe Bourdeau**
Université Libre de Bruxelles, 26, ave des Fleurs, B-1150 Brussels, Belgium (Chairman)

***Bernard Goldstein**
Professor and Chairman, Department of Environmental and Community Medicine, UMDNJ-Robert Wood Johnson Medical School, 675 Hoes Lane, Piscataway, NJ 08854, USA (Vice Chairman)

Gareth Green
Professor and Chairman, Department of Environmental Health Sciences, The Johns Hopkins University, School of Hygiene and Public Health, 615 North Wolfe Street, Baltimore, MD 21205, USA

***Miki Goto**
Professor, Department of Chemistry and Director, Institute of Ecotoxicology, Gakushuin University, 1-5-1 Mejiro, Toshima-ku, Tokyo 171, Japan

J. R. Hickman
Bureau of Chemical Hazards, Environmental Health Center, Department Health and Welfare, Tunney's Pasture, Ottawa, Ontario Canada KlA OL2

Vladimir Landa
Czechoslovak Academy of Sciences, Institute of Entomology, Branisovska 31, 370 05 Ceske Budejovice, Czech Republic

***Aly Massoud**
Professor and Chairman, Department of Community, Environmental and Occupational Medicine, Ein Shams University, Abbassia Cairo, Egypt

***Michel Mercier**
Director, International Programme on Chemical Safety, World Health Organization, 1211 Geneva 27, Switzerland

Thomas Odhiambo
Director, International Centre for Insect Physiology and Ecology, P.O. Box 30772, Nairobi, Kenya

***Blanca Raquel Ordonez**
Environmental Advisor of the Minister of Health, School of Public Health, Lopez Cotilla 739, Col. del Valle, 03100 D.F., Mexico

Dennis V. Parke
Head, Department of Biochemistry, University of Surrey, Guildford, Surrey GU2 5XH, UK

David B. Peakall
Monitoring and Research Centre, Old Coach House, Campden Hill, London, UK

Jerzy Piotrowski
Lodz Medical Academy, Institute of Environmental Research, Narutowicza 120A, 90-145 Lodz, Poland

David P. Rall
5302 Reno Road, NW, Washington, DC 20015, USA

***Robert G. Tardiff**
Vice President for Health Sciences, EA Engineering, Science, and Technology, Inc., 8401 Colesville Road, Silver Spring, MD 20910, USA (Editor)

*Member of the Executive Committee

Participants of the Workshop

Amos Banin
The Hebrew University of Jerusalem, P.O. Box 12, 76100 Rehovot, Israel

Larry W. Barnthouse
Environmental Sciences Division, Oak Ridge National Laboratory, P.O. Box 2008, Oak Ridge, Tennessee 37831-6036, USA

George C. Becking
Interregional Research Unit - MD A206, International Programme on Chemical Safety, World Health Organization, P.O. Box 12233, Research Triangle Park, NC 27709, USA

N.P. Bochkov
Presidium of Russian Academy 14, Solyanka Str. 109801, Moscow, Russia

Maurice Bonneau
INRA/Centre de Recherche de Nancy, Station de Recherche sur les Sols, la Microbiologie et la Nutrition des Arbres Forestiers, 54280 Champenoux, France

Philippe Bourdeau
Université Libre de Bruxelles, 26, ave des Fleurs, B-1150 Brussels, Belgium

Joanna Burger
Environmental and Occupational Health Sciences Institute, 681 Frelinghuysen Road, P.O. Box 1179, Piscataway, NJ 08855-1179, USA

Patrick Calow
University of Sheffield, Department of Animal and Plant Sciences, P.O. Box 601, Sheffield S10 2UG, UK

Marie-Madeleine Coûteaux[1]
Centre d'Ecologie Fonctionnelle et Evolutive, CNRS, BP 5051 Route de Mende, 34033 Montpellier Cedex 1, France

T. Damstra
National Institute of Environmental Health Sciences, P.O. Box 12233, Research Triangle Park, NC 27709, USA

Reinhard Debus
Fraunhofer-Institut für Umweltchemie und Ökotoxicologie, D-5948 Schmallenberg, Germany

Michael H. Depledge
Ecotoxicology Group, Institute of Biology, Odense University, DK-5230 Odense M., Denmark

Francesco di Castri
Centre d'Ecologie Fonctionnelle et Evolutive, CNRS, BP 5051 Route de Mende, 34033 Montpellier Cedex 1, France

Jürgen Elsner
Institut für Toxicologie ETH, CH-8603 Schwarzenbach, Switzerland

Bernard Goldstein
Department of Environmental and Community Medicine, UMDNJ–Robert Wood Johnson Medical School, 675 Hoes Lane, Piscataway, NJ 09954, USA

Gareth M. Green
Harvard School of Public Health, 677 Huntingdon Avenue, Room 814, Boston, MA 02115, USA

R.H. Groves
CSIRO Biological Control Unit, 335, Avenue Paul Parguel, 34000, Montpellier, France

S.P. Hopkin
Department of Pure and Applied Zoology, School of Animal and Microbial Sciences, University of Reading, P.O. Box 228, Reading, RG6 2AJ, UK

John N.R. Jeffers
Glenside, Oxenholme, Kendal, LA9 7RF Cumbria, UK

Werner Klein
Fraunhofer-Institut für Umweltchemie und Ökotoxicologie, D-5948 Schmallenberg, Germany

Vladimir Landa
Institute of Entomology, Czech Academy of Sciences, Branisovska 31, 370 05 Ceske Budejovice, Czech Republic

Rick A. Linthurst[1]
Environmental Protection Agency, AREAL (MD-75), Research Triangle Park, NC 27711, USA

W. Lyle Lockhart
Department of Fisheries and Oceans, 501 University Crescent, Winnipeg, Manitoba, Canada R3T 2N6

Aly Massoud
Department of Community, Environmental and Occupational Medicine, Ein Shams University, Abbassia Cairo, Egypt

David A. Mouat
Biological Science Center, Desert Research Institute, P.O. Box 60220, Reno, Nevada 89506, USA

Blanca Raquel Ordonez
Environmental Advisor of the Minister of Health, School of Public Health, Lopez Cotilla 739, Col. del Valle, 03100 D.F., Mexico

David B. Peakall
17 St. Mary's Road, Wimbledon, London, SW19 TB2, UK

Jerzy Piotrowski
Lodz Medical Academy, Institute of Environmental Research, Narutowicza 120A 90-145, Lodz, Poland

Peter Principe
AREAL (MD-75), US Environmental Protection Agency, Research Triangle Park, NC 27711, USA

William H. Queen
Institute for Coastal and Marine Resources, East Carolina University, Greenville, NC 27858-4353, USA

F. Ramade
Laboratoire d'Ecologie et de Zoologie, Bâtiment 442, Université de Paris-Sud, 91405 Orsay, Cedex, France

Heyle E. Roda
Environmental Research, P.O. Box 447, Lakefield, Ontario, Canada KOL 2HO

T. Savinova
Murmansk Marine Biological Institute, Academy of Sciences of Russia, 17 Wladminskaya St., Murmansk 183023, Russia

Patrick Sheehan
Chemrisk Division, McLaren/Hart, 1135 Atlantic Avenue, Alemada, CA 94501, USA

Donald W. Stanley
Institute for Coastal and Marine Resources, East Carolina University, Greenville, NC 27858, USA

William A. Suk
National Institute of Environmental Health Sciences, Division of Extramural Research and Training, P.O. Box 12233, Research Triangle Park, NC 27708 USA

Robert G. Tardiff
Vice President for Health Sciences, EA Engineering, Science, and Technology, Inc., 8401 Colesville Road, Silver Spring, MD 20910, USA

Herman A. Verhoef
Vrije Universiteit, Department of Ecology and Ecotoxicology, Amsterdam, The Netherlands

J. Verneaux
Hydrobiology-Hydroecology Laboratory and the Institute of Environmental Sciences and Technology of the University of Franche-Comté, Besançon, France

Bernard Weiss
Department of Environmental Medicine, University of Rochester Medical Center, Rochester, NY 14642, USA

Xu Xiao-Bai
Environmental Chemistry Division, Organic Analytical Research Center for E-E Sciences, Academia Sinica, P.O. Box 2871, 000085 Beijing, People's Republic of China

1. Chairman

Part 1

JOINT REPORT

1 Introduction, General Conclusions, and Recommendations

1.1 INTRODUCTION

Ecological policy and management decision makers require sound scientific information upon which to base and justify their decisions; therefore, environmental research and monitoring must provide the full range of information required by decision makers. Historically, studies of the effects of chemicals on flora, fauna, and the environment have focused almost exclusively on identifying how those effects benefit or harm humans. As a result, environmental decision makers have selected natural resources that are particularly valued for food, shelter, medicine, recreation, or other uses. Only in the last few decades have relationships been recognized between these critical natural resources and the many components of the environment not of direct use to humans. As knowledge of the interconnectedness of the Earth's ecosystems increases, so does the recognition of the importance of obtaining comprehensive information about the effects of chemicals on all components of ecosystems. The success of concepts such as "sustainable development" is inconceivable without an improved understanding of the function and integrity of ecosystems and of how they are affected by human-induced stresses. This growing awareness of the importance of "healthy" ecosystems is coupled with a heightened interest in the ethical questions related to managing the uses of our natural environment.

Recognizing that all ecosystems are of indisputable importance to our future, useful approaches to make decisions to protect them must be developed and disseminated. Because the cost of effective environmental protection can be significant (e.g., the expense of developing and using alternative methods in manufacturing, proper waste treatment and disposal methods, recycling; installing pollution controls and monitoring their effectiveness), efforts to control the introduction of chemicals to the environment must be justifiable. Decisions to incur the cost of protecting the environment are influenced by factors such as:

1. the value of the ecosystem in terms of societal and cultural priorities, economic considerations, and ecological functions;

Methods to Assess the Effects of Chemicals on Ecosystems
Edited by R. A. Linthurst, P. Bourdeau, and R. G. Tardiff

2. the certainty that an effect will occur at expected or observable levels;
3. the geographic extent and magnitude of the effect;
4. the certainty of the cause of an observed effect; and
5. the management options available to reduce the risk of adverse effects.

This volume presents approaches and methods to study the effects of chemicals on ecosystems in order to collect information that promotes risk-based decision making. Chapter 2 provides a discussion of the conceptual approaches to conduct ecological research and monitoring that includes characterizing "healthy" ecosystems, setting and maintaining objectives for ecosystem management, and using ecological risk assessment as a model for studying the effects of chemicals on ecosystems. Chapter 3 provides an overview of available methods to study the effects of chemical stressors in aquatic and terrestrial ecosystems. Chapter 4 focuses on the importance of large geographic-scale (regional) environmental monitoring and assessments.

The remaining chapters present papers contributed for review by the participants of SGOMSEC 10 Workshop, held in Montpellier, France, in March 1993. These papers describe and evaluate available methods to monitor chemical contamination and their effects on ecosystems.

1.2 ECOLOGICAL RISK ASSESSMENT: A TOOL FOR DECISION MAKING

Ecological risk assessment is useful to provide a framework to obtain information needed to make management decisions. Ecological risk assessments help to identify existing problems, anticipate the risks of planned actions, establish regulatory program and research priorities, and provide a scientific basis for regulatory actions. Most importantly, environmental risk assessment provides the basis to prioritize management and regulatory efforts because objective comparisons of a variety of situations can be made. Using risk as a common denominator allows the decision maker to select the most effective management actions. Following implementation of management actions, environmental risk assessment can also help to measure the effectiveness of those actions.

Before any data are collected or any research takes place, the researcher must consider how the environmental risk assessment will be carried out and what specific questions must be answered. The process of conducting an environmental risk assessment requires carefully identifying assessment objectives (environmental values to be protected), deciding the appropriate scale and level of biological organization, assembling multidisciplinary data collection and assessment teams, rigorously interpreting results using both quantitative and qualitative methods, and communicating the results in a manner that facilitates risk management. Chapter 2 presents a framework for environmental risk assessment drawn from several recent sources that present unified principles for assessing the ecological risks of toxic chemicals and other stresses. The components of the suggested framework are illustrated in Figure 1.1. This risk assessment framework is intended to facilitate integration of human health and ecological risks of toxic chemicals and other environmental stresses, and, as far as the US is concerned, was developed by combining essential features of the National Academy of Sciences/National Research Council (NRC, 1983) framework for health risk assessment and the US Environmental Protection Agency (USEPA, 1992) framework for ecological risk assessment. This proposed framework is discussed within the contributed papers. This framework has four components: (1) hazard identification; (2) exposure assessment and exposure-response assessment; (3) risk characterization; and (4) risk management.

1.2.1 HAZARD IDENTIFICATION

This step determines whether a particular danger exists, if the effects associated with the hazard are sufficiently significant to warrant further study or immediate management action, and the kinds of data required to determine the level of risk. This systematic planning step establishes the goals and focus of the assessment, and identifies the major factors to be considered. A major factor to be considered at this step is the selection of ecologically based endpoints relevant to the ultimate decisions to be made. Information used for hazard identification include short-term or screening toxicity tests and reviews of existing information that characterize the potentially affected ecosystems and the contaminants in question.

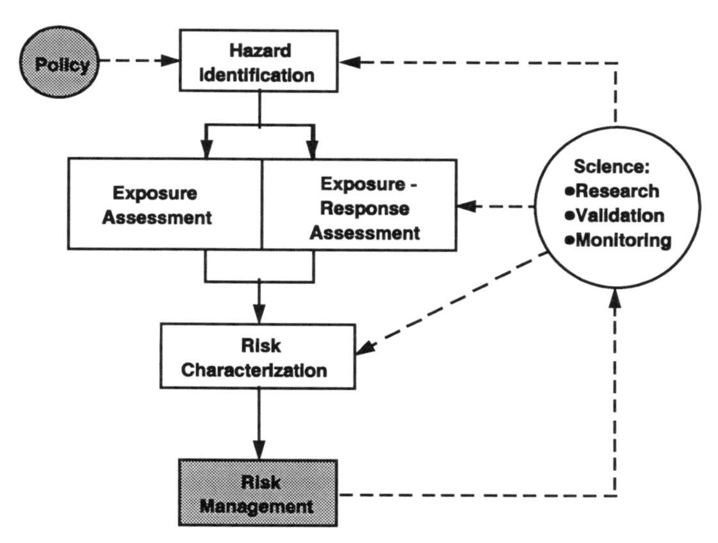

Figure 1.1. Risk assessment framework that integrates human health risk (NRC, 1983) and ecological risks of chemicals (USEPA, 1992)

1.2.2 EXPOSURE ASSESSMENT

This step is the determination of exposure to the hazardous agent in question. This process includes measurement or prediction of movement, fate, and partitioning of chemicals in the environment. This step is typically accomplished through chemical analysis of site media or ecological receptors and/or mathematical modelling.

1.2.3 EXPOSURE-RESPONSE ASSESSMENT

This process is the determination of the relation between the magnitude of exposure and the probability of occurrence of the expected effects. Information useful in this step includes toxicity data (chronic toxicity, mode of action, sensitivities of particular species), mesocosm or field test data, field surveys comparing exposed and unexposed sites, and population or ecosystem modelling.

1.2.4 RISK CHARACTERIZATION

This step involves describing the nature and magnitude of risks, including the inherent uncertainties, expressed in terms understandable to decision makers and the public. This step integrates information from the previous steps, and communicates it to decision makers in a manner relevant to the decisions being made. Because the purpose of risk assessment is to support decision making, communication with decision makers is a critical step in an environmental risk assessment.

1.2.5 RISK MANAGEMENT

This process is the one whereby decisions are made about whether an assessed risk needs to be managed, and the means for accomplishing it, for the protection of public health and environmental resources. Managing risks involves making decisions based on the information collected in the previous steps of the risk assessment along with a consideration of social and cultural values, economic realities, and political factors.

Monitoring and research are used to provide the data necessary to conduct an environmental risk assessment. Important aspects to consider when planning monitoring for a risk assessment include the assumptions involved in determining which environmental values are to be protected by the risk management decisions; which levels of biological organization must be studied to obtain the data necessary to answer the pertinent questions; and the temporal and spatial scales relevant to the study. Each aspect is discussed in greater detail throughout this volume.

1.3 ECOSYSTEM OBJECTIVES

The first steps in performing an environmental risk assessment are the identification of the resources at risk and the definition of desired or acceptable conditions of those resources. Ecological "values" to be protected must be identified clearly to help define the monitoring objectives and the assessment endpoints of concern. Social, cultural, political, economic, and ecological values must contribute to the selection of assessment endpoints. Frequently, risk assessments focus only on resources of recognized economic value to human society. This narrow definition of the resources at risk ignores the critical components of an ecosystem that support and preserve the resources traditionally valued, because ecological values cannot be quantified easily or in terms that are meaningful to environmental policy makers. Risk assessments based solely on a narrow economic definition of valuable resources ultimately may fail to detect and quantify the risks that threaten those resources indirectly through effects on other components of the ecosystem. Furthermore, management efforts that neglect ecological values ultimately may be ineffective at protecting economically valuable resources. An approach is needed to translate the ecological value of ecosystem components that are not valued by human society into information that contributes to management decisions or can be used to justify these decisions. Such an approach must account for the fact that definitions of ecosystem values will always be influenced primarily by the needs and desires of human society and provide guidance to relate ecologically important components of an ecosystem to direct and indirect human uses of the system.

Once the values of an ecosystem are defined, its condition or degree of "health" must be characterized relative to some standard of condition in order to assess risk. Defining a "healthy" ecosystem provides the standard against which to evaluate its current condition and to predict the potential changes in that condition over time (i.e., trends) in response to a stress. Unfortunately, the current state of ecological science is insufficient to enable ecologists to define the optimum condition of most ecosystems. In most cases, defining the optimum level of species diversity within an ecosystem or to compiling a taxonomic list that precisely represents the optimal condition of a community is impossible. Several properties or attributes have been proposed to determine whether an ecosystem is in optimal condition of "health:" homeostasis, the absence of disease, diversity or complexity, stability or resilience, vigour or "scope for growth;" and the balance between ecosystem components.

All of these have shortcomings that limit their utility. For example, homeostasis, stability, and resiliency cannot be expected in ecosystems in early stages of succession. Vigour also depends on the successional stage, and the balance between ecosystems is difficult to evaluate without a reliable reference. Characterizing the condition of an ecosystem on the basis of diversity assumes a relationship between diversity and stability that cannot be generalized confidently. Despite these limitations, evaluation of the condition of an ecosystem should be possible by studying the structural and functional properties demonstrated to be essential in determining the condition of ecosystems.

Chapter 2 describes an approach for ascribing value to ecosystems known as the "Ecological Benefits Paradigm," discusses the advantages and difficulties of using the approach, and includes suggestions for appropriate assessment endpoints and related indicators. Reasonable approaches are suggested to define the condition of aquatic and terrestrial ecosystems based on the current state of ecological knowledge. The utility of defining quality objectives for ecosystems is also discussed. Quality objectives quantify desirable characteristics of ecosystems, and can be used to gauge the magnitude of risk and the efficacy of management strategies. Chapter 2 concludes with a discussion of the importance of effectively communicating risk to all concerned with the condition and protection of an ecosystem ("stake-holders").

1.4 METHODS FOR STUDYING EFFECTS

Once the objectives and relevant scales of an environmental risk assessment are determined, a data collection plan must be developed. To acquire useful information, such a plan involves identification of methods for both field and laboratory studies and of the tools available to assist an environmental manager in measuring environmental conditions. These tools include chemical analyses, toxicity tests, field surveys and assessments, and special studies such as biomarkers and simulated ecosystem studies (i.e., mesocosms or microcosms). Any of these tools may provide the information needed to assess environmental conditions; however, the use of several of these approaches in combination may increase the level of confidence in the conclusions.

After determining the study objectives and implementing a design, measurement endpoints or indicators must be selected. Indicators provide data about the assessment endpoints determined to be most relevant to the study objectives. Methods to measure indicators can be grouped into several categories: field-ecological surveys of the contaminated site; chemical analyses of samples from the site; and toxicity tests of site media (water, soil, or sediments) either in the laboratory or on site. Special studies, such as bioaccumulation tests, and simulated ecosystem studies, such as microcosms or mesocosms, may also be used.

A choice that must be made in developing an evaluation strategy is whether the question is best answered by field or by laboratory studies. Laboratory studies permit better control of naturally varying environmental factors that can confound interpretation of biological response, and may provide an improved means to address specific questions about pathways or modes of action. Field studies provide an opportunity to translate the potential for impact into an estimate of extent of alterations. Mesocosm studies represent an intermediate approach that can potentially bridge the gap between these two extremes. The nature of the exposure and the type of information needed as input to the risk assessment will determine which is the most fruitful approach.

An important issue involved in the selection of an indicator to evaluate the significance of effects of chemicals is the determination of what level of biological organization is most appropriate to the ecosystem objective being evaluated. Biota can be sampled at the cellular, organismal, population, or community levels. Generally, the higher levels of organization address questions concerning extent and magnitude of impact. Indicators associated with the lower levels of organization tend to measure exposure to stress or contamination, and are generally more useful in identifying specific causes of population and community modifications. The various levels are complementary, and the selection of indicators at alternative levels depends on the questions posed by, and objectives of, the risk assessment.

Quality assurance and quality control are important parts of any data collection plan, but are particularly important in the context of risk assessment. Risk assessment often involves comparing populations in space or time. Failure to standardize methods or quality of the data produced could lead to the conclusion that one population is less affected than another, when the result only reflects differences in accuracy among investigators. The extent of quality assurance necessary depends largely on the scale of the study. Studies with larger spatial scales generally involve a greater number of participating organizations, and, therefore, require more attention to standardization. Studies encompassing longer temporal scales require quantification of the extent to which methodological improvements over time alter results.

Chapter 3 provides an overview of currently available monitoring methods for aquatic and terrestrial systems, and discusses appropriate uses of these methods within the ecological risk assessment framework. Many of these methods are discussed in more detail in the contributed papers.

1.5 MEASURING EFFECTS ON LARGER GEOGRAPHIC SCALES

Once the values of interest have been defined and methods are being considered, the risk assessment must be focused on the proper spatial and temporal scales. To focus the assessment, the geographic extent of the resource at risk and what portion of the resource has already been compromised must be known. Increasingly, scientists and decision makers are becoming aware of environmental problems that occur at large geographic scales: acid deposition, non-point source pollution, and loss of biodiversity. For this reason, environmental managers and decision-makers should focus not only on effects on the local scale, but on the multimedia impacts that occur at larger scales. Most environmental monitoring is limited to the local environment, so that specific pollutants can be linked to the source and controlled. Regrettably, this type of monitoring provides little information about the effects of chemical releases on entire large ecosystems. Such information would allow an enhanced evaluation of the risks of action or inaction.

Defining risk typically involves collecting and interpreting monitoring data that characterize ecological processes. Ecological processes occur on many different scales, and the spatial scale of the assessment is an important consideration. For instance, most point source contaminant inputs to aquatic systems occur on local scales, and the biological assemblages at risk may be limited to the immediate depositional zone. Nonpoint source inputs of contaminants, such as pesticides from agricultural land use practices, may affect assemblages at the watershed level, or over entire regions having similar farming practices and land use patterns. Atmospheric inputs typically affect areas that include multiple watersheds, and may involve global scales, as in the case of ozone depletion. Although environmental risk assessment requires investigations on these large scales, 90 percent of biological studies have been estimated to be conducted on spatial scales of less than 1 m^2 (Levin *et al.*, 1990). Investigators should consider conducting studies on the scale at which decisions are to be made, or use statistical and mathematical models to extrapolate from scientifically valid scales to levels more useful to managers.

Part of defining the risk may involve estimating the degree and rate to which the status of the resource is changing, in which case the temporal scale of the assessment must be addressed. The appropriate temporal scale may be related to the timeframe over which the effect manifests itself or to the timeframe over which corrective actions occur. For example, cessation of particulate inputs may resolve smog in less than a year, whereas groundwater contamination may take many years to clear after the cessation of inputs. Assessing groundwater condition based on a single year of data may provide a deceptive, snapshot impression that conditions are bad, when in fact the inputs have stopped and the effects are lessening, but the decrease will be observable only on the time scale of decades.

The spatial scale and temporal scales over which a risk assessment is to be conducted will determine the methods available to document and evaluate effects of chemicals. Some indicators appropriate at local scales may be inappropriate as biogeographic boundaries are crossed. While local scale and site specific approaches are effective assessment methods for certain types of stressors and ecological resources, a larger scale perspective permits a broader, more comprehensive view of environmental problems that may impact large geographic areas. Chapter 4 describes spatial scales frequently used for such assessments and discusses their applicability to selected ecosystem objectives, and also discusses how data collection options are affected by changes in scale, and how assessment techniques may vary.

1.6 GENERAL RECOMMENDATIONS

1. Ecological risk assessment is a useful framework to make decisions and to design monitoring programs that contribute to environmental policy and management decisions. Identifying ecosystem values, reference conditions, and scales of interest are important first steps in the environmental risk assessment process that will guide the selection of monitoring methods.
2. No single best method exists for comprehensive evaluation of a contaminated ecosystem. The most appropriate methods vary according to the nature of the

ecosystem and the contaminant in question, cost and time constraints, and political and social constraints. However, development and employment of standardized methods are essential to perform replicate analyses and to implement quality assurance and control plans that ensure that data are reliable, valid, and comparable.
3. Investigations at all levels of biological organization and geographic scales are important to understand and document the effects of chemicals on ecosystems.
4. Further development, validation, and standardization of appropriate methods to assess the ecological risks of chemical agents are required. Statistical designs for ecological monitoring must be improved. Dose–response curves should be established at the community and ecosystem levels. Studies of synergistic and antagonistic interactions among chemicals must be expanded. Research is also needed for biomarkers to identify exposures and toxicity in non-temperate biomes.
5. Accidental discharges of chemicals should be recognized as opportunities to improve our understanding of an ecosystem's inherent resiliency and capacity for recovery. Information gained from accidents can be especially valuable in developing countries where ecosystems are generally poorly characterized and where the fate and effects of pollutants are not well-documented.
6. Long-term monitoring should continue in order to enable estimation of natural variability, as well as to establish a baseline against which to evaluate the effects of disturbances. Characterization of ecosystems in developing countries is essential. Standard environmental risk assessment approaches (i.e., species diversity and the use of keystone species) may not be useful, if the components of the ecosystem in question have not been determined. Further research to characterize pristine ecosystems is needed to enable scientists to identify and quantify deviations from natural conditions.
7. The importance of time scales should be recognized, especially in the rates of contaminant release and in the resulting effects on biota.
8. Methods to assess chemical contaminant levels in biological samples involve two strategies: examination of the properties of the ecosystem under study, and examination of biota from the ecosystem for contaminant effects. Either test, used alone, provides an incomplete picture of contamination; using both approaches is recommended to obtain a complete examination of contaminated ecosystems.
9. Microcosm and mesocosm tests are useful intermediates between laboratory bioassays and ecosystem monitoring. Microcosm tests are generally small-scale laboratory tests that involve a few species, whereas mesocosms are relatively large, constructed in the field, and involve most or all species within an ecosystem. Mesocosm tests are most often conducted in aquatic systems. Whole-lake manipulations have been conducted on the landscape scale and involve intentional manipulation of entire aquatic systems.
10. Mathematical modelling should continue to be used routinely to evaluate the cycling of materials such as nutrients or metals between biotic and abiotic components of an ecosystem. The simplest models are to be used to estimate chemical loading and chemical mass balances within aquatic systems.

1.7 REFERENCES

Levin, S.A., Harwell, M.A., Kelly, J.R., and Kimball, K.D. (Eds.) (1990) *Ecotoxicology: Problems and Approaches*. Springer Verlag, New York, 547 pp.

NRC (National Research Council) (1983) *Risk Assessment in the Federal Government: Managing the Process*. National Academy Press, Washington, D.C., 191 pp.

USEPA (US Environmental Protection Agency) (1992) *Framework for Ecological Risk Assessment*. Report No. EPA/630/R-92/001. US Environmental Protection Agency, Risk Assessment Forum, Washington, D.C.

2 A Conceptual Approach for Ecological Research, Monitoring, and Assessment

A critical first step in studying the effects of chemicals in ecosystems is making a clear statement of the ecosystem values to be protected. Historically human societies have placed value on ecosystems based on the ability of the ecosystem to provide products and services deemed necessary by the society. These values have varied between societies and over time, but all have had an economic basis. An awareness of the historical emphasis on economic benefits and a desire to communicate the wide range of benefits provided by ecosystems in meaningful ways have resulted in the development of the Ecological Benefits Paradigm. This paradigm, which classifies ecosystem benefits or values as use versus non-use benefits, is introduced in Section 2.1 and discussed in greater detail in Chapter 19.

This paradigm facilitates environmental management decisions by providing a means of comparing and prioritizing essentially dissimilar environmental situations, and the efforts and resources available to manage and protect these ecosystems.

To maintain the availability of the desired ecological benefits, the conditions of an ecosystem that is capable of supplying the benefits must be described. Environmental managers must be able to characterize the desired condition (health) of the ecosystem in a way that will allow assessment of nominal (desirable or acceptable ecological condition) versus subnominal (undesirable or unacceptable ecological condition) states, so that they will know when a problem develops that deserves attention. Section 2.2 describes methods to characterize ecosystems. Environmental managers must also be able to clearly state the ecosystem quality objectives (Section 2.3) to be achieved so that the desired benefits may be maintained.

To protect the ecosystem and maintain the desired benefits, full analysis of the historical, current, or potential environmental contamination threatening the ecosystem is necessary. The steps in conducting comprehensive environmental assessment studies are presented in Section 2.4. Ecological risk assessment is offered as a tool that environmental managers may use to compare environmental risks and, thereby, more effectively prioritize actions and the allocation of resources to protect or restore ecosystems.

If risks are not translated into familiar or relevant terms for use by policy makers, little may be done to manage or reduce them. The importance of effectively communicating the results of environmental monitoring and assessment and the decisions based on environmental studies is discussed below.

Methods to Assess the Effects of Chemicals on Ecosystems
Edited by R. A. Linthurst, P. Bourdeau, and R. G. Tardiff

2.1 THE ECOLOGICAL BENEFITS PARADIGM

Degradation of ecosystems is generally agreed to be undesirable. Many reasons exist to preserve an ecosystem, and they vary according to the needs of the interested parties. Some of the numerous motivations to preserve the integrity of ecosystems, discussed in Chapter 1, are social and cultural, aesthetic and ethical, and the traditionally important economic motivations. Terrestrial and aquatic ecosystems provide a wide array of goods and services for human use; historically, the desire to maintain the flow of these goods and services has been the impetus for action to preserve and protect ecosystems. Because most environmental policies and management actions are initiated on the basis of perceived degradation or loss of these goods or services, ascribing value to ecosystems based on their ability to provide such benefits affords a useful framework to evaluate the analytical methods available to study the effects of chemicals on ecosystems. Historically, most evaluations of ecological effects of chemical contamination have related impacts to effects on the supply of products and services of importance to human cultures. Recently, a framework was developed that attempts to classify ecosystem values in terms of the benefits they provide to humans as *use* versus *non-use* benefits. This framework, known as the Ecological Benefits Paradigm, is discussed in greater detail in Chapter 19 of this document. Some benefits can be quantified easily, because they relate to the production of commodities and provision of services and amenities. Examples of these uses that are directly related to market benefits are food crops, water resources, and construction materials. Use benefits that do not provide marketable commodities or services, such as recreational opportunities, are referred to as non-market use. The non-use benefits, such as those that relate to ecological quality (e.g., biodiversity) are more difficult to quantify; in the Ecological Benefits Paradigm, they are known as "neglected benefits." Increasingly, the protection of ecosystems for "neglected benefits" is being recognized by scientists and environmental managers. Table 2.1 presents the types of benefits assigned to ecosystems by the Ecological Benefits Paradigm.

The Ecological Benefits Paradigm is useful to environmental managers by providing a method to prioritize ecosystem benefits, and, thereby, help to allocate the limited resources available for monitoring, managing, and protecting ecosystems. The relative value of protecting an ecosystem depends, in part, on the needs of the local human population; often the benefits that one group considers important are not valued by another group. These varying value systems may result in conflicting priorities for environmental management. For example, in some countries, the expanding industrialized populations seeking recreational opportunities may place demands on the same undeveloped areas that other, less industrialized indigenous groups have relied on for subsistence. Another example is seen in the situation that occurred in Minimata Bay, Japan. Protecting Minimata Bay from mercury contamination was important to the local population who relied heavily on fish from the bay for food. If the people living near Minimata Bay had obtained their food from other sources, they would not have been affected by mercury contamination in the bay, and the relative value of monitoring and protecting the ecosystem may not have been considered worth the cost based solely on the market benefits of the ecosystem.

Table 2.1. Benefits of the Ecological Benefits Paradigm

Use benefits	Non-use benefits
Market benefits	*Neglected benefits*
1. Food	1. Existence values
2. Live animals (non-food)	2. Historical, heritage, cultural, spiritual values
3. Animal materials	3. Philanthropic values
4. Non-animal commercial inputs	4. Bequest values
5. General water provision	5. Intrinsic values
6. Fossil fuels	6. Intergenerational equity
7. Other fuels (biomass)	7. Non-human/habitat benefits
8. Wood materials (other than fuel)	8. Preservation of genetic diversity (biodiversity)
9. Livestock forage	9. Environmental infrastructure maintenance cycles
10. Pollination	10. Climatic effects
Non-market use benefits	11. Contaminant/pollutant effects
1. Recreational uses	12. General scientific and research value
2. Tourism	13. Scarcity/uniqueness

Ecological, as well as market benefits, provided by an ecosystem should be considered in developing environmental policy. Although these so-called neglected benefits have not been neglected or misunderstood by ecologists, they have been virtually ignored in the development of national policies, partially due to the difficulty in characterizing or quantifying them. Furthermore, these non-use benefits can be quantified only by indirect economic methods (e.g., economic benefits of a recreational area), and cannot be related easily to small changes in ecological condition. If environmental policy makers wish to emphasize benefits that can be affected by small or incremental changes in ecological condition, they must focus on benefits for which the most direct measures are available. Unfortunately, this situation could lead analysts to continue to ignore ecological benefits that are difficult to quantify. Considering quantifiable benefits exclusively, however, is acceptable only for preliminary analysis because other benefits could easily be so important as to significantly alter the outcome of the analysis.

Although applying the Ecological Benefits Paradigm pushes ecological science closer to what some consider an undesirable tendency to value ecosystems in a purely economic manner, its use could improve environmental policy making and management strategies by considering all ecological goods and services. This framework would allow ecologists to describe ecological benefits to policy and decision makers in an easy to understand format, and would help clarify policy makers' preferences by stating them in terms of scientifically defensible data and not as subjective values.

An example of how the use of the Ecological Benefits Paradigm might facilitate decisionmaking by policy makers is when considering the management decision to harvest trees from a hypothetical old-growth forest. The authors point out that the greatest benefits from a virgin old-growth forest are realized from non-market benefits. Once the decision has been made to harvest the old-growth trees, most of the non-market benefits have been lost, but the market benefits from timber sales have been realized. Continuing the analysis provides an important insight: after the trees are harvested, not only are the non-market benefits lost, so are the market benefits. The resulting benefits profile of the harvested forest is much smaller than the benefits profile of the intact forest.

2.2 CHARACTERIZING THE CONDITION OF ECOSYSTEMS

2.2.1 DESCRIBING THE CONDITION OF ECOSYSTEMS

To evaluate an ecosystem's ability to provide the desired benefits, ecologists, managers, and policy makers must have a reliable method to characterize ecosystem condition. Defining a "healthy" ecosystem provides a baseline against which to evaluate the current condition of an ecosystem and to predict the potential effects of changes in that condition over time (i.e., trends in resources).

Some ecologists have attempted to describe ecosystem "health" in terms of human health. However, this is difficult or even impossible, because the kinds of analytical tools used in medical science do not exist in ecological science. However, two distinct features of ecosystems, stability and resilience, may be potentially useful in describing a measure of "health" in ecosystems (Pimm, 1984; Holling, 1986). Costanza (1992) identified six major concepts derived from the studies of ecosystem stability and resilience that could provide a framework for characterizing ecosystem health. These concepts define ecosystem health as: homeostasis, absence of disease; diversity or complexity; stability or resilience; vigour or "scope for growth;" and balance between system components.

As Costanza and Principe state, these concepts, while useful, are limited in their ability to describe complex ecosystems. They do, however, provide a basis to develop a more practical definition of ecosystem "health" that can be described as: a system free from "distress syndrome" (defined as the irreversible processes of system breakdown), stable, and self-sustaining. A "healthy" ecosystem is one that is active, maintains its autonomy over time, and is resilient to stresses (Haskell *et al.*, 1992).

2.2.2 STRUCTURAL AND FUNCTIONAL CHARACTERISTICS TO EVALUATE THE CONDITION OF ECOSYSTEMS

To characterize ecosystems, the structural and functional properties of ecosystems must be differentiated, particularly in the context of evaluating the effects of chemical or other stresses on an ecosystem. To properly characterize an ecosystem, both structural and functional parameters must be examined. Structural parameters (e.g., productivity, species composition, and demographic descriptors) and functional parameters (e.g., nutrient cycling and trophic structure) are interdependent, but have distinct characteristics. Structural parameters describe the parts that make up an ecosystem, that is, they simply tell what is there. Functional parameters describe the actions or inner-workings of a system and how the components work. Structural characteristics, descriptive of ecosystem components, generally are easier to assess (with the exception of underground microbiology), because fewer parameters can be measured over a shorter time period. Structural properties correlate with functional properties, that are expressed as rates, and must be measured in a more complicated manner. Measuring a functional property such as decomposition of plant matter on a forest floor requires taking multiple, complex measures over time; yet describing microbial composition is all the more difficult.

Functional redundancies can buffer ecosystems from the effects of perturbations within specific populations. Thus if groups of organisms are killed, their functions may be assumed by other groups of organisms. The overall effect is that the particular ecosystem functions remain unchanged, while the structure of the ecosystem changes substantially. Structured populations are able to resist stress-induced changes through compensatory alterations in growth and reproduction, which also may affect the structural characteristics of the ecosystem. These mechanisms for mediating stress have limitations, and, if the magnitude or duration of the stress exceeds these limitations, the condition of the ecosystem may decline. Environmental management strategies often give higher priority to sustaining ecosystem function than to preserving ecosystem structure; however, an ecosystem in which essential functions are performed by a severely reduced community may be more sensitive to stress in the future.

Several functional and structural properties of ecosystems can be monitored to evaluate ecosystem condition. Examples of functional parameters proven to be valuable indicators of stress are primary and secondary productivity, decomposition, and mineralization. Primary productivity, the rate at which green plants convert light energy into organic matter, provides the energy essential for an ecosystem to operate, and is dependent on abiotic factors (e.g., light, temperature, humidity, soil structure), biotic factors, and trophic interactions. Secondary productivity, the rate

Table 2.2. Structural and functional characteristics of aquatic ecosystems

↓Characteristic	↓Streams	↓Rivers	↓Lakes	↓Estuaries	↓Near-coastal	↓Ocean	↓Ease of measurement[a]
Structural							
Biodiversity	H	H	H	L	L	–	1
Relative density	L	L	–	–	–	–	1
Dominance	–	–	–	M	M	–	5
Food web characteristics	L	L	H	H	M	L	8
Genetic diversity	H	H	H	H	H	H	10
Functional							
Primary productivity	H	H	H	H	M	L	1
Decomposition	L	L	M	H	M	L	1
Energy flow	–	–	M	M	M	–	5
Mineral cycles	–	–	H	H	H	H	8
Nutrient cycles	H	M	–	–	–	–	8
Keystone species	–	–	H	M	H	–	5
Index of biotic integrity	H	H	–	–	–	–	10

L = low; M = medium: H = high, – = unimportant
[a] On a scale of 1 to 10, where 10 is the most difficult

at which herbivores utilize the energy from green plants and, in turn, supply energy for carnivores, is dependent on primary productivity, but is affected by other factors as well. Primary and secondary productivity may be influenced both directly and indirectly by chemical stresses. Decomposition of organic matter depends on climate, chemical composition of the substance, and type and quantity of decomposer organisms. The rate of decomposition appears to be a valuable indicator of any kind of ecosystem stress as any chemical stressor affecting these factors will alter decomposition rate. Mineralization rates of nitrogen, sulphur, and phosphorus compounds are also critical functional indicators of ecosystem health. These functional parameters are discussed in greater detail elsewhere in this document.

Structural properties of ecosystems may also be useful indicators of ecosystem conditions. These may include: changes in species composition and abundance (i.e., increase in stress-tolerant species or decrease in native species with a corresponding increase in introduced or exotic species); reduced biodiversity; shorter, less complex food webs; reduced population density; and reduced genetic diversity.

Tables 2.2 and 2.3 summarize some structural and functional characteristics that are useful indicators of the condition of some major aquatic and terrestrial systems. These parameters may be monitored to evaluate changes in ecosystems or to assess trends that may suggest the need for concern.

Table 2.3. Structural and functional characteristics of terrestrial ecosystems

↓Characteristic	↓Forests	↓Grasslands	↓Deserts	↓Tundra	↓Agro-ecosystems	↓Ease of measurement[a]
Structural						
Species composition	M	M	M	M	–	5
Biodiversity	M	M	M	M	–	5
Food web complexity	M	M	L	L	L	8
Relative density	M	M	M	M	–	2
Genetic diversity	L	L	L	L	–	10
Functional						
Primary productivity	H	H	L	L	H	1
Secondary productivity	M	M	M	M	–	3–7
Decomposition	M	L	L	L	L	1
Mineralization	M	M	L	L	H	3

L = low; M = medium; H = high; – = unimportant
[a] On a scale of 1 to 10, where 10 is the most difficult

2.3 SETTING AND MAINTAINING QUALITY OBJECTIVES FOR ECOSYSTEMS

Having defined the values of an ecosystem and identified its characteristics, environmental managers can set ecosystem quality objectives or quantitative goals for acceptable conditions. Quality objectives can be described as the specific ecosystem conditions that must be achieved or maintained to ensure that the desired state of an ecosystem is maintained, and that the desired ecosystem benefits are available. Examples are habitat continuity and maintenance of species parameters.

Such objectives provide a standard with which to compare the effects of chemical stress and to evaluate the effectiveness of corrective actions.

A wide variety of natural and anthropogenic stressors can influence ecosystem quality objectives. While this book focuses on chemical and physical stressors, the impact of naturally occurring and non-chemical stressors should not be underestimated. Habitat destruction is perhaps the most important and devastating ecosystem stressor; but from the point of view of this book, non-chemical stressors are important only to the extent that they interact with chemical stressors and to the extent that the effects of chemical and non-chemical stressors must be separated from one another to select appropriate regulatory and remedial strategies. Table 2.4 summarizes some common stressors that influence ecosystem quality objectives.

Table 2.4. Stressors that can influence ecosystem quality objectives

Natural stressors	Nonchemical stressors	Chemical stressors
Temperature variability	Land reclamation	Heavy metals
Oxygen variability	Solid waste	Organohalides
Salinity variability	Shipping	Oils
pH variability	Oil drilling	Detergents
Desiccation	Coastal restructuring	PAHs
Wave action	Dredging	Dioxins
Solar radiation	Dumping ash	PCBs
Flooding	Dam construction	Radionuclides
Predation	Irrigation	Excess nutrients (C,P,N)
Competition	Water extraction	Drilling muds
Weather	Paddy fields	Acidification
Forest fires	Introduced exotic species	Organotins
	Aquaculture	CFCs
	Agriculture	Ozone
	Deforestation	Gaseous air pollutants (SO_2, NO_x)
	Restriction of migration patterns	"Greenhouse" gases
	Overharvesting	Pesticides

Synergistic interactions between anthropogenic and natural stressors or among pollutants may increase the difficulty of identifying the principal cause of injurious effects. Many examples of interactions between non-chemical and chemical stressors exist. Generally, an ecosystem already weakened by stress is much more susceptible to the effects of additional stress. For example, a forest stressed by drought is more likely to be affected by either acid rain or sulphur dioxide (SO_2) than one that has been receiving an adequate amount of water. Experiments have proven that Norway spruce seedlings are quite sound after several weeks of exposure to high levels of SO_2 (200 ppb), whereas they are severely stressed if they experience drought at the end of the exposure period. Seedlings exposed to SO_2 recover from drought slower than those that endure drought without being exposed to SO_2. In the Vosges region of northeastern France, a seven-year field assessment of mineral nutrition in spruce and fir growing in very desaturated soils showed that magnesium content is generally low, but is much lower in dry years. Furthermore, terrestrial mammals and birds stressed with high body burdens of organochlorines have been shown to be more susceptible to disease and to have a large proportion of deformities.

2.3.1 SETTING QUALITY OBJECTIVES

Identifying and establishing quality objectives for an ecosystem are more easily accomplished by evaluating the baseline conditions of the pristine ecosystem. If historical data are unavailable to provide precontamination information, relying on data collected at carefully selected reference sites may be necessary to compare against the impacted study areas. These areas should be as similar to the study area as possible to eliminate variables in the extrapolation of data from the reference site to the impacted sites. However, some uncertainty will always exist with this approach.

Local conditions must be considered when setting quality objectives. Local conditions at one location may result in that ecosystem responding at a different rate than at another, or may result in reduced or increased resilience. For instance, the type of underlying bedrock present may provide some buffering of the effects of acid rain surrounding one lake and not at another.

Departures from ecosystem quality objectives can be defined as effects, which may vary in magnitude and occur over different time scales. Furthermore, some effects may be reversible, while others may persist for long periods. Environmental managers must decide how great an effect needs to be over a specific area, and how long it must persist before regulatory and remedial actions become necessary. Accidental releases of chemicals can be useful study exercises to observe effects over a gradient of exposures in a way that would normally be impossible. The opportunities for research on the effects of petroleum products on coastal ecosystems, birds, and mammals provided by accidental oil spills are discussed elsewhere in this volume.

2.3.2 MAINTAINING QUALITY OBJECTIVES

Identifying the potential effects of chemical contaminants before a release occurs has become vital in maintaining ecosystems. A thorough evaluation of a chemical prior to its commercial use has become a common regulatory policy in many countries, and is designed to prevent damage to ecosystems. In addition, efforts are being extended to implement waste-limiting technologies to minimize industrial discharges. The introduction of between 1000 and 2000 new chemicals yearly necessitates evaluation of potential environmental toxicity of these chemicals (Xu and Pang, 1992). Prior to industrial-scale production, new chemicals are tested to varying degrees to define potential environmental effects. Such information can help to prevent environmental contamination. Often, a tiered structure of decision-making is designed to define criteria for higher level testing.

Because of the enormous volume of chemicals requiring evaluation, test procedures must be cost-effective, rapid, reproducible, and ecologically relevant. Current methods to evaluate ecotoxicity are largely laboratory-based, and include quantitative structure-activity relationship (QSAR) approaches, acute toxicity testing with selected species, multiple species testing, and life-cycle analyses (Xu and Pang, 1992).

Identifying potential effects of chemicals prior to release to the environment is founded on two principal methodologies: (1) controlled studies to evaluate the potential effects on specific ecosystems and (2) methods to provide an early warning of adverse effects in an ecosystem (i.e., monitoring programs). Controlled studies range from those of simple, acute toxicity to the more complex ones with microcosm and mesocosm. Attempts to predict more ecologically relevant changes frequently involve the use of microcosms and mesocosms, in which effects on food chains and ecosystem structure and function can be evaluated. The use of microcosm and mesocosm studies to evaluate contaminant effects on aquatic and terrestrial ecosystem is discussed in Chapter 3. Biomonitoring strategies (i.e., the bioavailability of chemicals is a measure of the extent to which a chemical accumulates in an organism) may serve as early warning signals of significant chemical contamination of an ecosystem. The use of biomarkers can also provide an indication of the need for regulatory action, and can assist in evaluating the effectiveness of remedial action. Biomonitoring strategies and biomarkers are discussed in Chapter 3.

A policy approach to maintaining environmental quality objectives is to define "exclusion zones" around a chemical release or discharge. Such areas permit limited deviations from quality objectives that are deemed to be acceptable (Peakall, 1992). Their definition provides a balance between the sometimes conflicting needs of industrial production and protecting the ecosystem. The economic benefits of allowing an exclusion zone are weighed against protecting the ecosystem, and sometimes the economic benefits can be determined as outweighing the damage to the ecosystem.

The time scales of chemical contamination are also important, particularly for persistent pollutants, because several years may be needed for contaminants to be reduced to nominal levels (Burger and Gochfeld, 1992). In some instances, ecosystem recovery to within quality objectives can be accelerated by restoration procedures. For example, the River Thames (England) has historically been negatively impacted by stormwater carrying untreated sewage, which led to reduced dissolved oxygen levels and to frequent dieoffs of aquatic organisms, leading subsequently to drastically reduced biodiversity in the river. Eventually, only a few species remained that are known to be tolerant of low oxygen. By controlling the stormwater surges to the river, the biodiversity in the river has been gradually increasing (i.e., pollution-tolerant organisms are being replaced by organisms less tolerant).

2.4 STUDYING THE EFFECTS OF CHEMICALS

2.4.1 ECOLOGICAL RISK ASSESSMENT AS A TOOL

Ecological risk assessment is a structured approach with four components that provide a systematic method to assess the effects of chemical contaminants on ecosystems. The four components of ecological risk assessment are:

1. Problem formulation: to identify and describe the ecological damage and resources potentially at risk.
2. Exposure assessment: to describe the magnitude, duration, frequency, and routes of contaminant exposure to potential receptors.
3. Ecological effects assessment: to identify the nature of the hazards associated with the contaminant(s) and to quantify the relationship between exposure stress and receptor characteristics.
4. Risk characterization: to integrate the exposure and hazard information by estimating the likely incidence of an effect under the conditions described.

The ecological risk assessment process is most useful to obtain the information necessary to identify the potential hazards from contamination, determine the relative risks associated with contamination, and facilitate management of risks in a manner that best protects social, cultural, political, and economic benefits associated with the area. The process allows the comparison and prioritization of various essentially dissimilar local situations by focusing on the potential risk to biota.

Ecological risk assessment is an approach to guide the process of evaluating chemically contaminated areas. This section provides a discussion of essential steps of any ecological study to evaluate chemical contamination of an area. These generalized steps are: problem formulation, study design, and data analysis using statistical and modelling applications.

2.4.2 PROBLEM FORMULATION

2.4.2.1 Establishment of goals

The first step in diagnosing the cause of an environmental problem is to state clearly the questions to be answered. A clearly focused statement of goals assists in the selection of test biota, sampling sites, sampling and analytical methods, etc. Generally, the questions centre on identification of contaminants in the area, whether their levels are elevated above background levels, their bioavailability, and their potential to injure biota.

Table 2.5. Characteristics of an area to plan an ecological risk assessment

General characteristics	Examples
Area History	1. Historical land use 2. Industrial activities
Abiotic Features	1. Physiography 2. Geology 3. Hydrology 4. Meteorology
Contamination	1. Types of chemicals 2. Quantities and dispersal 3. Chemical and physical properties 4. Toxic effects
Biotic Features	1. Habitats present (a) Habitat types present (b) Availability of habitats (patchiness) (c) Critical or sensitive habitats present 2. Species present (a) Soil biota (b) Plant species (c) Animal species (d) Trophic webs and species interactions (i) Potential bioaccumulation in higher animals (ii) Potential human exposure

2.4.2.2 Identification and characterization of chemical stressors

Chemical stressors must be identified and characterized early in the study design process. An understanding of the most likely pathways of movement of contaminants through the ecosystem helps guide the sampling process. Chemical characteristics will sometimes help predict a contaminant's fate and transport and its temporal and spatial movement through site media. Knowing where the chemical is most likely to be compartmentalized and whether it is likely to bioaccumulate in biota helps to determine the most appropriate sampling and analysis methods.

2.4.2.3 Characterization of the ecosystem potentially at risk

Thoroughly characterizing an area before conducting an ecological assessment can result in substantial savings in time and resources. Relevant characteristics of an area to be evaluated before deciding which assessment methods to address the desired endpoints and yield relevant data are listed in Table 2.5. Narratives discussing the importance of each characteristic are available in other references; this section provides only a listing of the specific characteristics.

2.4.2.4 Identification of relevant endpoints

The concept of ecological endpoints, described as expressions of the values or benefits to be protected, provides a useful method to guide the study design (Suter, 1989). A clear statement of the endpoints helps focus the study, and ultimately facilitates communication of the study results. Assessment endpoints are described as formal expressions of the actual ecological values to be protected, referred to as environmental characteristics at risk from contamination. Characteristics of good assessment endpoints are: social relevance, biological relevance, unambiguous descriptor, measurable or predictable; capable of being damaged by contaminant, and logically related to the final decision (adapted from Suter, 1989, 1993). Examples are species diversity and reproductive integrity of a valued species.

Measurement endpoints are quantitative expressions of the assessment endpoints, that is, the actual measurements of the characteristics that are to be protected. Criteria of good measurement endpoints are: corresponding to or predictive of the assessment endpoint; readily measured; appropriate to the scale, exposure pathway, and temporal dynamics of the study area; low in natural variability; diagnostic; broadly applicable; and standardized. They also have existing data sets (Suter, 1989, 1993). Examples of measurement endpoints to quantify the specific assessment endpoint of species diversity are species abundance and composition. An example of a measurement endpoint for reproductive integrity of a valued species is age/size class structure of the population.

2.4.3 STUDY DESIGN

2.4.3.1 Types of sampling designs

Sampling locations may be determined quantitatively by random or non-random methods. Conclusions from random methods are the more reliable. Common types of random sampling designs are simple random and stratified random. The simple random method is most useful in areas that are highly uniform in habitat patterning, such as grasslands; however, sites not uniform in contaminant concentration, or other characteristics, make purely random sample site selection less efficient. Cluster sampling is generally used at a chemically contaminated study site.

Stratified random sampling is a modification of random sampling in which similar habitats within the larger study area are selected, and sites in those similar habitats are selected randomly. Stratified random sampling provides the opportunity to conduct some statistical analysis while increasing sampling efficiency. Examples of stratified random sampling are studies of the different levels in a forest ecosystem and studies of aquatic systems where the area is first stratified by salinity then sampled randomly.

If quantitative data are not required for statistical analysis, non-random sampling may be conducted, in which specific study sites are located along a transect or on a predetermined grid within the area. While efficient, non-random sampling may not provide data suitable for statistical analyses, thereby introducing immeasurable uncertainty in the conclusions.

2.4.3.2 Test method selection

No single type of test can provide all of the information necessary to determine the extent and magnitude of environmental contamination. Therefore, several types of tests should be employed. Test methods typically include field surveys, chemical analysis of media samples, toxicity tests (laboratory and field applications), and bioaccumulation studies. Additional studies such as biomarkers and microcosm and mesocosm studies may also be included. The advantages and requirements of numerous test methods for aquatic and terrestrial ecosystems are discussed in Chapter 3. Only together as part of a carefully designed field or laboratory study can these tools provide the information necessary to establish linkages between contamination and adverse effects in biota.

2.4.3.3 Quality assurance procedures

Any environmental data collection program is only as good as the data it collects. Thus, any program must be designed with an adequate level of quality assurance (QA) to ensure that the data collected are of the quality and quantity needed to answer the questions for which the study was designed. The first step in ensuring quality is to develop data quality objectives (DQOs) as part of the design process. DQOs are qualitative or quantitative statements of the level of uncertainty that a decision maker is willing to accept in decisions made with environmental data.

Development of clearly focused DQOs very early in the study design process can save time and money, while producing the kind of data that address the major issues of a study. The logical process for DQO development is described by Neptune and Blacker (1990). Their process, presented in Figure 2.1, was modified for use in chemically contaminated areas, and in the US contains six major steps (USEPA, 1993a).

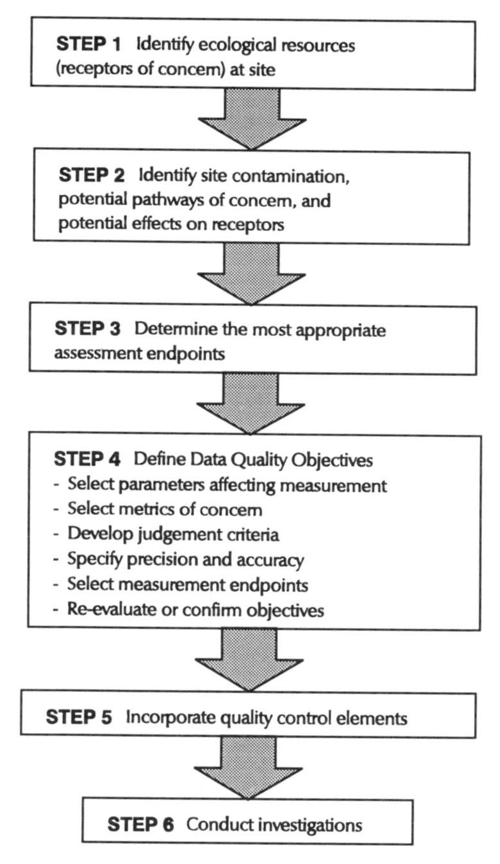

Figure 2.1. The modified logic process to develop DQOs (USEPA, 1993a)

After DQOs have been developed and a study has been designed to meet these data objectives, the study details need to be developed. Typically, the DQOs are produced via a quality assurance project plan (QAPP). QAPPs provide a link between the DQOs and the specific steps to be taken to achieve the data objectives. QAPPs are required by the USEPA in all of their data collection activities (USEPA, 1984). Elements generally found in USEPA's QAPP are:

1. description of the goals and objectives of the study;

2. organization of the project and quality assurance responsibilities;
3. DQOs and sampling strategy and design;
4. sampling procedures;
5. sample custody, storage and shipping procedures;
6. calibration procedures;
7. analytical procedures;
8. procedures for data management (i.e., data reduction, validation, and reporting);
9. internal quality control checks;
10. quality assurance audits;
11. quality assurance reports;
12. preventative maintenance procedures;
13. data assessment procedures; and
14. corrective actions.

Other tools to ensure the quality of a data collection project include the use of performance evaluation samples (i.e., analysis of a sample with a known concentration of a substance), and development of standard operating procedures for the specific procedures to be used routinely in the project (i.e., sample collection and analysis procedures). Very large studies may employ other quality tools such as Quality Management Plans, data quality audits, and management and performance audits by outside parties. These approaches are generally applicable to most data collection projects, and help to ensure, along with proper quality control measures, that the data generated are of the quality and quantity needed to answer the questions posed by the study initiator and decision maker.

2.4.4 STATISTICAL ANALYSIS AND MODELLING

Statistical analysis is a necessary step to determine the importance of the data collected and to evaluate whether the data are sufficient to answer the questions posed by the study. Integration and synthesis of the data collected into assessments of environmental condition are critical. The use of standard parametric procedures for analysis of normally distributed populations may be used in many applications. However, environmental data sets are frequently not normally distributed as is usually found with the life sciences. As such, special non-parametric (distribution-free) statistical procedures are commonly required. Environmental data also are frequently characterized by large measurement errors (random and systematic), missing or questionable data points, data near or below the detection and quantitation limits, and complex spatial and temporal trends and patterns (Gilbert, 1987). Thus, to minimize these factors as much as possible, statistical analyses should be planned carefully before the study is initiated.

Due to the complexity of environmental data, the presentation of results is critical. Graphical methods to display data should be employed wherever possible. Cumulative distribution functions (CDFs) are a recommended method to display data. CDFs display information on central tendency (mean, median) and data range in a graphical format. An example of a CDF is provided in Figure 2.2.

The use of geographic information systems (GISs) is also recommended, because of its ability to display spatial and temporal data using a mapping format. Geographic information systems are especially useful to monitor a regional scale, and are discussed in Chapter 4.

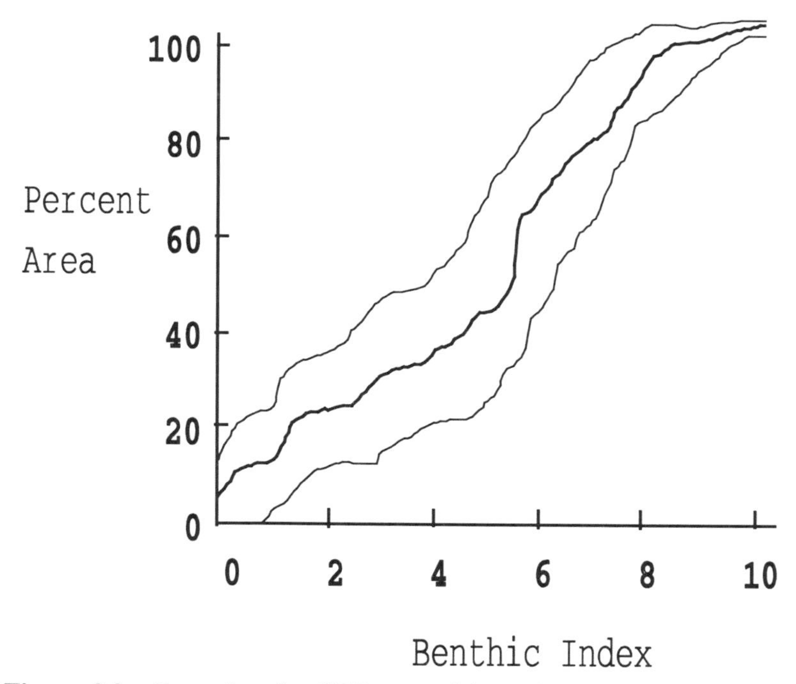

Figure 2.2. Example of a CDF: a confidence interval of 90 percent for an estimated benthic index for the Virginian Province (adapted from USEPA, 1993b)

As an integral component of the assessment process, mathematical models are used to integrate data or to generate missing data. They may be either statistical, such as an analysis of variance, or mechanistic (a quantitative description of the relationship between certain parameters). Mechanistic models are most often thought of when the "model" is used. These models are applied to describe for a substance an organism's exposure, its toxicity, or its transport and fate. Examples of their use in ecological assessment include establishing links between the measured responses in receptors of concern to population-level responses and to information on the spatial extent of contamination. Models may also be used to incorporate the effects of chemicals on ecosystem structure and processes and to predict the likely fate and transport of chemical contaminants. Models can be categorized according to their structure. Some of the more complex models attempt to depict the cycling of materials such as nutrients or metals among biotic and abiotic components of the ecosystem, and are often associated with chemical loadings and mass balance studies or studies of the ability of ecosystems to process chemicals. Another category of models emphasizes individual processes, such as the transfer of chemicals between the different media or the assimilation of toxicants or nutrients by the biota. Another advantage of models is that they allow scientists to verify their assumptions about the way an ecosystem works. The use of mathematical modelling to characterize risk at the regional scale is discussed in Chapter 4.

2.5 COMMUNICATING RISK

The effective communication of environmental data informs interested parties in relevant and understandable terms about the scientific data, analysis, interpretation, and recommendations that resulted in the environmental decision. Communicating an environmental decision is just as important to the success of an initiative as the ecological study on which the decision is based. Attempts to maintain ecosystem quality objectives have a much greater chance of success if the decision and the reasons behind it are effectively conveyed to interested parties. Communicating environmental risks brings the management strategy full circle by linking the assessment results to the efforts to protect the ecosystem benefits of concern.

Each society has a characteristic decision-making process: from a highly focused and hierarchical to broadly-based requiring active public participation. Since target audiences are different, communications pathways must be tailored to fit each. Although the following discussion focuses on a broadly-based process, the principle of matching the communications medium and the form of the message to the audience and the specific level of decision making is applicable to all situations.

2.5.1 TARGET AUDIENCES

Communicating scientific information effectively requires identifying target audiences and using language and media that the audiences can most easily understand. Audiences for environmental monitoring and assessment information are varied, and can include:

1. environmental managers, administrators, and regulators who require information to assess both risk and the effectiveness of regulations and management practices;
2. political leaders and government agency officials who set public policies and allocate financial resources;

3. industrial leaders and private sector professionals who advise and determine the choices of technology, raw materials, and plant location for economically productive activities;
4. environmental organizations, community groups, and private citizens who influence decisions in the public and private sectors through advocacy, voting procedures, and lobbying activities;
5. university faculty and students, public school administrators, and teachers who determine the curricula that can be a means of educating the public in environmental issues; and
6. researchers in the scientific community who determine the needs for additional scientific endeavour.

The diversity of audiences and their differing needs for scientific, technical, or interpreted information mandates a varied communications initiative.

2.5.2 CONTENT

Information must be provided in a context relevant for the level of the decision making expected of the target audience. The need to collect data, perform analyses, or derive conclusions should be explained clearly within the framework of the decision-making process. The specific data collected, their organization and analysis, and the conclusions and interpretation can be presented at a level of technical detail appropriate for the audience and its decision-making responsibilities. The sources and types of data, their strengths and weaknesses, significance of the findings, and alternative interpretations provide a basis to evaluate the reliability and conclusiveness of the findings.

Although most communication is presented through textual material, it is enhanced in clarity, persuasiveness, and understanding by the use of well-designed maps, charts, graphs, and other visual methods. For example, the presentation of the results of mathematical models, although highly technical in content, is an effective method to convey the dynamic nature of ecological functions and structures.

2.5.3 METHODS

The method of communicating data should vary with the audience being addressed. Scientific and technical data are initially entered, analysed, and stored in computerized data bases. Computer programs can also be developed to organize, structure, and present the data in a multilayered format for use at variable levels of technical complexity by audiences with different needs and skills. In addition, the data and interpretations may be presented in regular scientific and technical journals for peer review, critique, and validation, and through agency technical reports for program assessment, regulatory needs, and archiving. For wider use, data summaries and interpretations may be published in professional journals read by the relevant users. Scientific background articles in the general press and subscription periodicals and monographs, edited volumes, books, and textbooks inform the public, and serve as educational material for schools and universities. News reports appearing in the popular press, television, and radio are effective to achieve public awareness, but not to convey the full depth of technical issues of environmental problems. The accuracy of these initial reports, however, is critical in forming public attitudes; unfortunately, misinformation or misinterpretation of information is often very difficult to correct. Finally, oral presentations of reports to scientific societies, professional meetings, community groups, legislative hearings, and court proceedings are important additions to more formal communications. The various methods for communicating ecosystem data and the audiences for which the methods are appropriate are presented in Table 2.6.

Table 2.6. Methods to communicate ecosystem data to various audiences

↓Target audience	↓*Computer information systems*	↓*Professional meetings, briefings, and hearings*	↓*Professional publications*	↓*Periodicals, books*	↓*Textbooks*	↓*Press, TV, radio*
Scientists	X	X	X			
Managers/ administrators	X	X	X	X		
Political leaders			X	X		X
Environmental organizations	X	X	X	X		
Educational institutions	X		X	X	X	X
Public				X		X

2.5.4 TIMING OF COMMUNICATIONS

Proper timing to communicate monitoring and assessment information is vital to the overall success of management decisions. Input to databases for regulatory and assessment purposes is ongoing and not time sensitive. Status reports to agencies responsible for management are periodic as required by the program, and are time sensitive, but not urgent (i.e., years). Publication in scientific journals is appropriate to research schedules that have a long timeframe (e.g., months). Background articles in the public media are ongoing, but more effective when placed in the timeframe of pending public decisionmaking.

The timing of communication is essential when viewed in the context of critical decisionmaking, such as research planning and financing, pending legislation, annual budget authorizations, and appropriation cycles (when lobbying efforts are maximal and effective), in relation to current or planned tort proceedings, in local or regional economic planning and development hearings and decisions, and in the course of formulating environmental impact assessments. Ensuring that monitoring information is communicated in a proper and timely manner is an essential component of any research and monitoring program.

2.6 REFERENCES

Burger, J., and Gochfeld, M. (1992) Trace element distribution in growing feathers: additional excretion in feather sheaths. *Arch. Environ. Contam. Toxicol.* **23**, 105–108.

Costanza, R. (1992) Toward an operational definition of ecosystem health. In: Costanza, E., Norton, G.G., and Haskell, B.D. (Eds.) *Ecosystem Health: New Goals for Environmental Management*, pp. 236–253. Island Press, Washington, D.C.

Gilbert, R.O. (1987) *Statistical Methods for Environmental Pollution Monitoring*. Van Nostrand Reinhold, New York, 320 pp.

Haskell, B.D., Norton, B.G., and Costanza, R. (1992) What is ecosystem health and why should we worry about it? In: Costanza, E., Norton, G.G., and Haskell, B.D. (Eds.) *Ecosystem Health: New Goals for Environmental Management*, pp. 1–18. Island Press, Washington, D.C.

Holling, C.S. (1986) The resilience of terrestrial ecosystems: local surprise and global change. In Clark, W.C., and Munn, R.E. (Eds.) *Sustainable Development of the Biosphere*. Cambridge University Press, Cambridge, Melbourne, New York.

Neptune, D. and Blacker, S. (1990) Applying the data quality objective (DQO) process to research operations. In: Hart, D. (Ed.) *Proceedings of the Third Annual Ecological Quality Assurance Workshop*, pp. 5–22. Environment Canada, Burlington, Ontario, and US Environmental Protection Agency, Environmental Monitoring Systems Laboratory, Las Vegas, Nevada.

Peakall, D. (1992) *Animal Biomarkers as Pollution Indicators*. Chapman and Hall, London, England, 291 pp.

Pimm, S.L. (1984) The complexity and stability of ecosystems. *Nature* **307**, 321–326.

Suter, G.W., II (1989) Ecological endpoints. Chapter 2 in: Warren-Hicks, W., Parkhurst, B.R., and Baker, S.S., Jr. (Eds.) *Ecological Assessment of Hazardous Waste Sites*. Report

No. EPA 600/3-89/013. US Environmental Protection Agency, Environmental Research Laboratory, Corvallis, Oregon.

Suter, G.W., II (1993) *Ecological Risk Assessment.* Lewis Publishers, Chelsea, England, 538 pp.

USEPA (US Environmental Protection Agency) (1984) *Policy and Program Requirements to Implement the Mandatory Quality Assurance Program.* Order No. 5360.1. U.S. Environmental Protection Agency, Office of Administration and Resources Management, Washington, D.C., 3 April.

USEPA (1993a) *Biological Assessment at Hazardous Waste Sites: An Overview of Issues and Methods.* Proceedings of Seminar Series, 10 February 1993. Prepared for US Environmental Protection Agency, Office of Solid Waste and Emergency Response, Toxics Integration Branch, Washington, D.C., by Tetra Tech, Inc., Baltimore, Maryland.

USEPA (1993b) *Virginian Province Demonstration Report: EMAP-Estuaries: 1990.* Report No. EPA/620/R-93/006. US Environmental Protection Agency, Office of Research and Development, Washington, D.C.

Xu, X., and Pang, S. (1992) Briefing of activities relating to the studies on the environmental behavior and eco-toxicity of toxic organics. *J. Environ. Sci. (China)* **4**(4), 3–9.

3 Methods to Study Chemical Effects

Once the objectives and relevant scales of an environmental risk assessment have been determined (see Chapter 2), the data collection plan must be designed in a manner that will help establish linkages between contamination and observed adverse effects. Many tools are available to assist the environmental manager in assessing environmental condition. These include chemical analyses, toxicity tests, field surveys and assessments, and special studies such as biomarkers and simulated ecosystem studies (i.e., mesocosms or microcosms). Any one of these approaches may provide the information needed to assess environmental conditions; however, the use of several of these approaches in combination may increase the level of confidence in the assessment. This chapter presents an overview of the approaches that may be used to assess the ecological conditions of both aquatic and terrestrial ecosystems.

3.1 METHODS SELECTION

After the study objectives have been determined and a design is in place, measurement endpoints or indicators must be selected. Indicators provide data about the assessment endpoints determined to be most relevant to the study objectives. Methods to measure indicators can be grouped into several categories: field-ecological surveys of the contaminated site; chemical analyses of samples from the site; and toxicity tests of site media (water, soil, sediments) either in the laboratory or on site (*in situ*). Special studies such as bioaccumulation tests and simulated ecosystem studies (e.g., microcosms or mesocosms) may also be used.

Field surveys can evaluate changes in populations or communities of plants or animals that may be the result of contamination. Preliminary field studies conducted before an entire study is undertaken are recommended, as they may provide information about variability within the study area, thereby helping to define the study design. They may also provide general information on contamination and impacts to biota. Such information helps guide the selection of sampling sites, ecological receptors, and sampling methodologies. Chemical analyses can help determine the extent and quantity of contaminants at the site, while toxicity testing may provide a direct indication of the effects of contamination on site biota. Models can be used to predict the probable fate and transport of

Methods to Assess the Effects of Chemicals on Ecosystems
Edited by R. A. Linthurst, P. Bourdeau, and R. G. Tardiff
 Published by John Wiley & Sons Ltd

chemical contaminants, and, thereby, help determine the most useful plant and animal species for toxicity testing and other special studies. Models and other approaches may also be used to link measured responses in receptors of concern (toxicity test data) to population-level responses (field surveys), and to obtain information on the spatial extent and chemical components of contamination (chemical analyses). The use of each type of test and the careful integration of the data generated help to establish linkages between site impacts and adverse effects observed in biota.

3.2 OVERVIEW OF METHODS

Methods applicable to the evaluation of chemical effects in both terrestrial and aquatic systems are discussed in the following sections.

3.2.1 CHEMICAL ANALYSES OF SAMPLES

Chemical analyses are an integral component of any chemical effects study. Once preliminary studies have identified the sites to be sampled, chemical analyses of samples (soil and sediment, water, air, or biota) can be conducted to determine concentrations of various contaminants. In some cases, these data may be compared with those collected from an uncontaminated reference site. Chemical analyses are the first step in establishing linkages between adverse effects observed in biota and specific contaminant levels. Information about the total and bioavailable elemental concentrations, the concentrations of organic and inorganic ionic species, and the concentration of molecular organic species added to the environment by various human activities (technological, agricultural, recreational) must be collected.

Significant advances in analytical chemistry over the past 20 years have provided scientists with numerous methods to characterize environmental contamination. These methods enable researchers to perform rapid and accurate multielemental analyses at a reasonable cost. Similarly, modern organic analytical methods such as gas chromatography (GC), mass spectrometry (MS), high performance liquid chromatography (HPLC), and various combinations of these technologies have enabled chemists to detect very low levels (as low as parts per billion or trillion) of organic environmental contaminants routinely. Some more commonly used analytical methods are presented in Table 3.1.

The level of complexity of chemical analysis varies with the medium being sampled. Water is perhaps the simplest medium to analyse for chemical contaminants. Even though the concentrations of contaminants in a water sample may be low, the sample can be concentrated before analysis to improve detection and quantification. Analysis of soil and sediment samples may add another level of complexity, along with the difficulty of assuring homogeneity of the sample.

Analysis of biological tissue (plant or animal) for chemical residues is frequently the most complicated and most expensive task because of the complexity of the tissue matrix. The information obtained from a residue study, however, may be very valuable in ascertaining whether contaminants are bioavailable to biota. Residue or bioaccumulation studies are discussed in greater detail below.

Table 3.1. Major analytical approaches to detect common environmental chemical stressors

Stressor Class	Major Methodologies
Elemental	Atomic Absorption Spectroscopy (AAS) Inductively Coupled Plasma–Atomic Emission Spectroscopy (ICP-AES) Inductively Coupled Plasma–Mass Spectrometry (ICP-MS)
Ionic	Ion Selective Electrodes (ISE)
Molecular organics	Gas Chromatography (GC) Liquid Chromatography (LC) High Pressure Liquid Chromatography (HPLC) Gas Chromatography–Mass Spectrometry (GC-MS) High Pressure Liquid Chromatography–Mass Spectrometry (HPLC-MS) Gas Chromatography–Fourier-Transform Infra-Red Spectroscopy) (GC-FTIR) Super-Critical Fluid Chromatography (SFC) Nuclear-Magnetic Resonance (NMR)
Mineral	X-ray Diffractometry (XRD)
Polymer	Mass Spectrometry (MS) Liquid Chromatography–Mass Spectrometry (LC-MS)

If contaminants of interest are likely to be dispersed through the air, they should be monitored. Airborne contaminants include volatile organic compounds, gaseous compounds, and particulates that adhere to airborne particles. However, in many settings including some industrial settings, airborne contaminants are typically present only in low concentrations, thus requiring that large volumes of air be sampled over long periods of time.

Proper procedures to collect, handle, and store samples are also critical to obtain useful analytical data. Samples must be collected with the appropriate containers, and stored correctly to prevent contamination or loss of the contaminant by leaching or binding to the container and to ensure that metabolic processes have been halted.

3.2.2 TOXICITY TESTS

Toxicity tests provide a direct measure of the bioavailability of toxicants, and, when combined with chemical analyses and field surveys, can help establish linkages between site contamination and adverse ecological effects. Toxicity tests evaluate acute, subchronic, and chronic exposures and measure biological endpoints such as mortality, reproductive performance, growth, and behavioural changes. Also by using toxicity tests, the relative toxicity of a mixture of chemicals can be assessed by taking into account synergistic or antagonistic interactions among chemicals.

The major disadvantage of traditional laboratory-based toxicity tests is the difficulty in extrapolating effects observed in the laboratory to those observed in the wild. *In situ* (literally, "in position") toxicity assessments can help to address this problem. Numerous toxicity tests can be adapted for field use to evaluate the exposure of test organisms in site media, under "normal" conditions encountered in the field. Although *in situ* toxicity assessments are not as standardized as are laboratory toxicity tests, they are increasingly prominent in the ecological assessment process.

Conducting parallel toxicity tests in the laboratory and in the field may provide stronger inference of linkages between toxicity and contaminant exposure and reduce the problems associated with laboratory-to-field extrapolations of toxicity data. If the same effects are observed in both types of tests, a stronger case can be made for a causal relationship between the two.

Toxicity tests may be used in both aquatic and terrestrial habitats; however, procedures for aquatic toxicity testing are more developed than those for terrestrial toxicity testing. Traditional toxicity tests involve a single species, and measure endpoints such as mortality, growth, and reproduction. Numerous types of organisms are used in toxicity testing, including vertebrates (rodents, fish, and birds), invertebrates (plankton, amphipods, and insects), microbes, and plants (aquatic and terrestrial; vascular and non-vascular). Environmental matrices that can be tested include water, sediment, and soil. Test exposures may be either acute, subchronic, or chronic. Aquatic toxicity tests for acute effects typically require about four days (representing a small portion of the lifetime of the organism); tests for chronic and subchronic effects range from 7-30 days (representing a much larger portion of the lifetime of the organism). Terrestrial toxicity tests range from hours to days for acute effects to weeks and months for chronic and subchronic effects. Examples of specific aquatic and terrestrial toxicity tests are found below.

3.2.3 FIELD ASSESSMENTS

Field surveys of terrestrial and aquatic habitats can complement chemical analyses and toxicity testing, and may decrease the uncertainty in the assessment process by providing direct measures of impacts on site biota. Field surveys of contaminated sites provide information about the extent and patterns of contamination, and may help identify sites to sample for chemical analysis. Surveys can indicate the presence of sensitive plant and animal species that may be affected by contamination and also help identify potential species for further study (i.e., toxicity testing). Field surveys also provide information about the effects of contaminants on the structure and function of populations and communities at a site when data from a contaminated site are compared with data from carefully selected reference sites.

Community-level field studies generally provide the most information about the biological integrity of a system, while permitting examination of individual taxa within the community. Communities may be assessed from either a structural or functional perspective. An assessment of community structure defines biotic characteristics (e.g., abundance, diversity, and species composition) at a specific point in time, whereas an assessment of community function measures the rate of biological processes (e.g., species colonization rates and nutrient cycling) of the ecosystem.

The use of community-level studies in environmental monitoring is normally performed from a structural perspective, because structural studies take less time, are technically less complex, and facilitate comparisons with data from other kinds of studies. Community-level studies, however, are frequently difficult to interpret, and the causal links with contaminant exposure may not be readily apparent. Contamination often is not the only factor influencing community structure. Natural environmental factors such as temperature, moisture, pH, nutrient availability, and predator–prey relationships also affect community structure. Examples of specific field assessment methods for the aquatic and terrestrial environments are presented below.

3.2.4 SPECIAL ANALYSES

Special analyses such as bioaccumulation studies, biomarkers, or simulated ecosystem studies provide valuable information for an ecological assessment. Although these kinds of studies are not typically used in ecological risk assessments, they can be valuable when combined with information from other parts of the assessment process. Examples of some special analyses include bioaccumulation testing of dredged material for ocean disposal or mesocosm studies for pesticide registration.

3.2.4.1 Bioaccumulation studies

Bioaccumulation studies evaluate the net accumulation of a chemical in an organism through consumption of food or water containing the chemical. Bioaccumulation occurs when the rate at which an organism ingests a chemical exceeds the rate at which it excretes the substance. Bioaccumulation tests measure the actual uptake of a contaminant by organisms, and are especially useful when the contaminants at a site have high bioconcentration factors, such as PCBs or dioxins. Bioaccumulation studies may involve residue studies (i.e., measuring chemical residues in the tissues of organisms from the site) or a controlled study such as measuring residues in organisms exposed to a contaminated medium for a specific length of time, (i.e., a laboratory toxicity test or *in situ* bioconcentration study).

Bioaccumulation studies can provide direct measurements of contaminant bioavailability, whereas chemical analyses of soil and sediment cannot. They can also be used to evaluate potential human health risks when used to analyze flora or fauna that may be consumed by humans. Such studies can be difficult to interpret, however, because body burdens of a chemical residue frequently are not directly correlated with adverse effects. Other challenges encountered in bioaccumulation studies include the natural variability between individuals and within a population, interaction between contaminants, and biotransformation within the organism.

3.2.4.2 Biomarkers

A biomarker is a biological measure of an organism's response to a contaminant. Biomarkers are measurements of biological tissues, fluids, or cells that can be used to determine if an organism has been exposed to a contaminant or if a contaminant has caused biological changes. Biomarkers have been used for many years to assess human exposure and effects but only recently have they been applied to ecological assessments.

Biomarkers used to assess aquatic or terrestrial organisms generally evaluate physiological, histological, or biochemical parameters. They may provide information not available from residue studies, chemical analyses, toxicity tests, or ecological community studies. By measuring an actual exposure or biological response to a contaminant, biomarkers integrate the temporal exposures of the organism and the multiple pathways of exposure. Also, because most biomarkers provide an estimate of exposure or alterations at organismal or suborganismal levels, the responses observed are recorded after a compound has been metabolized or transformed within the organism, thereby providing a more realistic measure of exposure or changes.

Presently, few biomarker protocols are widely accepted for use with aquatic or terrestrial organisms, partly because clear causal relationships have not been established for many classes of contaminants. Furthermore, many responses to biomarkers may be produced by environmental influences such as seasonality; and, as with any biological population, intra-specific variability is always a confounding factor. Further discussion of the use of biomarkers in aquatic and terrestrial environments is provided below.

3.2.4.3 Microcosm and mesocosm studies

Microcosm and mesocosm tests are useful intermediates between bioassays and ecosystem experiments. They provide controlled experimental conditions in the laboratory or the field to study changes at any level (population or community or ecosystem) of a chemical or other stressor. Microcosm studies are generally small, contain a few species, and are conducted indoors, whereas mesocosm tests are relatively large, contain most or all the species from an ecosystem, and are usually conducted in outdoor settings. Given the expense and effort required to establish and maintain mesocosms, microcosms are often used when only one or a few species are required for a test.

Microcosm studies offer several advantages over mesocosm studies and field surveys. Multispecies microcosm studies provide greater ecological "realism" than single-species tests or basic multispecies tests. Microcosm tests are more space-efficient than mesocosm or field studies, and are easier to maintain under uniform conditions essential to replicate and standardize experimental procedures. Furthermore, the chemical, biological, and physical effects of a substance can be determined in one test system, rather than several. Unlike in field tests, microcosms eliminate the chance of contaminating the natural environment. Most importantly, microcosms enable researchers to observe the integrated effects of contaminants on community and ecosystem functions and pathways.

Data obtained in microcosm studies must be used with caution, however, due to the limitations of such studies. Most limitations result from the fact that a microcosm is an intentionally simplified representation of an ecosystem. Specific limitations include (1) extrapolating observations to the broader environment; (2) the absence of selected components of an ecosystem (i.e., lack of atmospheric deposition); and (3) the use of small population sizes, which may lead to chance extinctions.

Mesocosm tests are simulated field studies conducted in controlled environments such as artificial ponds or streams, large outdoor tanks, or littoral enclosures in a natural water body. Mesocosm tests also may be conducted in terrestrial habitats to evaluate the effects of chemical contaminants on vegetation. Mesocosm tests employ fully functional ecosystems to predict ecological effects or environmental fate processes and, as such, may be the best possible method to obtain information on how a stressor reacts to actual environmental conditions. A disadvantage of mesocosm studies is the difficulty in discriminating between effects caused by the stressor of interest and the natural variability of the ecosystem or community in question. Mesocosm studies are quite expensive and time consuming; consequently, their usefulness in the ecological risk assessment process may be limited.

3.3 ASSESSMENT OF CHEMICAL IMPACTS TO AQUATIC ECOSYSTEMS

A summary of the methods currently available to measure, diagnose, and quantify the effects of chemicals on aquatic (freshwater, estuarine, and marine) systems is provided below.

3.3.1 TOXICITY TESTS

Aquatic toxicity tests have been used extensively for hazard assessment for over 20 years. Many standardized methods have been developed, and published by the US Environmental Protection Agency (EPA), the US Army Corp of Engineers (COE), the Organization of Economic Cooperation and Development (OECD), and the American Society for Testing and Materials (ASTM). Tables 3.2 to 3.7 list commonly used protocols for acute and chronic toxicity tests for use with fish and invertebrates from marine and freshwater environments.

Table 3.2. Some freshwater acute toxicity tests

Species	*Ceriodaphnia dubia, Daphnia pulex and Daphnia magna*, fathead minnow, rainbow trout
Endpoint	Mortality
Duration	24, 48, or 96 hours
Temperature (°C)	20 or 25 for *Daphnia* and minnow; 12 for trout
Conditions	Static non-renewal and renewal, flow-through
Level of effort	Low
Citation	USEPA, 1991b

The most frequently used laboratory toxicity tests relate the concentration of the chemical in water to the time of death or some other observed manifestation in the test organism. Organisms used in aqueous aquatic toxicity tests include many species of fish, invertebrates, and algae. Commonly used fish species include: fathead minnow (*Pimephales promelas*) in freshwater systems, and the sheepshead minnow (*Cyprinodon variegatus*) or silverside (*Menidia* sp.) in marine and estuarine systems. Commonly used invertebrates include *Daphnia* sp. or *Ceriodaphnia dubia* in freshwater and the mysid shrimp (*Mysidopsis bahia*) in marine systems. Amphipods, such as *Hyalella azteca* in freshwater and *Ampelisca abdita* in saltwater, are frequently used to assess sediment toxicity.

Table 3.3. Some estuarine and marine acute toxicity tests (USEPA, 1991b)

Species	Mysid shrimp (*Mysidopsis bahia*), sheepshead minnow (*Cyprinodon variegatus*) and silverside (*Menidia* sp.)
Endpoint	Mortality
Duration	24, 48, or 96 hours
Temperature (°C)	20 or 25
Conditions	Static non-renewal, static renewal, and flow-through
Level of effort	Low

Table 3.4. Some freshwater chronic toxicity tests (USEPA, 1989)

Species/test	1. Fathead minnow larval survival and growth test 2. Fathead minnow embryo larval survival and tetratogenicity test 3. *Ceriodaphnia dubia* survival and reproduction test 4. Algal (*Selenastrum capricornutum*) growth test
Duration	7 days for tests 1, 2, and 3; 96 hours for test 4
Temperature (°C)	25
Conditions	Static renewal for tests 1, 2, and 3; static non-renewal for test 4
Level of Effort	Low

The duration of typical toxicity tests with aqueous samples ranges from four days for acute effects with an endpoint of mortality to 7 to 30 days for chronic and subchronic effects on survival, growth, or reproduction. Life-cycle tests are also needed, but are not frequently used, because the duration is long and the cost is high. Test durations to assess sediment toxicity typically range from 10 days or less for acute effects to 30 days for chronic effects. Test endpoints may include survival, reproduction, or emergence.

3.3.2 FIELD ASSESSMENTS

Field assessments of aquatic habitats may include surveys of all populations in the aquatic community including microbes, periphyton, plankton, macroinvertebrates, fish, and macrophytes. Community-level studies provide the most information about the biological integrity of an aquatic system, and allow an evaluation of individual species within the community. Either community structure or function may be assessed; however, structural evaluations are conducted more often, because they take less time, are less complicated, and produce easily interpreted data. A major consideration when assessing community structure is the possible effect by normally variable environmental factors such as salinity, temperature, or shading, as well as by contamination.

Table 3.5. Some estuarine and marine chronic toxicity tests

Species/test:	1. Sheepshead minnow or Island Silverside larval survival and growth test 2. Sheepside minnow embryo/larval survival and tetratogenicity test 3. *Mysidopsis bahia* survival, growth, and fecundity test 4. Sea urchin fertilization test 5. Algal sexual reproduction test
Duration:	7 days for tests 1, 2, and 3; 1.3 hours for test 4; 7–9 days for test 4
Temperature (°C):	25 for tests 1 and 2; 26–27 for test 3; 20 for test 4; 22–24 for test 5
Conditions:	Static renewal for tests 1, 2, and 3; static non-renewal for tests 4 and 5
Level of Effort:	Medium for tests 1, 3, 4, and 5; high for test 4
Citation:	USEPA, 1988

Any combination of taxonomic groups (algae, invertebrates, or vertebrates) and level of biological organization (individual, population, community, or ecosystem) can be used to assess the health of an aquatic system. Benthic invertebrates are used often in aquatic community studies, because as a group they integrate the effects of present and past conditions; they are generally abundant, relatively immobile, and have relatively long life cycles; and as a group, their ecological relationships are well studied. In addition, sampling procedures are well developed, and a single sampling often collects a considerable number of species from a wide range of phyla. Examples of community structure and function parameters that are used to assess aquatic communities are provided in Table 3.8. Specific procedures to assess the structural or functional parameters of aquatic ecosystems can be found in most aquatic ecology methods manuals such as that of Wetzel and Likens (1990), of EPA (1989), or of ASTM E1383 (1993).

3.3.3 SPECIAL ANALYSES

3.3.3.1 Bioaccumulation studies

Chemicals may enter organisms via food and sediment uptake or by uptake from the water across external membranes and gills. The three basic processes by which contaminants accumulate in aquatic species are bioconcentration, bioaccumulation, and biomagnification. Herein, bioaccumulation is used to represent all three pathways of contaminant uptake.

A first-level bioaccumulation study is used to determine residues in the biota. This study is uncontrolled (i.e., previous contaminant exposure is unknown); however, the study can provide background information, indicating whether the chemicals of interest can be taken up by biota at the site. Its findings are useful in the planning future studies.

A more involved approach to study bioaccumulation is to conduct *in situ* studies such as stream cage studies. These studies typically involve fish or benthic macroinvertebrates enclosed by a cage either attached to the substrate, suspended in the pelagic zone, or floating. These tests have the advantage of providing more "real-world" conditions; that is, contaminant accumulation proceeds at its normal rate (i.e., impacted by biotransformation and other fate processes) under normal temperature, light, and other exposure parameters. However, the advantages to this type of study are also some of its disadvantages. Many variables cannot be controlled (e.g., pollution slugs, extremely high tides or flows, temperature, light, and food availability). These make the test a more reliable estimator of the real world, but also add additional covariates that in turn make the data more difficult to interpret. Also the potential for escape of test organisms or a predator somehow entering the test enclosure is present. The logistics and costs of these studies also may be quite high.

Controlled environments such as mesocosms, microcosms, and artificial streams also offer a more "real-world" environment, although not to the same degree as *in situ* studies. They have many of the same advantages and disadvantages as *in situ* studies. The degree of similarity of controlled ecosystem studies to "real-world" conditions is dependent on the scale of the study (from laboratory flask on up to farm ponds) and its design.

The simplest bioaccumulation study is a laboratory study that tests one species at a time in a defined medium. The simplest of these tests are bioconcentration tests with fish or invertebrates. These tests are typically conducted with single chemicals or well-defined mixtures; however, they may also be modified for contaminated site media such as water or sediment. For instance, the US EPA may require a fish or oyster bioconcentration test for registration of a new chemical under the Toxic Substances Control Act (TSCA) (USEPA, 1992b) or in the pesticide registration process under the Federal Insecticide, Fungicide and Rodenticide Act (FIFRA) (USEPA, 1993b). Data obtained from these tests are used to develop a bioconcentration factor (a unitless value that indicates the degree to which a chemical concentrates in tissue with respect to the concentration in the surrounding medium [water]). The basic premise behind these tests is to expose the test organisms to a known chemical concentration until a steady state is attained (when uptake and depuration are equal) or for a designated time period (ASTM, 1988).

Table 3.6. Some freshwater sediment toxicity tests (ASTM E1383, 1993)

Species:	1. Amphipod (*Hyalella azteca*) 2. Midges: *Chironomus tentans*, *Chironomus riparius* 3. *Daphnia magna* and *Ceriodaphnia dubia* 4. Mayflies (*Hexagenia* spp.)
Endpoints:	1. Number of young; survival, growth & development; reproductive capacity 2. Larval survival and growth, adult emergence 3. Survival and reproduction 4. Mortality, growth, burrowing behaviour, moulting frequency
Duration:	10–30 days for tests 1 and 2; 2–7 days for test 3; 7–21 days for test 4
Temperature (°C):	20–25 for test 1; 20-23 for test 2; 25 for test 3; 17-22 for test 4
Conditions:	Static for all tests; flow-through for tests 1 and 2; recirculating for test 4
Level of effort:	Medium for all tests

Testing of dredged material is another application to assess the bioaccumulation potential of a substance. These methods have been developed by the U.S. Army

Corp of Engineers (1991) and the US EPA to assess the environmental impacts of dredged material dumped into ocean waters (USEPA, 1991a). Methods have also been recently developed to determine the environmental impact of dredged materials dumped into inland and near-coastal waters (USEPA, 1993a). These tests are also applicable to bioaccumulation testing of sediments. In general, these methods consist of exposing an organism (generally a benthic macroinvertebrate) to a quantity of sediment for a designated time (generally 10 days for metals and 28 days for organics or organometallics). Steady state bioaccumulation can then be determined by taking samples at designated intervals from the test chambers (USEPA, 1991b, 1993a).

Table 3.7. Some marine and esturine sediment toxicity tests (ASTM E1383, 1993)

Species:	1. Amphipods 2. Fish, crustaceans, zooplanctons, or bivalves 3. Infaunal amphipods, burrowing polycheates, mollusks, crustaceans, or fish
Material:	1. Whole sediment 2. Dredged material (elutriate) 3. Dredged material (whole sediment)
Endpoints:	1. Mortality, emergence, renurial 2. Mortality 3. Survival
Duration:	10 days for tests 1 and 3; 2 days for zooplancton and fish larvae in test 2 and 4 days for bivalves and crustaceans in test 2
Temperature (°C):	20–25 for test 1; 20–23 for test 2; 25 for test 3; 17–22 for test 4
Conditions:	Static for all tests; flow-through for tests 1 and 2; recirculating for test 4

3.3.3.2 Biomarkers

Biomarkers are biochemical, physiological, or cellular responses to a stressor. They may provide an indication of exposure to a stressor or of its consequences. They may be used in both field and laboratory settings; and, because the stress is sublethal, they are environmental monitors that can provide early warnings of environmental undesired consequences. Biomarkers such as metallothionein (metal binding proteins) or induction of the cytochrome P_{450} system (mixed function oxidases) have been used as research tools for several years to assess aquatic habitats. However, biomarkers have been employed only recently as environmental monitors.

The measurement of tissue residues (via bioaccumulation by biota) is the most direct biomarker of exposure. However, because some chemicals do not bioaccumulate, this measurement does not take into account toxic effects on the organism, excretion, or biotransformation. Also, the observed effects may not have biological significance. Most other "typical" biomarkers are indirect measures of exposure or stress that describe either a biological response or exposure that is of biological or toxicological significance. Examples of some commonly used biomarkers in the assessment of aquatic habitats are provided in Table 3.9.

Table 3.8. Community parameters assessed by aquatic field assessment methods (USEPA, 1989)

Structural parameters	Functional parameters
Abundance	Growth rates
Species richness	Metabolism
Biomass	Nutrient cycling
Indicator species	Primary production
Indices	
Guild structure	

Biomarkers can provide an added level of detail to an assessment of the aquatic environment, and can more closely infer a causal relationship, thereby decreasing the uncertainty in the conclusions. However, many of these biomarkers are still in the research stage, and, thus, must be used cautiously; a baseline uncontaminated site for comparison may also be required.

3.3.3.3 Controlled ecosystem studies

Controlled ecosystem studies have been used for many years to assess the impacts to aquatic environments, although the focus has been primarily on single chemical impacts. For instance, the US EPA Office of Pesticide Programs previously required simulated field testing (i.e., mesocosm studies) for many pesticides in the registration process. Simulated environment studies of aquatic environments include, in order of size and complexity, microcosms, mesocosms, artificial streams and ponds, and limnocorrals. As size is increased, the more ecosystem components can be added, and a closer replication of a natural system is attained. However, the more complex the model, the more covariants are obtained, and hence the greater the difficulty in interpreting the data. Also, as the complexity of the model increases, the cost and amount of effort involved increases. The ability to replicate systems also becomes increasingly more complex as one moves up the scale in size. However, even with the potential problems involved with the use of the large systems, the information obtained from them can be invaluable in the hazard and risk assessment processes.

Table 3.9. Examples of biomarkers (adapted from Huggett *et al.*, 1992; McCarthy and Shugart, 1990; and USEPA, 1989)

TYPE	EXAMPLES	CHEMICAL
Enzyme & Protein Inhibition	Mixed function oxidases	PAHs, PCBs, petroleum hydrocarbons
	Metallothioneins	Metals
	Cholinesterases	Carbamates, organophosphates
	δ-aminolevulinic acid dehydrase	Lead
Gross Indices	Skeletal abnormalities	Most chemicals
	Histopathology	Most chemicals
Genotoxicity	DNA adduct formation	PAHs
	Strand breakage	PAHs
	Micronucleus formation	DDT, PCBs, PAHs
	Oncogene activation	Carcinogens
	Mutations	Mutagens

The simplest controlled ecosystem study is a microcosm. These studies can range from flask size to aquarium size, and are generally conducted in a laboratory. They range from single species to multispecies systems; but, because they are one step up from a single-species standard laboratory bioassay, their resemblance to the real world is limited. However, recent research has increased the level of complexity in aquatic microcosms.

The standardized aquatic microcosm (SAM), developed by Frieda Taub and associates at the University of Washington, is a gnotobiotic (all organisms in the system are known), mixed flask culture system. This type of system differs from many other microcosms in that a specific ecosystem is not being replicated; rather, general ecosystem processes and functions are being tested. The idea behind the development of this test is to derive a "white rat" model that can be used by investigators worldwide, and can yield easily reproducible results that can be compared and then used in the hazard assessment process (Giesy and Allred, 1985).

To have a standardized aquatic microcosm, the test must be site independent and reproducible, and the water, organisms, and facilities must be available to all (Taub, 1985). This test uses distilled water to which a growth medium and laboratory cultured organisms are added (Taub, 1985). The organisms are easily distinguishable at ×40 magnification, have short lifespans, and have been previously used as toxicity testing organisms (Taub, 1985). The test uses fixed numbers of 10 algal species, cladocerans, amphipods, ostracods, protozoans, and rotifers as well as bacteria used as food for the algae and other microorganisms introduced with the zooplankton cultures (Taub and Crow, 1980). The logic is that by initiating microcosms with ample nutrients and small numbers of organisms, the spring–summer behaviour of a temperate aquatic community is being simulated (Taub, 1985). Re-inoculation of organisms is also employed to prevent random extinctions and to mimic immigration (Taub *et al.*, 1986). The test is conducted in gallon jars (24 jars with four treatment groups of six replicates each) under a standard light and temperature regime (Taub, 1985).

The test is run for 56 days after a seven-day assimilation period (Taub, 1985). The biology of the system has been developed to the point that all untreated containers show nitrate depletion, followed by algal increase that is terminated by a *Daphnid* population increase and an eventual crash of the Daphnid populations (Taub, 1985). Population dynamics as well as ecosystem level variables are monitored. However, measurements of community metabolism such as dissolved oxygen and chlorophyll-A are less variable than the organism counts (Crow and Taub, 1979). Inter- and intra-laboratory reproducibility, relying primarily on copper compounds, has been high (Giesy and Allred, 1985; Taub, 1985; Taub *et al.*, 1986).

The use of SAM as a screening microcosm, not simulating any specific ecosystem, has been supported by other investigators (Haque, 1980). However, it also has been criticized, because of its cost (Shannon *et al.*, 1986) and the fact that artificial communities used in a SAM may not be representative of "natural, co-adapted species assemblages" and thus may not be reliable for studies of ecosystem-level properties (Hammons, 1981).

Mesocosm or simulated field studies are large multispecies controlled ecosystem studies. They may include large pond systems, large outdoor tanks, artificial streams, and enclosures in the pelagic or littoral zone of an aquatic system. Mesocosms are functioning ecosystems, albeit on a smaller scale, and, therefore, provide a measure of both direct and indirect toxicity. However, the disadvantages of mesocosm tests are high logistics to cost, the ability to replicate, and, due to the high numbers of endpoints, difficulty in interpretation of findings (Graney *et al.*, 1994). Mesocosm tests have been used quite extensively in the four-tier FIFRA hazard evaluation process for pesticides.

Tier-4 testing consists of mesocosm (simulated field studies) and actual field studies, and is required on a case-by-case basis depending on the exposure potential and toxicological hazard of the pesticide (Bascietto, 1990; USEPA, 1982). Guidance criteria for registrants on conducting mesocosm testing of pesticides are available (Touart, 1988); however, no standardized methods for mesocosm testing exist, due to the complexity of mesocosms and the specific requirements of each testing situation.

3.4 METHODS TO ASSESS CHEMICAL IMPACTS ON TERRESTRIAL ENVIRONMENTS

Chemically contaminated terrestrial habitats are evaluated in much the same manner as aquatic habitats. After definition of the study objectives, three types of data must be obtained and integrated: those from chemical analyses of samples, toxicity tests (both in the laboratory and in the field), and field or ecological surveys of biota. In addition, incorporation of biomarker studies or controlled ecosystem studies may be appropriate to help diagnose the causes of observed effects on terrestrial habitats and to establish the causal relationships linking contamination to observed changes. The special challenges of studying ecosystems that experience climatic extremes are discussed in later chapters of this book. The most commonly used toxicity tests, field methods, and special analytical techniques for assessing chemical impacts in terrestrial habitats are discussed in this section. Relevant features of the most frequently used toxicity tests are presented in Table 3.10 (tests using vertebrate and invertebrate animal species and soil microbes) and Table 3.11 (tests using vegetation).

3.4.1 TOXICITY TESTS

Terrestrial toxicity tests may be conducted with vertebrates and invertebrates, vegetation, or soil microbes in both laboratory and field situations. Generally, the tests measure toxicity by directly exposing test biota to media samples collected from the site, or indirectly by exposure to eluates (water that has been filtered through soil or sediment samples to remove water-soluble constituents) or leachates prepared from site samples. Most often these tests evaluate acute toxicity on the population level; however, some tests using higher animals evaluate the effects of chronic toxicity on individuals. Many methods evaluate the condition of soils as an indication of chemical toxicity; because soils provide the essential foundation of terrestrial ecosystems, the integrity of the soil can provide an indication of the integrity of the entire ecosystem. References that discuss the methods in greater detail and standardized protocols, where available, are presented in Tables 3.10 and 3.11.

3.4.1.1 Invertebrate toxicity tests

Tests using invertebrates (earthworms and insects)

Due to their essential functions in ecosystems, soil invertebrates are useful targets to assess the ecological effects of chemicals. Invertebrate tests measure acute toxicity at the species level (earthworm tests) and the population and community levels (soil insect tests). Primary endpoints are survival, growth (measured as biomass), reproductive success, and behavioural changes. The tests can be used in most habitats, are applicable to a wide range of contaminants, can be conducted fairly rapidly, and offer a range of cost options. Table 3.10 provides more information on these types of tests.

Tests using soil microbes

Soil bacteria and fungi have critical roles in the cycling of carbon, nitrogen, sulphur, and phosphorus, and make substrates available in forms that higher plants and animals can utilize. Because of their unique role in stable ecosystems, the physiological functioning of soil microbes can be very effective in studying the effects of chemicals on terrestrial ecosystems. Toxicity tests using soil microbes measure changes in microbial metabolism, respiration, and nitrogen cycling of soil bacteria and fungi after exposure to contaminants. Short-term microbial toxicity tests are technically simple, rapid, and relatively inexpensive procedures; standard protocols for some tests are available commercially. Table 3.10 provides more information on these tests.

3.4.1.2 Vertebrate toxicity tests

Vertebrate toxicity tests evaluate acute, subacute, and chronic toxicity of chemicals by describing survival and growth (amphibian tests), reproductive success (amphibian, small mammal, and avian tests), and body burdens (small mammal and avian tests) of the test species. All media and most chemicals can be tested by these tests that generally have longer exposure periods than the invertebrate tests previously discussed. The endpoints are easily understood, and are relevant to economically important higher animals. Feeding studies (small mammal and avian toxicity tests) are especially useful to determine the potential uptake of contaminants into food webs and potential human exposure route (if animals tested are representative of a possible human food source). Cellular level tests are available that provide an indication of chemical effects on immune function and genetic material of test species. These tests, while technically complex, provide information about the potential hazards to humans. Standard protocols, many adapted from veterinary medicine, exist for most tests.

Table 3.10. Vertebrate, invertebrate, and microbial test methods to assess the toxicity to terrestrial ecosystems

Test/species	Chemical sensitivity	References
Earthworm survival *Eisenia foetida, Lumbricus terrestris*	Water-soluble chemicals, metals, pesticides, organics, mixtures	Callahan *et al.*, 1985; Edwards, 1983; Goats and Edwards, 1982
Insect tests Ants, crickets, fruit flies, mites, beetles	Pesticides, chemical mixtures (not for metals or herbicides)	Gano *et al.*, 1985; OECD, 1984; James and Lighthart, 1990
Amphibian tests *Xenopus laevis*	Metals, pesticides, organics	ASTM E1439
Small mammal tests Rodents, voles, ferrets	Any substance capable of contaminating feed stocks	ASTM protocols: 552, 555, 593, 757, 758, 1103, 1163, 1372, 1373
Avian tests Bobwhite, quail, mallard, pheasant	Any substance capable of contaminating feed stocks	ASTM E857 and E1062
Vertebrate immunotoxicity Birds and mammals	Selenium, pentachlorophenol	Rose and Friedman, 1976; Oppenheim and Schechter, 1976; Gewurz and Suyehira, 1976
Invertebrate immunotoxicity Earthworms	PCBs	Stein and Cooper, 1988; Eyambe *et al.*, 1990; Rodriguez-Grau *et al.*, 1989
Chromosomal aberration tests Small mammals residing on site	Any known genotoxicant	Brusick, 1980; McBee *et al.*, 1987
Bacterial luminescence test *Photobacterium phosphoreum*	Metals, pesticides, herbicides, volatile and semi-volatile organics, hydrocarbons	Bulich, 1982, 1986; Ribo and Kaiser, 1987; Ahn and Morrison,1991
Soil biota metabolic activity Soil bacteria and fungi	Metals	Burns, 1986; Ladd, 1985; Nannipieri *et al.*, 1986a, 1986b
Soil biota respiration rates Soil bacteria and fungi	Metals and pesticides	Doelman and Haanstra, 1984; Dumontet and Mathur, 1989
Soil biota nitrogen cycling Soil bacteria and fungi	Insecticides, herbicides	Parr, 1974

Table 3.11. Vegetation toxicity test methods to assess chemical impacts to terrestrial ecosystems

Test/species	Chemical sensitivity	References
Seed germination test: Lettuce *Lactuca sativa*	Metals, insecticides, herbicides, volatile and semi-volatile organics, hydrocarbons	US Code of Federal Regulations, 1985; USFDA, 1987b; Gorsuch *et al.*, 1990; Linder *et al.*, 1990; USEPA, 1989, 1992
Root elongation test: Lettuce, *Lactuca sativa*	Metals, insecticides, herbicides, volatile and semi-volatile organics, hydrocarbons	US Code of Federal Regulations, 1985; USFDA, 1987b
Seedling growth tests: Purchased lettude seeds or site-specific collected seeds	Metals, insecticides, herbicides, volatile and semi-volatile organics, hydrocarbons	US Code of Federal Regulations, 1985; USFDA, 1987c; OECD, 1984
Whole plant toxicity tests: Purchased lettuce seeds or site-specific collected seeds	Highly mobile, water-soluble compounds	Pfleeger *et al.*, 1991
Vascular plant toxicity tests: Plants from purchased seeds (cress, mustard) or site-specific collected seeds	Water-soluble compounds only	Ratsch, *et al.*, 1986; Shimabuku *et al.*, 1991
Photosynthetic inhibition tests/ chlorophyll fluorescence assay: Terrestrial plants	Water-soluble compounds only (if using soil eluate); all types of substances evaluated in field	Judy *et al.*, 1990, 1991; Miles, 1990

3.4.1.3 Vegetation toxicity tests

The vegetation toxicity tests discussed herein are used to test chemicals on crop species. Using eluates, the tests evaluate acute and subchronic toxicity both directly and indirectly with variable exposure periods (5 to 90 days), and they can be used in most habitats. The primary endpoints are survival (seed germination test), growth (seedling growth test and root elongation test), reproduction success (vascular plant toxicity tests), and photosynthesis rates (chlorophyll fluorescence assay). The tests are applicable at several levels of organization, and can be applied in both the laboratory and the field to test all types of chemicals. Standard protocols have been adapted from agricultural science that are relatively inexpensive and simple to perform. Seeds of standard test species (e.g., *Lactuca sativa*, lettuce) are available from commercial sources; seeds of site specific plants can also be tested. These tests offer a wide range of cost options, with some being fairly inexpensive (seed germination and root elongation tests) and others requiring expensive test equipment (photosynthesis inhibition test). Growth conditions must be carefully monitored during these tests, because nutrient limitations can complicate interpretation of toxicity effects.

These tests evaluate chemical effects on all stages of plant development; however, different stages of plant growth exhibit varying sensitivity to chemical insults. For example, seeds are the least sensitive stage of a plant's growth; if seed germination reveals significant toxicity, the environmental consequences are probably severe. Unlike seeds, young roots and seedlings are relatively sensitive to chemical insult, making tests that evaluate those stages especially sensitive assays and particularly effective at demonstrating the effects of low contaminant concentrations and chronic toxicity. Table 3.11 provides detailed information about these tests.

3.4.1.4 *In situ* toxicity tests

Certain toxicity tests described above have been adapted for use in the field to evaluate *in situ* toxicity conditions. The most frequently used tests measure earthworm survival, amphibian viability, and seed germination. On-site tests conducted in parallel with laboratory analyses offer certain advantages. *In situ* toxicity tests can address methodological biases often associated with laboratory tests and the uncertainties associated with extrapolation from laboratory to field and from standard test species to site-specific species. The on-site version of the tests may often reduce cost and workload by eliminating collection, shipping, handling, and disposal of waste, thus permitting the collection and measurement of greater numbers of samples. However, when conducting on-site tests, the environmental conditions at the site, including fluctuation in temperature and moisture conditions, must be considered. To do so may require rejection of many data sets, and also careful planning to be present during optimal conditions. For more detailed descriptions of methods see: *earthworm survival test* (Callahan *et al.*, 1991; Marquenie *et al.*, 1987; USEPA, 1992); *amphibian test* (Linder *et al.*, 1991; USEPA, 1992); and *seed germination test* (USDA 1985; Gorsuch *et al.*, 1990; Linder *et al.*, 1990; USEPA, 1989, 1992; USFDA, 1987b).

3.4.2 FIELD ASSESSMENTS

Field surveys are useful to evaluate the effects of chemicals on terrestrial ecosystems (USEPA, 1989). Field assessments discussed here include remote sensing methods, direct observation or long-term monitoring of permanent plots, field surveys of populations and communities, and adaptations of toxicity tests to field applications. Changes in an ecosystem can be monitored and evaluated at the same site over a long period of time. This technique is especially useful in forest studies when the individuals under study are long lived and the area is less disturbed. Examples of field assessments used to evaluate effects of contaminants are discussed below.

3.4.2.1 Remote sensing methods

Remote sensing may be used in several ways to assess vegetation of chemically contaminated sites. The main sources of radiometric data are the Landsat Multi Spectral Scanner (MSS) in the US, the Thematic Mapper (TM) in the US, and the Systeme Probatoire d'Observation de la Terre (SPOT) in France (USEPA, 1989; Koeln *et al.*, 1994). Resolution for the three types of data are: MSS: 80 metres; TM: 30 metres; and SPOT: 20 metres. For improved resolution, infrared and conventional photography from fixed-wing aerial aircraft may be supplemented. Remotely sensed data offer the following advantages: relatively unlimited accessibility; safe, non-intrusive assessment and monitoring; and the existence of historical data (MSS since 1972; TM since 1982; SPOT since 1984), the opportunity to assess large-scale seasonal and annual vegetation patterns (USEPA, 1989). Data derived from remote sensing methods can be used to map vegetation boundaries, estimate photosynthesis rates and drought stress, detect the effects of natural pests epidemics, and assess forest decline due to air pollutants. Advantages of remote sensing methods are discussed in greater detail in the contributed chapters.

3.4.2.2 Direct observation or ground truthing of remotely sensed data

A primary use of field survey methods is "ground truthing" or verification of remotely sensed data by direct observation in the field to determine the vegetation types and habitats present. A semi-quantitative method like the Relevé method, which has been used for many years worldwide, is usually sufficient to develop a description and patterning of the plant species present (Braun-Blanquet, 1932). This method will provide information sufficient to plan additional studies of plant and animal species to evaluate the influence of chemicals on site biota.

3.4.2.3 Long-term monitoring of permanent assessment plots

Long-term monitoring of permanently marked plots is often useful to follow changes in a terrestrial ecosystem (Clarke, 1986; Bonham, 1989; USEPA, 1989). This technique has been used extensively in forest habitats by national resource management agencies such as the US Forest Service. Permanent forest plots evaluate parameters indicative of overall forest condition and habitat suitability, (e.g., vegetation demographics, soil surface conditions, primary productivity, and nutrient cycling). Such long-term monitoring may be costly. The data quality is very dependent on the selection of the representative sample location. Detailed descriptions of the various types of permanent assessment procedures are presented later in this volume.

3.4.2.4 Population surveys

Vegetation surveys—community structure and floristics studies

These methods provide quantitative or qualitative information to help establish the extent and magnitude of chemical impacts on vegetation. Plant community parameters (e.g., species density and percent cover) of existing site plant communities or of defined test plant communities grown in a representative mesocosm in test soil are used to identify effects of chemicals. Endpoints are species abundance, species dominance, community structure, age-class or size-class distributions, and species distribution patterns. Vegetation sampling data provide information that can be linked with site history and toxicological information to establish causation. Vegetation sampling methods have been used widely in basic and applied plant ecology for many years. Many field sampling techniques exist, ranging from pseudo-quantitative to quantitative methods relying on defined-area plots (quadrants) or various plotless sampling including lines, points, or variable radius methods. Detailed descriptions of these methods are presented by Kapustka *et al.* (1989), Pfleeger (1991), USEPA (1992), Weinstein and Laurence (1989), and Weinstein *et al.* (1991).

Animal population surveys

Surveys of animal populations present on a site, when compared to references sites, provide information about the impact of contamination on demographic parameters and ecological diversity (USEPA, 1989; Bookout, 1994). To accurately estimate if vertebrate and invertebrate populations have been adversely affected by site contamination, a census or estimation of population numbers of resident species, age and sex ratios, reproduction rates and rearing success, and survival and mortality rates must be conducted (Johnson, 1994). An understanding of the life cycles and behaviour of animal species expected to be found at the site, as determined by studying a reference site, helps to guide the sampling plan.

Two major approaches to estimate animal populations are direct counts and indirect methods of observing the animals present (Johnson, 1994). Direct counts of representative sample sites are more cost-effective than conducting a complete census. The most common direct methods are sampling along transects (Anderson *et al.*, 1976) and within quadrants, recording direct observations by field teams, or sampling with capture traps. Selection of sampling times and capture method is determined by evaluating reference sites as well as documentation in the literature. Field teams must be experienced in animal handling and identification. If captured animals are to be used in long-term studies or toxicity tests, the individuals must be handled and marked in a manner that does not cause injury or death.

When direct observational methods are impossible, indirect evidence of animal life must be utilized. The sampling design and statistical interpretations are the same as for direct observation or capture data. Commonly used types of indirect signs include: dens, burrows, or nests; tracks, faeces, calls; and counts of carcasses. The use of indirect signs alone is problematic, but can be used cautiously for interpretation (Davis and Winstead, 1980).

3.4.3 SPECIAL ANALYSES

Tests that will provide information unavailable from chemical analyses, toxicity tests, or field assessments of the chemically contaminated site are often desirable. Demonstrating that a chemical is present in elevated levels at the site or that the chemical is toxic to biota is at times insufficient, but the chemicals must be shown to be incorporated into the food chain. Residue studies can provide this kind of information, and give indications of possible routes of exposure to humans through their food sources. Soil microcosm studies provide a picture of the overall impact on soil ecosystem functioning that is important to evaluate chemical impacts on primary productivity of plant and animal resources of value to human society. Biomarkers may also provide an indication of actual contaminant exposure or effects at the organismal level. Biomarkers and soil microcosm studies used to evaluate contaminant stress in terrestrial organisms are described below.

3.4.3.1 Biomarkers

Biomarkers are increasingly incorporated into assessments of aquatic and terrestrial ecosystems. General discussions of the utilization of biomarkers can be found in McCarthy and Shugart (1990) and Huggett *et al.* (1992).

3.4.3.2 Soil microcosm studies

Soil microcosm studies evaluate ecological effects and environmental fate and transport of solid and liquid contaminants by measuring adverse effects on growth and reproduction of native vegetation or crop plants. The tests also measure the uptake and cycling of nutrients in the terrestrial ecosystem. Possible endpoints include: primary productivity; bioaccumulation and translocation of contaminants in plant tissue; and nutrients lost in leachates. Originally designed to test grassland or agricultural soils, the tests have been widely used in numerous systems, and are standardized through ASTM E1197 (1992). Tests can be readily adapted to site-specific conditions, and have been used to evaluate complex chemical wastes, hazardous wastes, and agricultural chemicals. Because of the ease of collecting leachate, these tests are especially suitable to monitor nutrient losses in leachates. Regrettably, the duration of these tests is relatively long (6–8 weeks), and few commercial laboratories currently conduct them. More detailed descriptions of such methods are found in USEPA (1992b) and Van Voris *et al.* (1985).

3.5 DIAGNOSIS OF ENVIRONMENTAL PROBLEMS AND ESTABLISHMENT OF CAUSATION

This chapter has presented some of the varied tools that enable scientists to define the condition of chemically contaminated sites and to establish linkages between site contamination and observed adverse effects. These tools include chemical analyses, toxicity tests, field surveys, special studies such as biomarkers and simulated ecosystem studies, and mathematical modelling to evaluate assumptions (Figure 3.1). Each tool provides specific kinds of data, which, if used alone, are certainly useful; however, the greatest contribution of each tool is in providing a unique portion of an entire environmental puzzle. Chemical analyses of site samples show the extent and quantity of contamination. Data from toxicity tests (laboratory and *in situ*) demonstrate that contaminants are capable of causing damage. Field observations indicate the extent and patterns of contamination and provide supporting evidence that site populations have been altered structurally and functionally.

Once ecotoxicity has been measured using the tools described above, the next step is to demonstrate that the contaminants caused the damage. In most cases, the extent and nature of the toxic effects (especially in a retrospective risk assessment) do not unequivocally demonstrate which agent was responsible for the injury. Correlations between a substance and a form of damage may be obvious, but causation is most difficult to prove. As with other fields of toxicology, the scientist and decision maker may need to rely on statistical associations, because causal relationships can be established only indirectly.

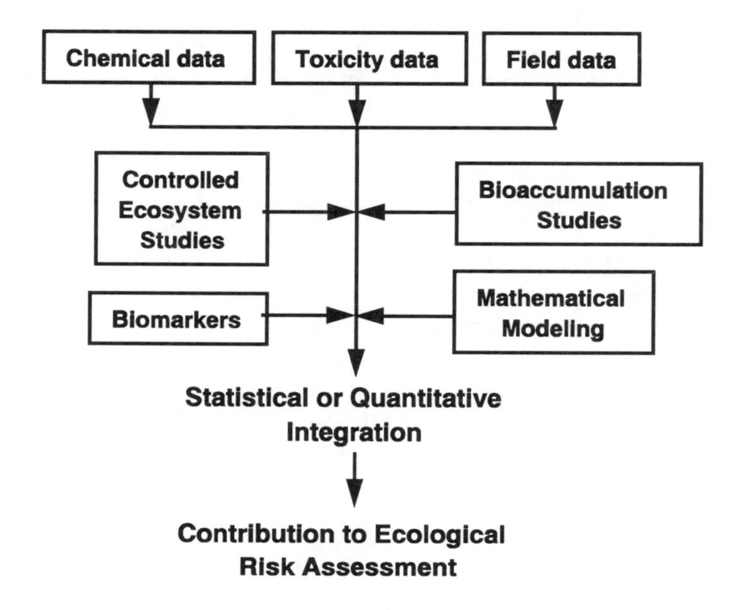

Figure 3.1. Types of data used in an ecological risk assessment

Scientists have used the criteria proposed by bacteriologist, Robert Koch, in the early 1900s when medical science was first struggling to understand cholera, typhoid fever, and tuberculosis, as a way to establish the weight of evidence for causation. These criteria can be adapted as guidelines to establish whether a chemical is causing (or has caused) the observed alterations; they are summarized as follows (as restated by Evans, 1976):

1. The effect is more pronounced in exposed biota.
2. The exposure is more frequent among biota exhibiting the effect.
3. The incidence of the effect is higher in exposed populations.
4. The appearance of the effect should follow exposure.
5. The responses should follow exposure along a gradient from mild to severe.
6. Exposure triggers a measured response, with a higher probability after exposure, and should not be seen in unexposed biota.
7. Experimental reproduction of the effect occurs more frequently in exposed biota.
8. Elimination of the suspected agent decreases the incidence of the effect.

If these criteria are met by the totality of the data, then causation can be claimed with a relatively high degree of certainty. The environmental manager must be aware of the many limitations inherent in ecological data when making management decisions. However, if the appropriate steps are followed (i.e., clear statement of objectives, sound study design, selection of appropriate assessment methodologies, careful data interpretation, and QA/QC integrated into the entire study process), the data will be sufficiently sound and comprehensive to make confident estimates of risk on which to base environmental decisions.

3.6 RECOMMENDATIONS

3.6.1 RECOMMENDATIONS FOR ENHANCING ASSESSMENTS OF AQUATIC AND TERRESTRIAL ECOSYSTEMS

1. Further research should be conducted to characterize pristine aquatic ecosystems so that deviations from the natural situation resulting from chemical contamination can be reliably quantified.
2. Monitoring programs should be implemented to estimate baseline data and to gain increased understanding of the natural variability within ecosystems. Long-term monitoring programs should be established as soon as possible to detect adverse effects of chemicals on ecosystems.
3. Additional research needs to be conducted using communities or organisms in micro- and mesocosms to establish dose–response relationships at levels above that of the single species. Often unclear is how change at one level of a community affects organisms at other levels in aquatic ecosystems. Therefore, mechanistic links must be established among effects at different levels of biological organization. Effects at the global level, in particular, deserve further attention.
4. Synergistic and antagonistic effects among chemicals need to be better understood in aquatic ecosystems.
5. Research should continue to identify useful biomarkers in nontemperate biomes, especially in polar and tropical regions. Uncertainties resulting from extrapolation of data generated in temperate biomes make these tests less useful than in other regions. What holds for a particular concentration of a pollutant in one ecosystem may not necessarily apply to all ecosystems due to differences in local conditions.
6. Accidental releases of chemicals should be viewed as an opportunity to increase understanding of the nature and severity of adverse effects and the efficacy of current monitoring techniques for chemical contaminants.
7. Recovery from accidental releases and chronic exposures, and the potential to manipulate the natural recovery process, should be studied. The robustness of ecosystems should be studied to increase understanding of the ability of a system to withstand chemical exposure without irreversible damage.
8. The importance of timeframes and time scales needs greater attention. The timeframe in which results are needed should be considered when selecting assessment and monitoring techniques. In addition, the nature of a chemical release and dispersal (i.e., patterns and rates; acute and chronic toxicity) should be considered when selecting methods.

9. Alternative methods should be developed that will more easily allow the translation of ecological characterizations into value or benefit characterizations.

3.7 REFERENCES

Ahn, B.K., and Morrison, G. (1991) *Soil Toxicity Screening Test with Photobacterium Activity in Environmental Site Assessment.* EcoTech, Inc., Irving, Texas.

Anderson, D.R., Laake, J.L., Crain, B.R., and Burnham, K.P. (1976) *Guidelines for Line Transect Sampling of Biological Populations.* Utah Cooperative Wildlife Research Unit, Logan, Utah, 27 pp.

ASTM E1383-93 (1993) Standard guide for conducting sediment toxicity tests with freshwater invertebrates. In: *Annual Book of ASTM Standards: Water and Environmental Technology*, Vol. 11.04, American Society for Testing and Materials, Philadelphia.

ASTM E552 (1992) Standard test method for efficacy of acute mammalian predacides. In: *Annual Book of ASTM Standards: Pesticides; Resource Recovery; Hazardous Substances and Oil Spill Responses; Waste Management; Biological Effects*, Vol. 11.04, pp. 235–238. American Society for Testing and Materials, Philadelphia.

ASTM E555 (1992) Standard practice for determining acute oral LD_{50} for testing vertebrate control agents. In: *Annual Book of ASTM Standards: Pesticides; Resource Recovery; Hazardous Substances and Oil Spill Responses; Waste Management; Biological Effects*, Vol. 11.04, pp. 242–243. American Society for Testing and Materials, Philadelphia.

ASTM E593 (1992) Standard test method for efficacy of a multiple-dose rodenticide under laboratory conditions. In: *Annual Book of ASTM Standards: Pesticides; Resource Recovery; Hazardous Substances and Oil Spill Responses; Waste Management; Biological Effects*, Vol. 11.04, pp. 262–267. American Society for Testing and Materials, Philadelphia.

ASTM E757 (1992) Standard test method for efficacy of a canine reproduction inhibitors. In: *Annual Book of ASTM Standards: Pesticides; Resource Recovery; Hazardous Substances and Oil Spill Responses; Waste Management; Biological Effects*, Vol. 11.04, pp. 428–431. American Society for Testing and Materials, Philadelphia.

ASTM E758 (1992) Standard test method for mammalian acute percutaneous toxicity. In: *Annual Book of ASTM Standards: Pesticides; Resource Recovery; Hazardous Substances and Oil Spill Responses; Waste Management; Biological Effects*, Vol 11.04, pp. 432–436. American Society for Testing and Materials, Philadelphia.

ASTM E857 (1992) Standard practice for conducting subacute dietary toxicity tests with avian species. In: *Annual Book of ASTM Standards: Pesticides; Resource Recovery; Hazardous Substances and Oil Spill Responses; Waste Management; Biological Effects*, Vol. 11.04, pp. 481–485. American Society for Testing and Materials, Philadelphia.

ASTM E1062 (1992) Standard practice for conducting reproductive studies with avian species. In: *Annual Book of ASTM Standards: Pesticides; Resource Recovery; Hazardous Substances and Oil Spill Responses; Waste Management; Biological Effects*, Vol. 11.04, pp. 706–716. American Society for Testing and Materials, Philadelphia.

ASTM E1103 (1992) Standard test method for determining subchronic dermal toxicity. In: *Annual Book of ASTM Standards: Pesticides; Resource Recovery; Hazardous Substances and Oil Spill Responses; Waste Management; Biological Effects*, Vol. 11.04, pp. 735–737. American Society for Testing and Materials, Philadelphia.

ASTM E1163 (1992) Standard test method for estimating acute oral toxicity in rats. In: *Annual Book of ASTM Standards: Pesticides; Resource Recovery; Hazardous Substances and Oil Spill Responses; Waste Management; Biological Effects*, Vol. 11.04, pp. 770–775. American Society for Testing and Materials, Philadelphia.

ASTM E1197 (1992) Standard guide for conducting a terrestrial soil-core microcosm test. In: *Annual Book of ASTM Standards: Pesticides, Resource Recovery; Hazardous Substances and Oil Spill Responses; Waste Management; Biological Effects*, Vol. 11.04, pp. 848–860. American Society for Testing and Materials, Philadelphia.

ASTM E1367 (1992) Standard guide for conducting 10-day static sediment toxicity tests with marine and estuarine amphipods. In: *Annual Book of ASTM Standards: Pesticides; Resource Recovery; Hazardous Substances and Oil Spill Responses; Waste Management; Biological Effects*, Vol. 11.04. American Society for Testing and Materials, Philadelphia.American Society for Testing and Materials, Philadelphia.

ASTM E1372 (1992) Standard test method for conducting a 90-day oral toxicity study in rats. In: *Annual Book of ASTM Standards: Pesticides; Resource Recovery; Hazardous Substances and Oil Spill Responses; Waste Management; Biological Effects*, Vol., 11.04, pp. 1107–1111. American Society for Testing and Materials, Philadelphia.

ASTM E1373 (1992) Standard test method for conducting a subchronic inhalation toxicity study in rats. In: *Annual Book of ASTM Standards: Pesticides; Resource Recovery; Hazardous Substances and Oil Spill Responses; Waste Management; Biological Effects*, Vol. 11.04, pp. 1112–1115. American Society for Testing and Materials, Philadelphia.

ASTM E1439 (1992) Standard guide for conducting the frog embryo teratogenesis assay—*Xenopus* (Fetax). In: *Annual Book of ASTM standards: Pesticides; Resource Recovery; Hazardous Substances and Oil Spill Responses; Waste Management; Biological Effects*, Vol. 11.04, pp. 1199–1209. American Society for Testing and Materials, Philadelphia.

ASTM STP 1091, American Society for Testing and Materials, Philadelphia.

Bascietto, J., Hinckley, D., Plafkin, J., and Slimak, M. (1990) Ecotoxicity and ecological risk assessment, regulatory applications at EPA. *Environ. Sci. Technol.* **24**(1), 10–15.

Bonham, C.D. (1989) *Measurements for Terrestrial Vegetation.* John Wiley, New York, 380 pp.

Bookout, T.A. (Ed.) (1994) *Research and Management Techniques for Wildlife and Habitats* (5th edn). The Wildlife Society, Bethesda, Maryland.

Braun-Blanquet, J. (1932) *Plant Sociology: The Study of Plant Communities.* McGraw-Hill, New York.

Brusick, D. (1980) Protocol 13: bone marrow cytogenetic analysis in rats. In: *Principles of Genetic Toxicology.* Plenum Press, New York.

Bulich, A.A. (1982) A practical and reliable method for monitoring the toxicity of aquatic samples. *Process Biochem.* March/April, 45–47.

Bulich, A.A. (1986) Bioluminescence assays. In: Bitton, G., and Dutka, B.J. (Eds.) *Toxicity Testing Using Microorganisms*, Vol. 1, pp. 57–74. CRC Press, Boca Raton, Florida.

Burns, R.G. (1986) Interaction of enzymes with soil mineral and organic colloids. In: Huang, P.M., and Schnitzer, M. (Eds.) *Interactions of Soil Minerals with Natural Organics and Microbes*, pp. 429–451. Soil Scientists Society of America, Madison, Wisconsin.

Callahan, C.A., Menzie, C.A., Birmester, D.E., Wilborn, D.C., and Ernst, T. (1991) On-site methods for assessing chemical impact on the soil environment using earthworms: a case study at the Baird and McGuire Superfund site, Holbrook, MA. *Environ. Toxicol. Chem.* **10**, 817–826.

Callahan, C.A., Russell, L.K., and Peterson, S.H. (1985) A comparison of three earthworm bioassay procedures for the assessment of environmental samples containing hazardous wastes. *Biol. Fertil. Soils* **1**, 195–200.

Chen, S.C., Fitzpatrick, L.C., Govern, A.J., Venables, B.J., and Cooper, E.L. (1991) Nitroblue tetrazolium dye reduction by earthworm (*Lumbricus terrestris*) ceolomocytes: an enzyme assay for nonspecific immunotoxicity of xenobiotics. *Environ. Toxicol. Chem.* **10**, 1037-1043.

Clarke, R. (Ed.) (1986) *The Handbook of Ecological Monitoring*. Oxford University Press, New York.

Crow, M.E., and Taub, F.B. (1979) Designing of microcosm bioassay to detect ecosystem level effects. *Int. J. Environ. Studies* **13**, 141–147.

Davis, D.E., and Winstead, R.L. (1980) Estimating the numbers of wildlife populations. In: Schemnitz, S.D. (Ed.) *Wildlife Management Techniques Manual* (4th edn), pp. 221–245. The Wildlife Society, Washington, D.C.

Doelman, P., and Haanstra, L. (1984) Short-term and long-term effects of cadmium, chromium, copper, nickel, lead, and zinc on soil microbial respiration in relation to abiotic soil factors. *Plant and Soil* **79**, 317–337.

Dumontet, S., and Mathur, S.P. (1989) Evaluation of respiration-based methods for measuring microbial biomass in metal-contaminated acidic mineral and organic acids. *Soil Biology and Biochemistry* **21**, 431–435.

Edwards, C.A. (1983) Report of the second stage in development of a standardized laboratory method for assessing the toxicity of chemical substances to earthworms. Report to the Commission of the European Communities, Rothamsted Experimental Station, Harpenden, Herts, England.

Evans, A.S. (1976) Causation and disease: the Henle–Koch postulates revisited. *Yale J. Biology and Medicine* **49**, 175–195.

Eyambe, G.S., Goven, A.J., Fitzpatrick, L.C., Venables, B.J., and Cooper, E.L. (1990) A non-invasive technique for sequential collection of earthworm (*Lumbricus terrestris*) leukocytes during subchronic immunotoxicity studies. *Laboratory Animals* **25**, 61–67.

Gano, K.A., Carlile, D.W., and Rogers, L.E. (1985) *A Harvester Ant Bioassay for Assessing Hazardous Chemical Waste Sites*. Report No. PNL-5434, UC-11. Pacific Northwest Laboratory, Richland, Washington.

Gewurz, H., and L. Suyehira (1976) Complement. In: Rose, N., and Friedman, H. (Eds.) *Manual of Clinical Immunology*, pp. 36–50. American Society of Microbiology, Washington, D.C.

Giesy, J.P., and Allred, P.M. (1985) Replicability of aquatic multispecies test systems. In: Cairns, J. (Ed.) *Multispecies Toxicity Testing*. Society for Environmental Contamination and Toxicology, Washington, D.C.

Goats, G.C., and Edwards, C.A. (1982) Testing the toxicity of industrial chemicals to earthworms. *Rothamsted Reports* 104–105.

Gorsuch, J.W., Kringle, R.O., and Robillard, K.A. (1990) Chemical effects on the germination and early growth of terrestrial plants. In: Wang, W., Gorsuch, J.W., and Lower, W.R. (Eds.) *Plants for Toxicity Assessment*, pp. 49–58.

Graney, R.L., Kennedy, J.H., and Rodgers, J.H., Jr (Eds.) (1994) *Aquatic Mesocosm Studies in Ecological Risk Assessments*. SETAC Special Publication Series. Lewis Publishers, Boca Raton, Florida.

Hammons, A.S. (Ed.) (1981) *Methods for Ecological Toxicology, A Critical Review of Laboratory Multispecies Tests*. Ann Arbor Science Publishers, Inc., Ann Arbor, Mich.

Haque, R. (1980) *Interlaboratory Evaluation of Microcosm Research.* Office of Environmental Processes and Effects Research. EPA-600/9-80-019. US Environmental Protection Agency, Washington, D.C.

Hornshaw, T.C., Aulerich, R.J., and Ringer, R.K. (1986a) Toxicity of *o*-creosol to mink and European ferrets. *Environmental Toxicology and Chemistry* **5**, 713–720.

Hornshaw, T.C., Ringer, R.K., and Aulerich, R.J. (1986b) Toxicity of sodium monofluoroacetate (Compound 1080) to mink and European ferrets. *Environmental Toxicology and Chemistry* **5**, 213–223.

Huggett, R.J., Kimerle, R.A., Mehrle, P.M., Jr, and Bergman, H.L. (1992) *Biomarkers: Biochemical, Physiological, and Histological Markers of Anthropogenic Stress.* Lewis Publishers, Boca Raton, 347 pp.

James, R.R., and Lighthart, B.. (1990) Bioassay for testing the lethal effects of bacterial pathogens on the predatory beetle *Hippodamia convergens* Gue. (Coleoptera: Coccinellidae). 600/3-90/090. US Environmental Protection Agency, Environmental Research Laboratory, Corvallis.

Jerne, N.K., and Nordin, A.A. (1963) Plaque formation in agar by single antibody-producing cells. *Science* **140**, 405.

Johnson, D.H. (1994) Population analysis. In: Bookout, T.A. (Ed.) *Research and Management Techniques for Wildlife and Habitats*, 5th edn, pp. 419–444. The Wildlife Society, Bethesda.

Judy, B.M., Lower, W.R., Miles, C.D., Thomas, M.W., and Krause, G.F. (1990) Chlorophyll fluorescence of a higher plant as an assay for toxicity assessment of soil and water. In: Wang, W., Gorsuch, J.W., and Lower, W.R. (Eds.) *Plants for Toxicity Assessment*, pp. 308–318. ASTM STP 1091. American Society for Testing and Materials, Philadelphia, Pa.

Judy, B.M., Lower, W.R., Ireland, F.R., and Krause, G.F. (1991) A seedling chlorophyll fluorescence toxicity assay. In: Gorsuch, J.W., Lower, W.R., Lewis, M.A., and Wang, W. (Eds.) *Plants for Toxicity Assessment*, pp. 146–160. ASTM STP 1115. American Society for Testing and Materials, Philadelphia, Pa.

Kapustka, L.A., LaPoint, T., Fairchild, J., McBee, K., and Bromenshenk, J. (1989) Vegetation assessment. In: Warren-Hicks, W., Parkhurst, B., and Baker, S. Jr (Eds.) *Ecological Assessment of Hazardous Waste Sites: A Field and Laboratory Reference*, pp. 8-40 to 8-72. EPA/600/3-89/013. US Environmental Protection Agency, Environmental Research Laboratory, Corvallis.

Kershaw, K.A. (1973) *Quantitative and Dynamic Plant Ecology*, 2nd edn. American Elsevier Publishing Co., Inc., New York.

Kochwa, S. (1976) Immunoelectrophoresis. In: Rose, N., and Friedman, H. (Eds.) *Manual of Clinical Immunology*, pp. 17–35. American Society of Microbiology, Washington, D.C.

Koeln, G.T., Cowardin, L.M., and Strong, L.L. (1994) Geographic information systems. In: Bookout, T.A. (Ed.) *Research and Management Techniques for Wildlife and Habitats*, 5th edn, pp. 540–566. The Wildlife Society, Bethesda, Md.

Ladd, J.N. (1985) Soil enzymes. In: Vaughan, D., and Malcom, R.E. (Eds.) *Soil Organic Matter and Biological Activity*, pp. 175–221. Martinus Nijhoff, Dordrecht, The Netherlands.

Linder, G., Greene, J.C., Ratsch, H., Nwosu, J., Smith, S., and Wilborn, D. (1990) Seed germination and root elongation toxicity tests in hazardous waste site evaluation: methods, development and applications. In: Wang, W., Gorsuch, J.W., and Lower, W.R. (Eds.) *Plants for Toxicity Assessment*, pp. 177–187. ASTM STP 1091, American Society for Testing and Materials, Philadelphia, Pa.

Linder, G., Wyant, J., Meganck, R., and Williams, B. (1991) Evaluating amphibian responses in wetlands impacted by mining activities in the western United States. In: Comer, R.D., Davis, P.R., Foster, S.Q., Grant, C.V., Rush, S., Thorne, O., and Todd, J. (Eds.) *Issues and Technology in the Management of Impacted Wildlife*, pp. 17–25. Thorne Ecological Institute, Boulder, Col.

Linder, G., and Richmond, M.E. (1990) Feed aversion in small mammals as a potential source of hazard reduction for environmental chemicals: agrichemical case studies. *Environmental Toxicology and Chemistry* **9**, 95–105.

McBee, K., Bickham, J.W., Brown, K.W., and Donnelly, K.C. (1987) Chromosomal aberrations in native small mammals (*Peromyscus leucopus* and *Sigmodon hispidus*) at a petrochemical waste disposal site: I. Standard karyology. *Archives of Environmental Contamination and Toxicology* **16**, 681–688.

McCarthy, J.F., and Shugart, L.R. (1990) *Biomarkers of Environmental Contamination*. Lewis Publishers, Boca Raton, 457 pp.

Marquenie, J.M., Simmers, J.W., and Kay, S.H. (1987) Preliminary assessment of bioaccumulation of metals and organic contaminants at the Times Beach confined disposal site, Buffalo, NY. Miscellaneous Paper EL-87-6. Department of the Army, Waterways Experiment Station, Corps of Engineers, Vicksburg, Miss.

Microbics (1992) Microtox™ Manual. Microbics Corp., Carlsbad, Calif. 619/438-8282.

Miles, D. (1990) The role of chlorophyll fluorescence as a bioassay for assessment of toxicity in plants. In: Wang, W., Gorsuch, J.W., and Lower, W.R. (Eds.) *Plants for Toxicity Assessment*, pp. 297–307. ASTM STP 1091. American Society for Testing and Materials, Philadelphia, Pa.

Nannipieri, P., Ciardi, C., Badalucco, K., and Casella, S. (1986a) A method to determine DNA and RNA. *Soil Biology and Biochemistry* **18**, 275–281.

Nannipieri, P., Grego, S., and Ceccanti, B. (1986b) Ecological significance of the biological activity in soil. *Soil Biology and Biochemistry* **6**, 293–355.

OECD (1984) *OECD Guidelines for Testing of Chemicals*. Director of Information, Organization for Economic Development (OECD), 2 rue André Pascal, 75775 Paris Cedex 16, France.

Oppenheim, J., and Schechter, B. (1976) Lymphocyte transformation. In: Rose, N., and Friedman, H. (Eds.) *Manual of Clinical Immunology*, pp. 81–94. American Society of Microbiology, Washington, D.C.

Parr, J.F. (1974) Effects of pesticides on microorganisms in soil and water. In: Guenzi, W.D., Ahlrich, J.L., Bloodworth, M.E., Chesters, G., and Nash, R.G. (Eds.) *Pesticides in Soil and Water*, pp. 315–340. Soil Science Society of America, Inc., Madison.

Pfleeger, T. (1991) Impact of airborne pesticides on natural plant communities. In: *Plant Tier Testing: A Workshop to Evaluate Nontarget Plant Testing in Subdivision J Pesticide Guidelines*. EPA/600/9-91/041. US Environmental Protection Agency, Environmental Research Laboratory, Corvallis.

Pfleeger, T., McFarlane, C., Sherman, R., and Volk, G. (1991) A short-term bioassay for whole plant toxicity. In: Gorsuch, J.W., Lower, W.R., Wang, W., and Lewis, M.A. (Eds.) *Plants for Toxicity Assessment*: Vol. 2, pp. 355–364. ASTM STP 1115. American Society for Testing and Materials, Philadelphia, Pa.

Ratsch, H.C., Johndro, D.J., and McFarlane, J.C. (1986) Growth inhibition and morphological effects of several chemicals in *Arabidopsis thaliana* (L.) Heynh. *Environmental Toxicology and Chemistry* **5**, 55–60.

Ribo, J.M., and Kaiser, K.L.E. 197. *Photobacterium phosphoreum* toxicity bioassay. I. Test procedures and applications. *Toxicity Assessment* **2**, 305–323.

Rodriguez-Grau, J., Venables, B.J., Fitzpatrick, L.C., and Cooper, E.L. (1989) Suppression of secretory rosette formation by PCBs in *Lumbricus terrestris*: an earthworm assay for humoral immunotoxicity of xenobiotics. *Environmental Toxicology and Chemistry* **8**, 1201–1207.

Rose, N., and Friedman, H. (Eds.) (1976) *Manual of Clinical Immunology*. American Society of Microbiology, Washington, D.C.

Shannon, L.J., *et al.* (1986) A comparison of mixed flask culture and standardized laboratory model ecosystems for toxicity testing. In: Cairns, J. (Ed.) *Community Toxicity Testing*. STP 920. American Society for Testing and Materials, Philadelphia.

Shimabuku, R.A., Ratsch, H.C., Wise, C.M., Nwosu, J.U., and Kapustka, L.A. (1991) A new plant life-cycle bioassay for assessment of the effects for toxic chemicals using rapid cycling *Brassica*. In: Wang, W., Gorsuch, J.W., and Lower, W.R. (Eds.) *Plants for Toxicity Assessment*, pp. 365–375. ASTM STP 1115. American Society for Testing and Materials, Philadelphia, Pa.

Stein, E.A., and Cooper, E.L. (1988) *In vitro* agglutinin production by earthworm leukocytes. *Developmental Comparative Immunology* **12**, 531-547.

Taub, F.B. (1985) Toward interlaboratory (round-robin) testing of a standardized aquatic microcosm. In: Cairns, J. (Ed.) *Multispecies toxicity testing*. Society for Environmental Contamination and Toxicology, Washington, D.C.

Taub, F.B and Crow, M.E. (1980) Synthesizing aquatic microcosms. In: Giesy, J.P. (Ed.) *Microcosms in Ecological Research*. US Department of Energy, Washington, D.C.

Taub, F.B. *et al.* (1986) Preliminary results of interlaboratory testing of a standardized aquatic microcosm. In: Cairns, J. (Ed.) *Community Toxicity Testing*. STP 920. American Society for Testing and Materials, Philadelphia, PA.

Touart, L.W. (1988) *Aquatic Mesocosm Tests to Support Pesticide Registrations*. Office of Pesticide Programs. EPA 540/09-88-035. US Environmental Protection Agency, Washington, D.C.

US Army Corps of Engineers (1991) *Evaluation of Dredged Material Proposed for Ocean Disposal*. Testing Manual. EPA 503/8-91/001.

USDA (US Department of Agriculture) (1985) *Rules and Regulations; Seed Germination/Root Elongation Toxicity Test*. US Code of Federal Regulations, 50 CFR 797.2750, Washington, D.C., 27 September.

USEPA (US Environmental Protection Agency) (1981) *Short Term Methods for Estimating the Chronic Toxicity of Effluents and Receiving Waters to Freshwater Organisms*. Report No. EPA 600/4-89/001. US Environmental Protection Agency, Office of Research and Development, Washington, D.C.

USEPA (US Environmental Protection Agency) (1982) *Pesticide Assessment Guidelines Subdivision E Hazard Evaluation: Wildlife and Aquatic Organisms*. Report No. EPA 540/9-82-024. US Environmental Protection Agency, Washington, D.C.

USEPA (US Environmental Protection Agency) (1988) *Short Term Methods for Estimating the Chronic Toxicity of Effluents and Receiving Waters to Marine and Estuarine Organisms*. Report No. EPA 600/4-87/028. US Environmental Protection Agency, Office of Research and Development, Washington, D.C.

USEPA (US Environmental Protection Agency) (1989) *Ecological Assessment of Hazardous Waste Sites: A Field and Laboratory Reference Document*. Report No. EPA/600-/3-89/013. US Environmental Protection Agency, Office of Research and Development, Corvallis, Oregon.

USEPA (US Environmental Protection Agency) (1990) *Short-Term Methods for Estimating the Chronic Toxicity of Effluents and Receiving Waters to Freshwater Organisms*. Report No. EPA 600/400-89/001. US Environmental Protection Agency, Office of Research and Development, Washington, D.C.

USEPA (US Environmental Protection Agency) (1991a) *Evaluation of Dredged Material Proposed for Ocean Disposal. Testing Manual*. Report No. EPA 503/8-91/001. US Environmental Protection Agency, Washington, D.C.

USEPA (US Environmental Protection Agency) (1991b) *Methods for Measuring the Acute Toxicity of Effluents and Receiving Waters to Freshwater and Marine Organisms*. Report No. EPA 600/4-90/027. US Environmental Protection Agency, Office of Research and Development, Washington, D.C.

USEPA (US Environmental Protection Agency) (1992a) *Evaluation of Terrestrial Indicators for Use in Ecological Assessments at Hazardous Waste Sites*. Report No. EPA/600/R92/183. US Environmental Protection Agency, Office of Research and Development, Washington, D.C.

USEPA (US Environmental Protection Agency) (1992b) *Toxic Substances Control Act*, Environmental Effects Testing Guidelines. US Code of Federal Regulations, 40 CFR 797, Washington, D.C., July 1.

USEPA (US Environmental Protection Agency) (1993a). *Evaluation of Dredged Material Proposed for Discharge in Inland and Near-Coastal Waters-Testing Manual (Draft). Inland Testing Manual*. Report No. EPA-000/0-93/000. US Environmental Protection Agency, Washington, D.C., August.

USEPA (US Environmental Protection Agency) (1993b) *Federal Insecticide, Fungicide, and Rodenticide Act, Wildlife and Aquatic Organisms Data Requirements*. US Code of Federal Regulations, 40 *CFR* 158.490, Washington, D.C., 1 July.

USFDA (US Food and Drug Administration) (1987a) Earthworm subacute toxicity. Section 4.12 in: *Environmental Assessment Technical Assistance Document*. US Department of Health and Human Services, Food and Drug Administration, Center for Food Safety and Applied Nutrition, Environmental Impact Section, and Center for Veterinary Medicine Environmental Staff, Washington, D.C.

USFDA (US Food and Drug Administration) (1987b) Seed germination and root elongation. Section 4.06 in: *Environmental Assessment Technical Handbook*. US Department of Health and Human Services, Food and Drug Administration, Center for Food Safety and Applied Nutrition, and Center for Veterinary Medicine, Washington, D.C.

USFDA (US Food and Drug Administration) (1987c) Seedling growth. Section 4.07 in: *Environmental Assessment Technical Book, Technical Assistance Document*. US Department of Health and Human Services, Food and Drug Administration, Center for Food Safety and Applied Nutrition, and Center for Veterinary Medicine, Washington, D.C.

Van Voris, P., Tolle, D., and Arthur, M.F (1985) *Experimental Terrestrial Soil-Core Microcosm Test Protocol. A Method for Measuring the Potential Ecological Effects, Fate, and Transport of Chemicals in Terrestrial Ecosystems*. Report No. EPA/600/3-85/047, PNL–5450. US Environmental Protection Agency, Office of Research and Development, Washington, D.C.

Walton, B.T. (1980) Differential life-stage susceptibility of *Acheta domesticus* [crickets] to acridine. *Environ. Entomol.* **9**, 18–20.

Weinstein, L.H., and Laurence, J.A. (1989) Indigenous and cultivated plants as bioindicators. In: National Research Council, Woodwell, G.M. (Chair) *Biologic Markers of Air Pollution Stress and Damage in Forests*, pp. 195–204. National Academy Press, Washington, D.C.

Weinstein, L.H., Laurence, J.A., Mandl, R.H., and Wälti, K. (1991) Use of native and cultivated plants as bioindicators and biomonitors of pollution damage. In: Wang, W., Gorsuch, J.W., and Lower, W.R. (Eds.). *Plants for Toxicity Assessment.* ASTM STP 1115. American Society for Testing and Materials, Philadelphia. Pa.

Wetzel, O.R.G., and Likens, G.E. (1990) *Limnological Analyses* 2nd edn. Springer-Verlag, New York.

4 Large Geographic Scale Environmental Monitoring and Assessment

This chapter provides a conceptual overview of the approaches and methods to conduct environmental monitoring and assessments of the effects of chemicals on geographic scales that cover entire regions. Most environmental monitoring is limited to the local environment so that specific pollutants can be linked to the source and controlled. Unfortunately, this type of monitoring provides little information about the effects of chemical releases on entire ecosystems. To better identify "systemic" environmental problems (e.g., acid deposition and loss of diversity) and the ecosystems at greatest risk, regional monitoring and assessment programs are being developed, such as the Environmental Monitoring and Assessment Program (EMAP) conducted by the US Environmental Protection Agency (Messer *et al.*, 1991; Kutz *et al.*, 1992). Regional programs like EMAP use large geographic scale monitoring to identify significant long-term changes in the condition of ecosystems.

In a manner similar to that used in Chapter 3, this chapter discusses important concepts that form the basis of large geographic-scale assessment methods, such as time scales, assessment endpoints, study design, examples of assessment methods, and methods to analyse data.

4.1 IMPORTANCE OF LARGE GEOGRAPHIC SCALE MONITORING AND ASSESSMENT

The effects of chemicals on ecosystems frequently extend beyond local boundaries. Increasingly, scientists and risk managers are becoming aware of environmental problems that occur on large geographic scales: acid deposition, non-point source pollution, and loss of biodiversity, to name a few. Consequently, environmental managers should focus not only on effects locally, but also on multimedia impacts that occur on larger scales. Ecosystems are interconnected, and disruption in one section may cause repercussions throughout the entire system in ways that could not be anticipated by evaluating chemical impacts on the system on only a local

Methods to Assess the Effects of Chemicals on Ecosystems
Edited by R. A. Linthurst, P. Bourdeau, and R. G. Tardiff

level.

A narrow focus on only the individual or community levels of an ecosystem is insufficient to make decisions and allocate resources for several reasons. First, releases of chemicals to the environment have impacts that are most often significant on large geographic scales. An example is the formation of acid precipitation in one area that results from fossil fuel combustion in distant areas. Second, the ecological resources of interest exist on large scales (i.e., fisheries and forests), and are affected by many complex climatic processes that themselves function on global levels. If environmental managers consider the complexities of the structural components of ecosystems and the functional processes that drive them, they can make more realistic and accurate evaluations of the ecological resources of interest. Such management decisions are often triggered by knowledge of the extent or magnitude of a problem. Recognizing that effects are widespread (i.e., a certain percentage of the ecological resource is degraded) may initiate corrective action to remedy degradation at the regional scale. While knowledge of the effect of chemicals at the organismal level is needed, recognizing that such an effect may be magnified at the population or community level is vital to managing effectively ecosystems of larger geographic scales.

Regions represent an intermediate hierarchical level between individual ecosystems (estuaries, forests, deserts, and agricultural areas) and the global biosphere. An understanding of regional scale processes is often required to explain or predict the behaviour of individual ecosystems. Regional-scale ecological characteristics also may be the most appropriate scale for addressing ecological effects of global scale processes. Examples of some environmental problems that may require a regional scale ecological assessment are provided below.

Regions are important units both to assess the effects of chemical releases and to manage biological resources. For example, the airshed is an appropriate unit to assess the success of controls on airborne chemical emissions, whereas the watershed is an appropriate unit for regional water quality. Similarly, forest resources are best managed on a regional scale; therefore, the effects of chemical exposures on these resources are best evaluated on a similar scale. While local scale and site-specific approaches are effective methods to ascertain the influence of some types of stressors and ecological resources, a larger scale perspective permits a more comprehensive view of environmental problems that may impact large geographic areas.

Programs to assess the adverse impacts of chemical pollutants on ecosystems can also be made considerably more effective if they include regional concerns, values, and characteristics. Making decisions in the overall context of reducing risk over large geographic areas may help to avoid contamination problems that can result from management actions that move pollutants from one medium to another. Employing a regional approach can enhance chemical pollutant control programs in the following ways:

1. Identification of specific chemical pollutants that are major environmental threats within a region, and specific areas within regional systems that are being threatened by the pollutants. Chemical pollutants enter regional environmental systems from numerous point and non-point sources, and their threat is dependent on many factors, including: toxicity, persistence in the environment, and bioavailability; the quantities discharged; and the relative sensitivity of organisms in the area of discharge. Complete assessments of the threats posed by chemical discharges must also consider numerous less obvious, yet equally important, factors, such as land use patterns and socioeconomic systems. Emphasizing a regional approach to environmental assessment should encourage researchers to assemble and integrate the existing data on these other environmental parameters, which are most often described in a regional context.
2. Establishment of regional research priorities, and the allocation of research funds to priority projects. Attempts to develop and implement regional pollution control programs should reveal information gaps that limit the effectiveness of these programs. These informational deficiencies should become research priorities for the region, and research funds should be channelled to these priority topics.
3. Development of monitoring programs and allocation of funds to program components. Monitoring programs are undertaken to identify changes in environmental systems triggered by biotic or abiotic stressors. A regional approach should include collection and organization of scientific data on the region, accurate interpretation of monitoring data, and the use of these data for resource management. A regional framework to address environmental problems should provide guidance on the development and structure of the monitoring program, parameters to be measured, and the sampling schedule.
4. Design of mitigation and restoration programs and allocation of funds to specific projects. Environmental problems resulting from the discharge of chemicals are frequently addressed by mitigation-restoration projects. A regional approach could identify the highest priority mitigation-restoration sites, thereby assisting risk managers to allocate the limited funds available for mitigation-restoration to the most cost-effective projects.
5. Justification of regulatory programs. Regulatory programs enjoy wider support by legislators, special interest groups, and the general public if rationales for these programs are clearly understood to be rational by all interested parties. The need to regulate an activity can be understood more readily if the regulation of the activity is explained in terms of the positive effects on the region as a whole rather than only on a specific site within the region.

4.2 DEFINING A REGION

Developing a regional ecological monitoring and assessment program requires defining the region of interest, which depends, in part, on the management objectives usually stated as questions. Individual ecosystems are normally defined operationally in terms of areas of relatively uniform physical conditions, biotic community composition, and land use (e.g., freshwater lake, cornfield, or estuary). Regions, in contrast, do not always have obvious natural boundaries, but, instead, are defined by spatiotemporal scale and human interest. A particular point on a map may belong to several different regions defined for different purposes: a 1000 km^2 watershed, a 10000 km^2 airshed, and a state or nation of very small or very large size.

Three relatively natural approaches can define regions: "ecoregion;" watershed (if large) or river basin; and airshed. Ecoregions are defined as areas with relatively uniform ecological characteristics (Omernik, 1987). The ecoregion approach can employ original ecosystem types believed to be present before human development, or it may rely on current land use (e.g., agricultural or urban, suburban, or rural). A watershed-based region, for example, is defined by selecting a particular body of water as a starting point and identifying the land surface area drained by all of the rivers feeding into that body of water. An airshed-based region is defined as an area of restricted air flow, such as an area bounded by mountains (e.g., the Los Angeles Basin in California) or an area within which plumes from individual sources become thoroughly mixed.

Such approaches are convenient for scientists, but are often impractical for environmental management, because management action is often constrained by political or socioeconomic boundaries. Regions, therefore, are often prescribed by district, state, or national boundaries that cross ecological boundaries or ecoregions. Suter (1993) has suggested that multiple definitions of a region should be employed for many regional assessment problems. Actually, a useful approach may be to first define regions by ecologically relevant boundaries, and then overlay this with political or socioeconomic boundaries. Discussion of three major approaches or paradigms used to define regions (landscape mosaic, watershed, and airshed) follow.

4.2.1 THE LANDSCAPE MOSAIC PARADIGM

Figure 4.1 depicts an idealized view of a region divided into different ecosystems and land-use types. This view of a region may be called the "landscape mosaic." A simplistic approach to assess the quality or condition of such a landscape is to evaluate the condition of each unit independently, and then to aggregate the results by summing the economic returns and other benefits from each unit; this is also accomplished by calculating an environmental quality index. This approach is severely limited in its ability to aggregate, because it cannot account for interactions between landscape units, and it ignores organisms such as birds and large mammals that require relatively large habitat sizes for survival or can move between landscape units. The frequency and severity of forest fires, for example, are influenced by the size and spatial distribution of vulnerable stands; the abundance of deer in managed forests is enhanced by the presence of clear-cut patches or forest edges containing palatable second-growth vegetation; and the persistence of spotted owls depends on the existence of large stands of undisturbed old-growth forest.

The new subdiscipline of ecological science called "landscape ecology" is devoted to studies of the ecological characteristics of habitat mosaics. Among the research topics pursued by landscape ecologists are the development of matrices that characterize landscape patterns (Turner, 1987), the development and testing of spatial simulation models that relate local-scale ecological processes to regional scale patterns (Dale and Gardner 1987; Costanza *et al.*, 1990), and the analysis of the influence of spatial distributions of ecosystem types on the persistence of populations (Pulliam, 1988) or the severity of pest infestations (Graham *et al.*, 1991).

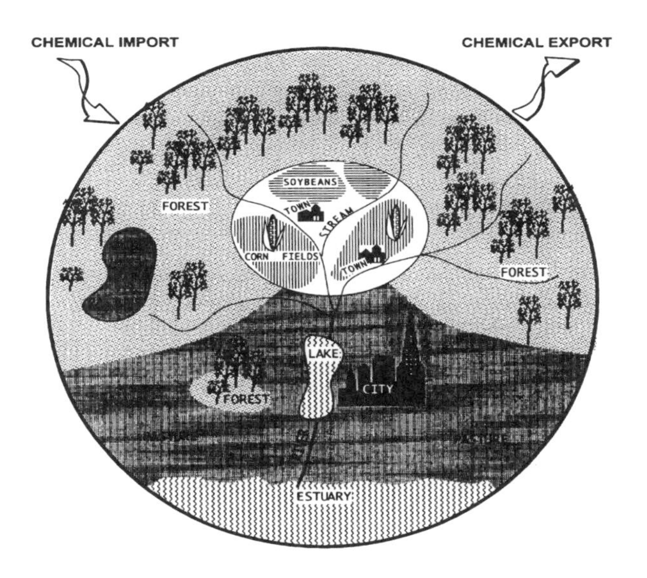

Figure 4.1 A representation of a landscape mosaic

4.2.2 THE WATERSHED PARADIGM

A region can also be characterized as a hydrologic network of connected streams, lakes, estuaries, and oceans. The ecological characteristics of any particular segment of this network are influenced by events occurring in upstream segments as well as those occurring within the segment. The watershed approach to regional assessment has been practised for several decades in water quality management.

Influences of multiple point sources of pollutants on downstream water quality can be calculated using available hydrologic simulation models (Thomann, 1972). Problems of major current interest involve interactions between terrestrial landscapes and watersheds. "Non-point-source pollution," including soils and chemicals contained in runoff from agricultural fields, feedlots, suburban lawns, and industrialized areas is a major source of many of the most persistent and ecologically harmful contaminants. A regional approach to protect or restore an estuarine ecosystem may include managing both the terrestrial and the aquatic components of upstream watersheds.

4.2.3 THE AIRSHED PARADIGM

Characterizing air quality only in terms of the behaviour of pollutant plumes and mixing zones within an airshed is inadequate to assess influences of atmospheric pollutants on regions. The airshed paradigm characterizes a region as a large volume of air that receives inputs from sources within the region and from other airsheds overlying other regions. Thus, an airshed is not a closed system, and an understanding of the rates, directions, and effects of interchanges with other discrete airsheds is essential to evaluate the regional effects of airborne substances. For example, combustion of fossil fuels in one region may release sulphur dioxide to a local airshed; the sulphur dioxide can then combine with atmospheric water vapour, to form sulphuric acid which lowers the pH of any precipitation. Commonly, although sulphur dioxide is released within a discrete local airshed, often it is carried on the wind to another airshed where it combines with water vapour forming acidic precipitation that falls in a region some distance from the original release.

4.3 SELECTING A TIME SCALE FOR REGIONAL MONITORING

Having identified spatial boundaries for a region, a time scale of interest must be specified. The scales of time and space are related yet different; and which scale is most relevant to the study objectives must be determined. The analyst must judge whether varying time scales or varying spatial scales provides the data needed to diagnose the problem. Spatial scales are important when evaluating the extent and magnitude of contamination or its effects on biota. Temporal scales are useful to discern contaminant-related trends or changes in populations or media with time. For instance, the same individuals may be tagged and repeatedly sampled with time to determine bioaccumulation of the substance. Tracking reproductive effects of chemicals is possible by evaluating population parameters (e.g., age and sex ratios or natality) of one population with time. In addition, changes in relative proportions of contaminant-sensitive and contaminant-tolerant species of one population will evaluate the effects of the chemical on population dynamics.

The temporal sampling approach employed by the US EPA's EMAP evaluates both local conditions and long-term trends in ecosystems by sampling on a rotating four-year sequential cycle (Overton *et al.*, 1990; Messer *et al.*, 1991). Sites are marked on a grid, and during the first year of sampling, one-fourth of the sites are sampled, with another (different) fourth of the sites sampled during the second, third, and fourth years. Starting with the fifth year, the sampling cycle is repeated with the first fourth of the site being revisited. This sampling design is well suited to detect persistent, gradual changes in local populations and to detect long-term changes that slightly affect the entire region.

Figure 4.2 illustrates the source of problems that can arise from the evaluation of an environmental problem on a limited, perhaps inappropriate, scale. When limited to those events that vary on a large scale (i.e., variability on a cycle of centuries), the assessment cannot discriminate events that vary on a much smaller scale (i.e., days); the converse is also true.

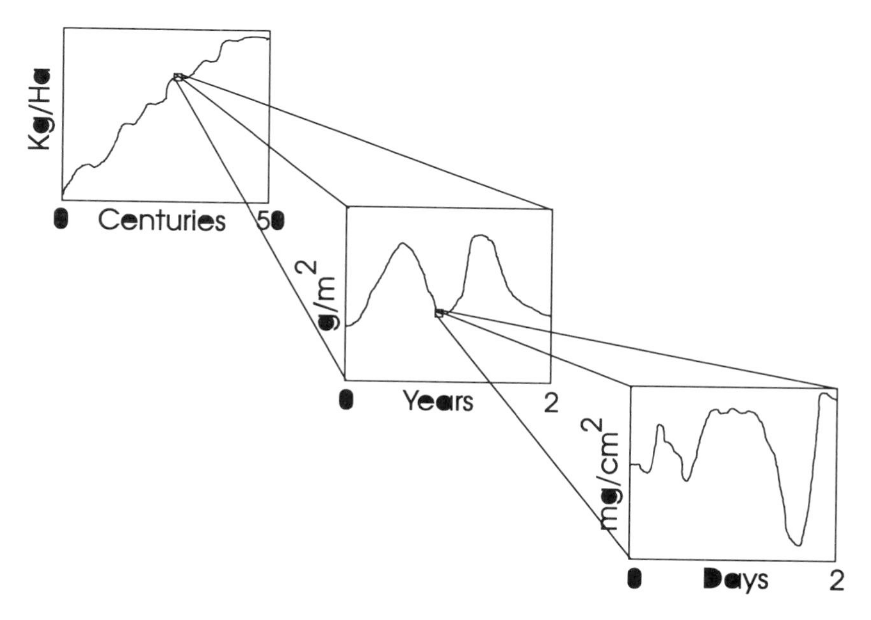

Figure 4.2. Illustration of the relevance of data at different time scales

For studies of regional effects of environmental pollutants, the relevant time scales are defined by those of pollutant releases and ecological responses. For assessments of regional effects of acid deposition on surface waters, for example, the US National Acidic Precipitation Assessment Program (NAPAP) chose a time scale of 30 years to ascertain reductions of atmospheric sulphur inputs and for the chemical composition of affected streams and lakes to reach equilibrium with the reduced loadings (NAPAP, 1991). When responses to anthropogenic stress are confounded by responses to natural climatic variability, time scales for assessments should include sufficient time to characterize climatic variability. Assessments of responses of forest ecosystems in the eastern United States to air pollution have been greatly complicated because the time scale of variation in temperature and precipitation in the eastern United States is roughly the same as that used to measure and predict changes in pollutant levels. The available time-series of climate records and forest condition data are too short to characterize this variation, and to separate its influence from the influence of pollutant deposition.

Clearly, the management time scale is also relevant to define time scales for regional assessments. Regulations must be reviewed according to prescribed schedules, and technologies for pollutant control cannot be implemented instantaneously. As with spatial scales, the natural time scales must first be defined for a regional assessment, and then overlaid on the appropriate management time scales.

4.4 REGIONAL RISK ASSESSMENTS

Suter (1993) described six general situations in which a regional-level ecological risk assessment may be the most appropriate approach. These are presented in Table 4.1, along with examples of each situation. A fuller discussion of these types of situations that may call for regional ecological risk assessments is provided in other chapters of this document.

Several issues associated with spatial and temporal scales must be considered when planning a regional ecological risk assessment. These include: (1) the spatial and temporal scales associated with various levels of biological organization at which effects may be assessed; (2) the relationship of landscape scales to the distributions of populations in an area, and the way in which they use the available resources; and (3) the spatial and temporal scales of potential chemical exposures. These issues are discussed here only briefly, but are presented in greater detail in a later chapter of this volume.

The effects of chemicals may be assessed at the biological organization levels of individual, population, community, ecosystem, region, or some combination of these levels. The spatial scale of a regional response to chemical exposure usually overlaps with, but extends beyond, that of individually exposed ecosystems (hundreds of kilometres to thousands of kilometres). The temporal scale of regional dynamics also overlaps that of the component ecosystems, but may be longer than for any one ecosystem in the region (tens of years to hundreds of years). Spatial and temporal scales associated with ecological effects at the various levels of biological organization are discussed by Sheehan (1984a, 1984b, 1992) and Suter (1993).

A second scale consideration for regional ecological risk assessments is the relationship between the scale of landscape patterns (including vegetation structure) and the distribution of animal species, and the scale of their use of resources (Holling, 1992). For example, at the regional scale, landscape responses to chemical stressors are often assessed in terms of changes in vegetation and vegetative structure, and little data are provided about the secondary effects of such changes in vegetation on animal populations. An understanding of the relationship between animal population dynamics and the structure of the landscape is essential to predict changes in animal populations that may result from changes in the vegetative structure of their habitat.

Table 4.1. Appropriate situations for a regional-level ecological risk assessment (Suter, 1993)

Assessment Concern	Examples
Local sources of toxic chemical releases that may have regional effects	Chernobyl nuclear reactor accident
Combinations of releases from individual sources, each within acceptable limits, may be unacceptable when combined	Fossil fuel combustion releasing SO_2 which combines with atmospheric water vapour producing H_2SO_4 and thereby lower the pH of precipitation; multiple effluent releases within a watershed
Regional scale processes that have an affect on the transformation and transport of airborne chemicals beyond that of local scale processes	Creation of photochemical smog
Pollutant emissions that have an effect on regional scales that do not occur on local scales	Depletion of stratospheric ozone by CFCs
Regional characteristics that are not present at local scales are more effectively protected by regional RA	Climatic processes
Success of regional regulatory and resource management programs can most adequately be assessed on a regional scale	Improvement in air quality within an airshed after emission regulations are implemented

The third consideration is the spatial and temporal scale of the chemical exposure itself. If the chemicals are widely distributed or have the potential to be, a regional approach is recommended. Broad-scale applications of pesticides, such as aerial spraying to control grasshoppers in the prairie regions of the United States and Canada (Sheehan *et al.*, 1987) and the spruce budworm in the forests of New Brunswick (Mitchell and Roberts, 1984), cover hundreds to thousands of kilometres, and may occur intermittently over years or decades. At the extreme, widespread aerial transport of photochemical oxidants (Skelly, 1980) and acids (Nash *et al.*, 1992) has contaminated large geographical regions (thousands to tens of thousands of kilometres). Similarly, multiple inputs of chemicals into river systems occurs over hundreds to thousands of kilometres of the watershed. Cumulative exposure to airborne and waterborne chemicals also takes place over long time periods (tens to hundreds of years).

4.4.1 ASCRIBING VALUES ON REGIONAL SCALES

As with any assessment, a clear statement of values and endpoints should be the first step in its planning. Regions possess the characteristics of their component ecological, socioeconomic, and geographic systems as well as unique characteristics associated with the interactions between these systems. The relationships among systems usually differentiate the region from a single ecosystem; therefore, the values associated with a region include the values associated with the individual component systems and the values associated with uniquely regional characteristics.

4.4.1.1 Values inherent in local systems within a region

Both the broad generic and more specific cultural values attributed to ecological systems are equally applicable to regions; therefore, the tables of market, non-market, cultural, and ecological values-benefits identified in Chapter 2 (Tables 2.1 and 2.2) are equally useful to analyse regional issues of chemical contamination. In fact, many of these values are best recognized at a regional scale (e.g., forage and timber production). A general goal of management at the regional scale is the conservation of all natural resources (including biological resources) for sustainable use (IUCN, 1980). Biological resources include: genetic diversity at the population-species level, biodiversity at the ecosystem-regional level of organization, and biological productivity (both primary and secondary) at the regional level.

4.4.1.2 Values unique to regions

The values unique to a region are the result of interactions between ecosystems and socioeconomic systems. Such interactions regulate ecological processes such as the cycling of water and nutrients vital to support sustainable resources. The International Union for the Conservation of Nature (IUCN, 1980) has ranked the conservation of fundamental ecological processes as a first priority objective. The basic components of this objective are:

1. The integrity of the hydrologic cycle is critical to the continued availability of water resources.
2. Disruption of the hydrologic cycles at the watershed-regional scale can result in dramatic changes in the quality and availability of water, and, in fact, has severe effects on human societies.
3. The cycling of nutrients, such as carbon, phosphorous, and nitrogen, is another regionally important ecological process that must continue uninterrupted to ensure the fertility of arable land on which agricultural production depends.
4. Climatic processes, that essentially control primary productivity, are an additional example of vital regional-level values resulting from interactions within the region. Changes in dependable climatic patterns severely disrupt agriculture and change the inherent liveability of certain areas.
5. Shifting rainfall patterns can turn a previously productive area into a non-productive area.

Table 4.2. Endpoints relevant to regional scale assessments

Assessment Endpoint	Illustrations
Traditional ecological endpoints	Changes in species numbers and diversity Loss of rare and endangered species Loss of primary productivity
Region-specific endpoints	Acid precipitation Degradation of air quality Habitat loss and fragmentation Enrichment of aquatic ecosystems by runoff of agricultural fertilizers Increased salinization of agricultural soils Disruption of nutrient cycles
Anthropocentric endpoints	Loss of economically important species Contamination and depletion of water resources Loss of aesthetically valued resources

Assessments of the impacts of chemicals on ecosystems should be planned with the consideration of the importance of these fundamental ecological processes that function on large geographic scales.

4.4.2 REGIONAL SCALE ASSESSMENT ENDPOINTS

After considering the target values, the endpoints should be chosen as the focus of any effects studies or monitoring. Although a number of local scale endpoints will also be important at the regional scale, for the ecological consequences of pollution, regional problems cannot be equated to the sum of local problems. In fact, the properties and characteristics of complex systems viewed on the regional scale are specific and qualitatively different from the properties of biological systems viewed from lower levels of organization.

Endpoints relevant to regional scale assessment can be classified into three general categories: traditional ecological endpoints; endpoints specific to the region; and anthropocentric endpoints (important to human cultures). Examples of these assessment endpoints that are important on regional scales are provided in Table 4.2. Certainly many of these examples are relevant to more than one assessment endpoint; for example, acid precipitation is an important regional-level endpoint and an anthropocentric endpoint.

4.5 DESIGN AND SAMPLING CONSIDERATIONS

Because regions are extensive in both time and space, subsamples are needed to represent the whole population of possible sample units. Site-specific point-based analyses of chemical constituents, for example, can be made at only a limited number of locations throughout the region and on a limited number of occasions to track the changes in those constituents over time.

Reliance on representative subsamples of the entire unit is used commonly in scientific studies. Dose-response tests are performed with only a limited number of organisms that represent an equally limited number of taxa. Even remote sensing techniques depend on reflected and emitted electromagnetic radiation, that is a sample of total radiation, both in terms of the patches of ground from which it comes and the parts of the spectrum over which it is measured. The approach to select sampling sites to be truly representative, therefore, is critical to the interpretation of the data collected for a regional risk assessment.

This section briefly discusses the study design considerations unique to regional scale ecological risk assessments. Chapter 2 provides a more detailed discussion of sampling design considerations that generally apply to most ecological risk assessments, and many excellent references are available that provide greater guidance on sampling design consideration (Green, 1979).

4.5.1 STATISTICAL INFERENCE

In a regional-ecological risk assessment, the desire to make statistical inferences about some defined region can be satisfied only when information that is representative of that region is available. In practice, two ways exist to ensure representativeness when selecting sampling sites. The sampling sites must be selected at random from all possible sites, or they must be selected by some systematic process that eliminates subjectivity. Systematic sampling is often easier in regional risk assessments; however, derivation of completely valid estimates of the precision of a characterization based on systematic sampling is impossible. Most forms of statistical analysis require an element of randomness in the sampling design; consequently, some form of random sampling is preferred to measure physical, chemical, or biological variables.

4.5.2 NUMBER OF SAMPLES

The number of samples required to characterize adequately the condition of, and changes in, an ecosystem on a regional scale depends on two major factors. The first is the natural variability of the characteristic of the ecosystem being measured. An initial task in developing a sampling design for a regional risk assessment, preliminary sampling (i.e., pilot studies), can help determine the inherent variation in characteristics to be measured. Alternatively, determination of variation from data collected for earlier studies in the same region may be possible.

The second factor determining the number of samples required is the desired precision of estimates, which can be expressed as the standard error of the mean of the sample determination. In practice, the number of samples collected for an assessment will be a compromise between the desired precision of the information and the cost of the research. Excellent references are available that offer more detail on determining sampling parameters (Green, 1979; Gilbert, 1987).

A most critical influence on the number of samples collected is the cost of monitoring and analysis. An approach to balance these costs against data uncertainty is the data quality objectives approach (DQO) developed by the US EPA (USEPA, 1991). This approach helps to ensure that the type, amount, and quality of data collected are adequate to meet project objectives in a cost-effective manner. The DQO process results in a quantitative statement of the level of uncertainty in the results that a data user is willing to accept.

The US EPA Environmental Monitoring and Assessment Program (EMAP) is a national ecological status and trends monitoring program that developed its sampling design based on the following DQOs. An example of the DQO for trend detection is that "EMAP should be able to detect a 2 percent change per year in a condition indicator over a decade with α of 0.2 and β of 0.3 (power of 0.7) for a resource class (e.g., wetlands or forests) on a regional scale." An example of EMAP's DQO for status assessment is that "for each indicator of condition and resource class on a regional scale, estimate the preparation of the resource in subnominal condition within 10 percent (absolute) with 90 percent confidence based on four years of sampling" (Kirkland, 1994).

4.5.3 STRATIFICATION

The precision of regional estimates can be improved by dividing the region of interest into multiple homogeneous units that are less variable than the region as a whole. In characterizing media chemistry, for example, the soil of a region can be stratified according to land use (including forestry, tillage, horticulture, rough grazing, etc.) and soil type, among other factors. Separate random samples from each stratum can be combined in various ways to obtain a more precise characterization of media chemistry than could be obtained by selecting sampling sites in a completely random fashion. Aquatic media may be stratified by salinity and temperature characteristics. For example, the EMAP has been conducting research on the indicators it uses to assess estuarine condition. These samples are stratified on the basis of salinity (marine, polyhaline, mesohaline, oligohaline-tidal, freshwater) to aid in assessing the reliability of using these indicators to reliably distinguish between polluted and unpolluted environments (Holland, 1990).

Linear features such as roads, hedgerows, and streams must be recognized as separate strata in regional assessments. Such linear features often have particular ecological mechanisms and adaptations that enhance biodiversity, heterogeneity, or productivity, and they usually need to be sampled separately. Ecotones, the interfaces between different habitats, also fall in this category of linear systems that should be regarded as separate strata when attempting to characterize a region.

4.5.4 MEASUREMENT OF CHANGE

Regional risk assessments often require measuring spatial or temporal changes in regional scale processes or ecological characteristics. Measuring the status of an ecosystem and detecting trends over time are competing interests: status is generally best assessed by sampling as much of the resource as possible at a given time, while trends are generally best detected by repeating measurements at the same locations or of the same individuals at regular time intervals (Overton *et al.*, 1990). Revisiting sampling sites at too frequent intervals, however, may provide insufficient time for recovery from measurement stress. To detect subtle trends in diffuse subpopulations, alteration-degradation of the sampling location or individuals must be minimized. Another useful approach is collecting measurements from different individuals within the same population over time, while still allowing for assessments of the regional population (Overton *et al.*, 1990). A difficulty of this approach is ensuring that repeat samples are truly from the same population and account for the natural variability within a population.

The points of measurement or the individuals (e.g., trees and animals) must be marked to ensure that repeat measurements are taken at the same place or from the same individuals. However, care must be taken to ensure that the marking does not, itself, introduce a bias. Unfortunately, marks of any kind in the environment often increase the chance of vandalism, leading either to loss of the marking or to a change in the environmental process being measured. The efficiency and precision of measurements of change in ecological systems or processes can be enhanced greatly by sampling with partial replacement.

4.5.5 IMPORTANCE OF PRELIMINARY SAMPLING

Developing an effective sampling design for a regional assessment requires preliminary knowledge of the region and its features to identify the appropriate size of the sampling units, the number of samples required to obtain the desired precision of estimates, and the possible presence of large-scale spatial patterns that necessitate stratification. Because these problems cannot be remedied by computation after the data are collected, preliminary or pilot studies to investigate such issues are highly recommended. Even though pilot studies are time consuming, and may not generate data that are directly useful in the final assessment, pilot studies cannot be circumvented without jeopardizing the validity of the ultimate assessment.

4.6 METHODS TO MONITOR AND ASSESS REGIONAL ENVIRONMENTS

The appropriate choice of methods to collect and analyse data at the landscape or regional level depends on the objectives of the assessment. For the purpose of discussion, these issues are assumed to have been determined, and an experimental design has been assumed to have been selected. Hypothetically, the issue to be addressed is the use of herbicides to control shrubs on extensive rangelands, and management decisions may be needed to assess the effect of this activity on a management unit scale. The spatial scale for management may be on the order of thousands of square kilometres. The experimental design will include the types of data to be collected, the size of the primary sampling unit, the numbers of samples to be collected, and the means to aggregate those samples at the regional level.

For this illustration, the effects of the herbicide on the soil and plants and on the region as a whole must be well understood. The effects can be considered at the molecular and cellular levels; the extent and distribution of the effects can be considered; and ecosystem and regional effects can be assessed. Clearly, different techniques are required at each level. Concentrations of chemicals must be measured in soil, soil-water, and plant tissue. The effects of the herbicide at the community level may need to be assessed. Methods for spatial characterization, as well as regional aggregation of physical, ecological, and socioeconomic effects, must be selected. Finally, methods to monitor trends in the effects of the chemicals and to monitor the effectiveness of management strategies regarding the use of the chemical must be identified and evaluated.

Even on the regional scale, the significance and importance of point (or sample-based) measurements and organism-level or mechanistic-level investigations cannot be underestimated. The organism is the smallest unit that interacts directly with the environment (Suter, 1993), and the organism is directly exposed to chemicals that may be toxic. The reproducing population (or, perhaps, the "canopy level" in forestry studies; Mooney *et al.*, 1991) is the smallest ecological unit that persists on a human time scale; hence, it is the lowest level that can be managed and protected. The aggregation of plant and animal populations into communities and their interaction at the ecosystem level form a basic unit of the landscape or region. The reader is referred to Chapter 3 of this report as well as to chapters on the effects of chemicals on forests, arid ecosystems, soils, aquatic systems, and others for more detailed information on these and other techniques. In the following, methods to conduct regional-level environmental monitoring and assessment are discussed, emphasizing ecosystem or landscape-level methods. Methods applied to the organism, population, and community levels are presented in Chapter 3.

4.6.1 REMOTE SENSING TECHNIQUES

Remote sensing techniques, perhaps most useful at the ecosystem and landscape levels, include systems that acquire data at more detailed levels, and then are "scaled up" or extrapolated in various ways. Aggregations of the French Système Probatoire de l'Observation de la Terre (SPOT) and Landsat (Multi-Spectral Scanner (MSS) and Thematic Mapping (TM)) pixels allow scientists and managers to classify or otherwise characterize land-use and land-cover information at the regional level. The relatively high temporal resolution of both systems allows researchers to monitor changes in the attributes of regions over time. Multidate imagery techniques, which allow quantitative assessments of change, are discussed in a subsequent chapter of this report. Developing techniques in landscape ecology allow assessment of fractal dimension, patch, connectivity, and other parameters from which landscapes can be assessed. METEOSAT and the National Oceanic and Atmospheric Administration's AVHRR (Advanced Very High Resolution Radiometer) may allow assessments of very large areas over short time intervals, and enable researchers to scale up information derived from higher resolution remote sensing, field, and laboratory measurements. AVHRR-derived vegetation indices allow for an assessment of biomass and net primary productivity at regional and landscape levels and, especially, for monitoring changes in those variables over time. Mooney *et al.* (1991) suggested that remotely sensed measurements of APAR might serve as an index for the combined effects of multiple stresses. Although extrapolations to higher levels of organizations can be a source of uncertainty, using remote sensing to assess stress at the regional level can be an effective tool that may lead to new insights about regional and global stressors and effects.

Several researchers have investigated relationships between ecosystem variables and indices derived from remote sensing data. Peterson *et al.* (1987) found strong relationships (R2 = 0.91) between Landsat TM near infrared/red ratios (Normalized Difference Vegetation Index, NDVI) and LAI (Leaf Area Index) (Peterson *et al.*, 1987) in closed-canopy, pure conifer forests in west central Oregon. Figure 4.2 (Peterson *et al.* 1987) illustrates relationships between LAI and NDVI and between NDVI and NPP (Net Primary Production) in conifer forests in Montana.

4.6.2 GEOGRAPHIC INFORMATION SYSTEMS

Analysing the effects of chemicals on the environment at regional levels typically requires compiling and integrating data from disparate sources. Most of these data sources are geographically referenced; that is, the information they contain is specifically referenced to geographic locations. Geographically referenced data sets can be integrated and manipulated using a geographic information system (GIS). A GIS allows spatial data to be compiled and integrated with additional data layers or non-spatial attributes. A GIS also allows the analyst to simulate impacts in the attributes of a given data layer or indicator in response to a hypothesized change in the environment that might stress that indicator in different ways. As a modelling tool, GISs have been used to estimate potential environmental impacts over large, spatially varying ecoregions. GIS techniques have allowed analysts to identify spatial and temporal trends based on ecosystem and landscape level data, as well as the projected impacts from current and future development. Consequently, GIS is an especially powerful tool for landscape and regional assessments. Excellent references exist on the concepts and applications of GISs (Ripple, 1987; Koeln *et al.*, 1994).

4.7 CHARACTERIZING RISK AT THE REGIONAL SCALE

4.7.1 STATISTICAL METHODS

A statistically-based design is important in any regional assessment of risk. This design will ensure that the data acquired during the program are of sufficient quantity and quality to answer the questions posed by the risk managers and regulators who initiated the study. The use of a sound statistical design and methods also provides a firm basis to deal with uncertainty in the risk analysis. In particular, statistical theory defines the conditions under which the scientist is able to make valid inferences about some defined region or landscape. Those conditions include the formal definition of the region in both time and space, and the examination of "fair" samples from locations, organisms, soils, water, and air from that "population." The statistician's definition of "fair," however, is strict, in that the samples must be derived objectively from the population and, in practice, such samples must be taken either systematically or at random. Inferences based on samples or case studies that are chosen subjectively by the scientists concerned with the risk analysis are not acceptable as a basis for valid inferences about the population (Jeffers, 1988b).

At an analytical level, statistical methodology provides a range of techniques for determining the level of uncertainty in any measurement or combination of measurements. Those techniques depend on replication and random sampling in the original measurements and greatly enhance the control of extraneous variation by the use of stratification, covariance, and ancillary measurements (Green, 1979). The dangers of extrapolation beyond the range of the original data remain a constant potential source of error, as does the accumulation of errors in the aggregation of data over space and time. However, the failure of basic assumptions in traditional statistical methods has become less important with the introduction of new computer-based techniques (Lunn and McNeil, 1991).

4.7.2 MODELLING

Models, defined here as the formal statement of relationships in physical or mathematical terms, play an integral role in regional environmental monitoring and assessment. A mathematical model may be used to assess the transport and fate of contaminants as well as effects at all ecological levels from organismal to regional. Models used in regional ecological risk assessments may be deterministic (not incorporating random events) or may be stochastic or probabilistic (incorporating random events or variables).

All wholly deterministic modelling formulations depend on estimation of the model parameters as a special case of statistical inference (i.e. from data obtained as fair samples from some defined population). All models used to synthesize data have specific assumptions that need to be checked before the model is used for making decisions. Indeed, verifying and validating models against new data sets are essential components of the systems analysis in which the model building is embedded. The sensitivity of models to small changes in their basic parameters, a process known as sensitivity analysis, should be conducted whenever they are used.

Ideally, modelling should be embedded in a formal process of systems analysis that enables comparison of alternative models and facilitates validation and refinement of models (Jeffers, 1978). The range of alternative models may include qualitative or semi-quantitative representations of the functional relationships between the landscape elements. More usually, it will include deterministic mathematical models of the basic ecological processes, expressed as differential or difference equations. The effects of airborne and waterborne pollutants on agricultural and forest ecosystems have all been expressed in this way, as have the consequences of alternative energy strategies on political and continental regions.

Increasingly, however, risk analysis and other disciplines employ stochastic (hence, non-deterministic) models in which probability distributions represent the uncertainty in the relationships between several, perhaps many, variables. Starfield and Bleloch (1986) and Tijms (1986) described the use of such models for a variety of applications, including wildlife conservation and management. Most stand simulation models used to evaluate the effects of pollutant levels, pest infestations, or hydrological processes on forest, agricultural, or wetland landscapes are of this type. Markov models, which depend on the probabilities of transitions to and from a limited number of states, are widely used to simulate ecosystem succession (Jeffers, 1988a), and play an important role in creating regional scale models for ecological risk assessment, particularly when combined with methods of pattern analysis (Howard, 1991). The methods of linear, non-linear, and dynamic programming that emerged from the operations research applications of World War II have long been adopted by environmental managers in the search for optimal control strategies, especially in the management of regional scale systems for multiple goals. An extension of game theory to meta-game theory provides an appropriate model for describing conflict and finding stable resolutions to conflict, and has been used widely to model and resolve conflicts at the regional scale, for example, conflicts between polluting industries and national or regional control agencies. A recent advance in modelling of systems of all sizes is the application of network and topological models (Barnsley, 1988; Higashi and Burns, 1991).

With increasing development of expert systems, namely computer systems aimed at offering the same advice with justification that would be given by a human expert, the role of the "rule bases" (on which such systems depend) is increasing in interest. Such "rule bases" typically consist of sets of logical statements, ideally in common language, to help the manager choose the appropriate option(s) to manage a system. Research on the development of rule-based systems as a basis for environmental decision support systems is currently proceeding (Guariso and Werthner, 1989). Appropriate computer-based methods to extract a "rule base" from data have also been developed and are currently being refined.

Regional scale models are already being used to synthesize regional-level data and make inferences for regional-scale management of environmental risks. The best-known examples are the models used to quantify impacts of acidic deposition on terrestrial and aquatic ecosystems in Europe and North America. Hordjik (1991) described the use of a regional scale acid deposition model in designing sulphur emission control strategies for western Europe. Wright *et al.* (1991) and Baker *et al.* (1990) described regional-scale models used to assess effects of acidic deposition on lakes and streams. Although these models have not been used extensively for management purposes, models of the effects of climate change on regional ecosystem characteristics are being developed and tested (Agren *et al.*, 1991).

4.7.3 SOURCES OF UNCERTAINTY

Variation is an inherent property of living organisms and their environment. The genetic processes of sexual reproduction result in a range of phenotypes that expressed similar morphological characteristics or reactions to stimuli or environmental conditions. Chemical and physical processes within the environment also result in heterogeneity in both time and space; consequently, every measurement of the environment is subject to variability that needs to be determined and, if possible, controlled to improve the ability to perform assessments.

This variability is further compounded by the fact that the accuracy with which the physical, chemical, or biological properties of the environment can be measured is at least partly determined by the cost and technical sophistication of the instrumentation available and the technical abilities of the scientist or technician performing the measurements. Errors are possible even with the most technically advanced instrumentation.

In any investigation of complex systems, hierarchy theory emphasizes the importance of working at the coherent level in the hierarchy (Allen and Starr, 1982). The perceived variability in environmental systems may be increased by attempts to link ecosystem processes to forcing functions, such as temperature, precipitation, and moisture content, at too high a level in the hierarchy. Similarly, attempts to characterize ecosystems on the basis of processes at too low a level in the hierarchy may replicate variability that is apparently unrelated to ecosystem function, for example, attempting to determine the response of forests to airborne pollutants on the basis of measurements of photosynthesis in individual leaves. Choosing the proper level in the hierarchy to characterize risk is a significant step in reducing uncertainty in any assessment.

Regional scale risk assessment does not end with the analysis and determination of uncertainty. The information derived from the assessment and uncertainty analysis must be presented to decision makers and risk managers in a clear and effective manner, because decisions are often made on those parts of the assessment that contain the highest degree of uncertainty.

4.8 RECOMMENDATIONS TO ENHANCE REGIONAL ENVIRONMENTAL MONITORING AND ASSESSMENTS

Encourage additional support for research on:

1. ecotoxicological and ecological research at the landscape level;
2. concentrations of organics and trace elements in the environment to establish baseline values and identify long-term trends;
3. interactions between landscape components;
4. sensitivity of various regional landscapes to chemicals;
5. steps in the assessment-management process that have the weakest foundations.

Support development of improved methods to:

6. evaluate exposures and effects at regional levels;
7. determine ecological and socioeconomic values at a regional scale;
8. develop methods to scale from point base to region;
9. develop better methods to organize and present regional data;
10. develop better methods to organize, structure, and present regional ecosystem data for multiuser access.

Encourage and enhance:

11. increased use of synthesis (modelling) in addition to analysis at all spatial and temporal scales in risk assessment;
12. incorporation of sensitivity analysis of the models used in a risk analysis;
13. improved statistical education for those who conduct regional risk assessments and for those who use them;
14. education in effective risk communication methods for environmental professionals.

4.9 REFERENCES

Agren, G.I., McMurtrie, R.E., Parton, W.J., Pastor, J., and Shugart, H.H. (1991) State-of-the-art models of production–decomposition linkages in conifer and grassland ecosystems. *Ecol. Appl.* **1**, 118–138.

Allen, T.F.H., and Starr, T.B. (1982) *Hierarchy: Perspectives for Ecological Complexity*. University of Chicago Press, Chicago.

Baker, J.P., Freda, J., Christensen, S.W., and Sale, M.J. (1990) Biological effects of changes in surface water acid–base chemistry. In: *Acidic Deposition: State of Science and Technology*. NAPAP Report 13. National Acid Precipitation Assessment Program, Washington, D.C.

Barnsley, M. (1988) *Fractals Everywhere*. Academic Press, London.

Costanza, R., Sklar, F.H., and White, M.L. (1990) Modeling coastal landscape dynamics. *Bioscience* **40**, 91–107.

Dale, V.H., and Gardner, R.H. (1987) Assessing regional impacts of growth declines using a forest succession model. *J. Environ. Man.* **24**, 83–93.

Dobson, S. (1985) Methods for measuring effects of chemicals on terrestrial animals as indicators of ecological hazard. In: Vouk, V.B., Butler, G.C., Hoel, D.G., and Peakall, D.B. (Eds.) *Methods for Estimating Risk of Chemical Injury: Human and Non-Human Biota and Ecosystems*, pp. 537–548. SCOPE 26. John Wiley & Sons, New York.

Gilbert, R.O. (1987) *Statistical Methods for Environmental Pollution Monitoring*. Van Nostrand Reinhold, New York.

Graham, R.L., Hunsaker, C.T., O'Neill, R.V., and Jackson, B.L. (1991) Ecological risk assessment at the regional scale. *Ecol. Appl.* **1**(2), 196–206.

Green, R.H. (1979) *Sampling Design and Statistical Methods for Environmental Biologists*. John Wiley & Sons, Chichester.

Guariso, G, and Werthner, H. (1989) *Environmental Decision Support Systems*. Ellis Horwood Ltd, Chichester.

Higashi, M., and Burns, T.P. (1991) *Theoretical Studies of Ecosystems: The Network Perspective*. Cambridge University Press, Cambridge.

Holland, A.F. (Ed.) (1990) *Near Coastal Program Plan for 1990: Estuaries*. US Environmental Protection Agency, Environmental Monitoring and Assessment Program, Washington, D.C.

Holling, C.S. (1992) Cross-scale morphology, geometry and dynamics of ecosystems. *Ecol. Monogr.* **62**(4), 447–502.

Hordjik, L. (1991) Use of the RAINS model in acid rain negotiations in Europe. *Environ. Sci. Technol.* **25**, 596–603.

Howard, P.J.A. (1991) *An Introduction to Environmental Pattern Analysis*. Parthenon Publishing, Casterton.

IUCN (1980) *World Conservation Strategy*. International Union for the Conservation of Nature, Gland, Switzerland.

Jeffers, J.N.R. (1978) *An Introduction to Systems Analysis: With Ecological Applications*. Edward Arnold, London.

Jeffers, J.N.R. (1988a) Statistical and mathematical approaches to issues of scales in ecology. In: Rosswall, T., Woodmansee, R.G., and Risser, P.G. (Eds.) *Scales and Global Change: Spatial and Temporal Variability in Biospheric and Geospheric Processes*. SCOPE 35. John Wiley & Sons, Chichester.

Jeffers, J.N.R. (1988b) *Practitioner's Handbook on the Modelling of Dynamic Change in Ecosystems*. SCOPE 34. John Wiley & Sons, Chichester.

Kirkland, L.L. (1994) *EMAP Quality Management Plan*. US Environmental Protection Agency, Washington, D.C.

Koeln, G.T., Cowardin, L.M., and Strong, L.L. (1994) In: Bookout, T.A. (Ed.) *Research and Management Techniques for Wildlife and Habitats*, 5th edn., pp. 540–566. The Wildlife Society, Bethesda.

Kutz, F.W., Linthurst, R.A., Riordan, C., Slimak, M., and Frederick, R. (1992) Ecological research at EPA: new directions. *Environ. Sci. Technol.* **26**, 860–865.

Lunn, A.D, and McNeil, D.R. (1991) *Computer-Interactive Data Analysis*. John Wiley & Sons, Chichester.

Meadows, D.H. (1991) *The Global Citizen*. Island Press, Washington, D.C.

Messer, J.J., Linthurst, R.A., and Overton, W.S. (1991) An EPA program for monitoring ecological status and trends. *Environ. Monitor. Assess.* **17**, 67–78.

Mitchell, M.F., and Roberts, J.R. (1984) A case study of the use of fenitrothium in New Brunswick: The evolution of an ordered approach to ecological monitoring. In: Sheehan, P.J., Miller, D.R., Butler, G.C., and Bourdeau, P.H. (Eds.) *Effects of Pollutants at the Ecosystem Level*. SCOPE 22. John Wiley & Sons, Chichester.

Mooney, H.A., Winner, W.E., and Pell, E.J. (1991) *Response of Plants to Multiple Stresses*. Academic Press, San Diego.

NAPAP (1991) *1990 Integrated Assessment Report*. US National Acidic Precipitation Assessment Program, Washington, D.C.

Nash, B.L., Davis, D.O., and Skelly, J.M. (1992) Forest health along a wet sulphate/pH disposition gradient in north-central Pennsylvania. *Environ. Toxicol. Chem.* **11**, 1092–1104.

Omernik, J. (1987) Ecoregions of the conterminous United States. *Ann. Assoc. Am. Geogr.* **77**(1), 118–125.

Overton, W.S., White, D., and Stevens, D.L. (1990) *Design Report for EMAP*. Report No. EPA 600/3-91/053. US Environmental Protection Agency, Corvallis.

Peterson, E.B., Chan, Y.-H., Peterson, N.M., Constable, G.A., Caton, R.B., Davis, C.S., Wallace, R.R., and Yarranton, G.A. (1987) *Cumulative Effects Assessment in Canada: An Agenda for Action and Research.* Cat. No. EN106-7/1987E. Minister of Supply and Services Canada, Ottawa.

Pulliam, H.R. (1988) Sources, sinks, and population regualtion. *Am. Natur.* **132**, 652–661.

Ripple, W.J. (Ed.) (1987) *GIS for Resource Management.* American Society for Photogrammetry and Remote Sensing, and American Congress on Surveying and Mapping, Falls Church, Virginia.

Sheehan, P.J. (1984a) Effects on individuals and populations. In: Sheehan, P.J., Miller, D.R., Butler, G.C., and Bourdeau, P.H. (Eds.) *Effects of Pollutants at the Ecosystem Level.* SCOPE 22. John Wiley & Sons, Chichester.

Sheehan, P.J. (1984b) Effects on community and ecosystem structure and dynamics. In: Sheehan, P.J., Miller, D.R., Butler, G.C., and Bourdeau, P.H. (Eds.) *Effects of Pollutants at the Ecosystem Level.* SCOPE 22. John Wiley & Sons, Chichester.

Sheehan, P.J. (1992) Ecotoxicological considerations. In: Cote, R.P., and Wells, P.G. (Eds.) *Controlling Chemical Hazards.* Unwin Hyman, Boston.

Sheehan, P.J., Baril, A., Mineau, P., Smith, K.K., Hartenist, A., and Marshall, W.K. (1987) *The Impact of Pesticides on the Ecology of Prairie Nesting Ducks.* Technical Report Series No. 19. Environment Canada, Canadian Wildlife Service, Ottawa.

Skelly, J.M. (1980) *Photochemical Oxidant Impact on Mediterranean and Temperate Forest Ecosystems: Real and Potential Effects. Effects of Air Pollutants on Mediterranean and Temperate Forest Ecosystems.* General Technical Report PSW-43. Pacific Southwest Forest and Range Experimental Station, Berkeley.

Starfield, A.M., and Bleloch, A.L. (1986) *Building Models for Conservation and Wildlife.* Macmillan, London.

Suter, II, G.W. (1993) *Ecological Risk Assessment.* Lewis, Boca Raton, Florida.

Thomann, R.V. (1972) *Systems Analysis and Water Quality Management.* McGraw-Hill, New York.

Tijms, H.C. (1986) *Stochastic Modelling and Analysis.* John Wiley & Sons, Chichester.

Turner, M.G. (1987) Simulation of landscape changes in Georgia: A comparison of 3 transition models. *Landscape Ecol.* **1**, 29–36.

USEPA (1991) *Planning for Data Collection: The Data Quality Objectives Process for Environmental Decisions* (Draft, 16 October 1991). US Environmental Protection Agency, Office of Research and Development, Quality Assurance Management Staff, Washington, D.C.

Wright, R.F., Cosby, B.J., and Hornberger, G.M. (1991) A regional model of lake acidification in Norway. *Ambio* **20**, 222–225.

Part 2

CONTRIBUTED PAPERS

5 Biodiversity in the Assessment of Freshwater Quality

J. Verneaux
Université de Franche-Comté, France

5.1 BACKGROUND

5.1.1 POLLUTION AND OTHER DEGRADATION FACTORS

Water pollution often represents a very complex set of factors that degrade aquatic systems. Intervention in the return of a river to its prepolluted condition can attack the mosaic of habitats; for example, the erection of a large dam can modify biological structures and reduce the consumer populations (invertebrates and fish) as much as the effects of industrial discharges to surface water. Estimating the degree of overall degradation of a site due to pollutant load requires targeted objectives that depend upon different methods of remediation.

5.1.2 MANIFESTATIONS OF POLLUTION DAMAGE

In a lake or stream, the consequences of various discharges depend on the flow of the river and the assimilation capacity of the receiving environment. This ability to transform and transfer allochtonous river drift depends on the physical, chemical, and biological characteristics of the system in question. Three possibilities exist:

1. The introduction and dissemination of toxic or inhibiting substances. When nutrients are unlikely to enter the trophic structure (due perhaps to toxic inhibitory substances), pollution causes a gradual and eventually complete disappearance of species in the biological structure, as has been demonstrated for some pesticides, cyanides, detergents, and metals.
2. Excessive amounts of nutritive substances (organic matter, including nutrients). When the river drift of nutritive substances is progressive yet still within the assimilation capacity of the system, eutrophication (i.e., complexation of the biological structure) may be only temporarily accelerated.
3. The amount of river drift presumed to be nutritive exceeds the assimilation capability of the receiving water. This condition results in the gradual

Methods to Assess the Effects of Chemicals on Ecosystems
Edited by R. A. Linthurst, P. Bourdeau, and R. G. Tardiff

development of a pollution condition characterized by the accentuated simplification of the consuming biological structure. The accumulation of unused substances implies a chemical modification of the environment of which several parameters first reach, and then exceed, the tolerance limits of an increasingly large number of species; the simplification of the consuming biological structure worsens this process, which accelerates exponentially.

Pollution of an aquatic system is manifested at the population level by three phenomena. Modification of the structure of the initial population results in the development of a few saprophage or euryoeces populations, such as Oligochetes or the *Hydropsychidae*, and the decrease in abundance of other more sensitive organisms, such as certain *Heptageniidae* or Plecopters. Simultaneously, progressive desertion of habitat by lentic facies in favour of lotic facies can be observed, whereby species of the initial population do not appear or disappear—a process termed "imposed habitat changes" (Verneaux, 1973). Appearance, then proliferation, of species elective of specific river drift occurs; for example, this may occur because of intense development of certain algae, bacteria, and fungi (Cladophoraceae, *Spirogyra, Sphaerotilus, Leptomitus, Fusarium,* Cellolobacteria, Ferrobacteria, and Sulphobacteria) downstream from organic or specific discharges. Gradual disappearance in a specified order of all or part of the initial population then follows. These phenomena emphasize the differences between the eu-functional and dysfunctional systems, regardless of their potential trophic level. Eutrophication corresponds to an increase in biodiversity, and is, therefore, the opposite of the adverse effect of pollution or of the general system degradation phenomena which produce a decrease in the biodiversity of consuming organisms. Thus, an environment can be polytrophic without being eutrophic; and, for numerous reasons, a polluted environment may become dystrophic.

5.1.3 SPECIALIZED AND PRACTICAL METHODS

Analysis of published findings unveils the contrast between the large number of biological analyses performed, whose variety justifies both the statement by Bartsch and Ingram (1966) that "there are as many methods as biologists working on this topic" and the observation that few proposals include a complete, precise, and standardized protocol used as a standard by others. Depending on the specialization of the analysts, the protocols show great variety in the types of organisms, sampling procedures, taxonomic units selected, and procedures for data analysis.

The great variety of techniques employed in biocenotic testing requires that quantitative data be representative of the taxokenoses, and underscores the perception that each investigator develops and uses custom-tailored methods based on biological and ecological peculiarities of the group of organisms in question, characteristics of the investigated biotopes, work scale, and objectives of the investigations.

When a method of comparative analysis is proposed, defining appropriate and specific sampling protocols is problematic, because the parts of the method are interdependent. The macrobenthos sampling techniques that are employed jointly during seminars organized by the European Commission on biological indicators have shown the superior efficacy of a differentiated sampling protocol, in which the number of readings and the habitat categories are predefined (Verneaux *et al.*, 1982)—in contrast to other techniques in which the types of habitats are either not defined or are subjected to unrestricted research for a given time duration (Mouthon and Faessel, 1978).

The general sampling methods should include monitoring of:

1. "drift" organisms directly in the water;
2. impact of periphyton "growth" and macrobenthos levels on the local nutrient supplies; and
3. micro-organisms (including periphyton), macrobenthos, and plants (such as mosses) that concentrate micropollutants (Mouvet, 1986) on the artificial supplies or substrates.

Although the number of comparative analysis protocols is rather small, the sampling techniques are apparently subjected to successive revisions (Downing and Rigler, 1984). Virtually all organisms, from unicellular ones to fish, have been used as indicators for a given component of freshwater quality. Studies pertaining to stagnant water (lakes, ponds) especially employ planktonic organisms and organisms related to streams, the benthic organisms. A distinction must be made between essentially fundamental biocenotic analyses, which require a determination of the species, and practical methods in which identification is restricted to units of taxonomic groups that can be identified by non-specialized operators.

Although methods using Diatoms (Descy and Coste, 1991), Mollusks (Mouthon, 1993), and the Oligochaetes (Lafont *et al.*, 1991) exist, the most frequently employed practical methods involve the simplified analysis of the macrobenthos of streams. In France, the dulcicole macroscopic invertebrates include approximately 150 families, 700 genera, and 2200 listed species (Ilies *et al.*, 1978).

Depending on the nature of the organisms employed and the objectives being pursued, two main trends can be distinguished in the assessment of water quality and the biogenic tendencies of systems:

1. assessments based on the presence of organisms considered indicators of a specific type of contamination (which includes analyses for bacterial contamination and those involving "biotic indices"); and
2. global approaches, based on the investigation of all or part of the aquatic populations, for which the absence of some organisms is as significant as the development of populations of other organisms.

5.1.4 PROCEDURES FOR DATA ANALYSIS

Based on fundamental parameters for the analysis of populations by specific variety (or taxonomic richness) and density (or abundance) of individual organisms, the criteria have evolved from a simple comparison of specific indicators to the interpretation of biological structures in relation to reference organizations (biotypology) established in the abstract space of mathematical analyses. Direct comparisons of species indices provide observations that are especially interesting when they allow specification of biocenotic evolution. No specific analytical procedure is required, and the value of the conclusions depends on the precision of the findings and on the biological and ecological knowledge of the operators.

Graphic procedures or simple formulas such as species deficit ("Artenfehlbetrag;" Kothé, 1962) allow a global or group-by-group visualization of the changes which have an impact on biocenoses. The advantage of such determinations is that they offer a most complete perspective of the population of a location at a given time, that can be used as a reference for subsequent studies. This situation mandates that choices regarding impact studies be made by a recognized expert to have some value. Furthermore, true impact studies must be provisional; for that reason, they require access to biological and parametric models. Such access is a very rare occurrence.

From the baseline parameters to study the populations formed by the taxonomic variety and abundance of individual organisms, authors have proposed numerous indices of diversity and regularity, of which the most frequently used are that of Shannon (1948) with incorporation of measurement of the regularity ("evenness") of taxonomic distribution, including "equitability," "diversity," and "redundance" (Lloyd and Gehlardi, 1964; Patten, 1962; Hurlbert, 1971). These techniques are discussed and critiqued in the works of Margalef (1974), Pielou (1975), and Legendre and Legendre (1979). The main deficiency of these indices seems to be that similar values are offered for situations that may be very different.

The structure of populations can be investigated by their distributions of variety and abundance. Applying the log-normal model of Preston (1948) to the analysis of populations of benthic Diatoms, Patrick *et al.* (1954) proposed to use the flattening of the Gaussian curve, a function of the number of species and the standard deviation, as the indicator of water quality. Without consideration for some distributions resulting in different curves, the examination of numerous readings shows that, according to the taxons in question and to sampling procedures and types of environments, highly divergent structures are obtained, and they are well suited for the adjustment of the most diverse functions (binomial, positive and negative, logarithmic, exponential, linear) when multiple peaks do not appear. However, modifications that involve the distributions may constitute a useful contribution to the formulation of interpretative hypotheses (Amanieu *et al.*, 1981).

The findings need to be compared with biological reference organisms. Whenever the environments under study have been previously investigated with exhaustive definition of the biological organizations in abstract mathematical terms, new findings can be analysed with the initial structure used as a reference (i.e., biotypology) (Verneaux, 1973). Whether a water flow is considered in isolation or as a hydrographic network, the replacement of one set of species with another throughout a water flow system is plotted into the first two axes of an AFC (Hill, 1974; Orlóci, 1975; Benzecri *et al.*, 1973) by an ecological continuum of species in the shape of a U ("Guttman effect"), to which other specific structure can be compared).

Within a biogeographic area of interest, the essential problem is loss of reference, or loss of the assurance that a sufficient number of control (sampling) stations (which might have degraded to varying degrees) exist for the establishment of a minimum reference structure. Such frameworks were established for 12 water flows through the Doubs hydrographic network on the basis of findings made between 1969 and 1972. All subsequent modifications are interpreted in relation to this structure.

The species of the organisms are not determined, and, to assess the water quality, the water-sediment interface (i.e., a wholistic aquatic environment) is characterized with the aid of simple formulas or standard tables, taking into account the nature of the taxons and the taxonomic variety. Since these different methods have been synthesized (Hellawell, 1986), two methods will be described.

5.2 ANALYSIS OF BENTHIC COMMUNITIES AND THE QUALITY OF WATERWAYS

Based on simplified analyses of the macrobenthos, biotic indices have been designed to allow numerous operators (taxonomic non-specialists given adequate training) to establish a balance and draw a large-scale map of the general status from a national hydrographic network (Verneaux and Tuffery, 1967). A notation table offers a direct index of the sampling station (0–10) as a function of the nature and variety of the fauna in relation to three benthic samples of 1/10 m^2 in running water (lotic facies) and three benthic samples in calm water (lentic facies). Despite its simplicity and its wide use in Europe in forms adapted to particular aspects of the networks, this method has low sensitivity. By contrast, the maximum index class (Ib = 9±1) represents suboptimal quality, explaining its use as a reference for, rather than an index of, the absence of pronounced degradation. Thus, the index is not a trend indicator.

Attempts to improve sensitivity and precision (Verneaux *et al.*, 1976) have revealed that these objectives can be achieved for jurassic-type pre-mountain water streams; however, limitations then become enhanced. For instance, the relative quality of the systems is underestimated with little slope or warmer systems that are naturally less well suited to develop such organisms as Plecopters or Heptageniidae. Nevertheless, these tests have contributed to the definition of a more exact sampling protocol.

In parallel, Chandler (1970) proposed a similar method, relying on a score of

0–100 obtained from a table combining the nature and abundance of the fauna (five estimated classes) with specific identification of most indicator groups. From practical experience, the family constitutes the basic unit, and the abundance criteria are discarded (Armitage *et al.*, 1983; Hellawell, 1986). Similar tests (experimental protocol Cb2) have shown that the continual presence of numerous families is inadequate for them to be selected as indicator taxons (Verneaux *et al.*, 1981). These processes led to the proposal of a simplified protocol (Verneaux *et al.*, 1982).

The analysis of large amounts of data defined the following characteristics:

1. the minimum practical size of a sample as 1/20 m^2;
2. the required and sufficient number of readings as 8;
3. the precise sampling protocol that circumscribes the mosaic of habitats;
4. the listing of the taxons used (135, of which 38 served as indicators based on frequency and fidelity); and
5. the table of standard index values (0–20) according to the nature and taxonomic variety of the benthic fauna collected by the proposed protocol.

The classification of indicators is accomplished using two series of factor analyses of the distribution of families in sampling stations with little or no degradation, and then with the Rhithron (a stream with predominantly Cyprinides) altered in various ways (Benzecri *et al.*, 1973). The classification of the taxons according to their relative general tolerance was performed in a manner similar to that for fish (Verneaux, 1981). This relative ranking differs considerably from those hierarchies of the "score systems" (Armitage *et al.*, 1983), in which the positions of some taxons seem to be the result of the fact that several readings were made in environments that had already been degraded considerably (i.e., "loss of reference").

5.3 BIOLOGICAL GENERAL QUALITY INDEX

This standard describes a method to determine the standardized global biological index published by Verneaux *et al.* (1982), which was promulgated under the name IBGN by AFNOR (1992). IBGN assesses the biogenic tendency of a waterstream station from the results of a macrofauna analysis that is considered to produce a comprehensive expression of the general quality of a waterstream station under otherwise constant conditions. When applied to an isolated site, the method determines the global biological quality within a range of parameters, whose maximum value corresponds to the optimal combination of set variety with the nature of the benthic macrofauna. When applied comparatively (e.g., upstream and downstream from a discharge), the method evaluates the effect of a disturbance on the receiving environment within the limits of its sensitivity.

The benthic macrofauna samples (diameter > 500 μm) are taken at each station, according to a sampling protocol that takes into account the different types of flexible net (diameter = 500 mm), using: retractable panels, removable base (1/20 m^2), sampler "Surber" position, rack, metal frame, cutting blade, sampler "Shrimpnet" position, and habitats defined by the support structure and the flow speed.

Table 5.1. IBGN values according to the nature and the taxonomic variety of the macrofauna

Variety class→		14	13	12	11	10	9	8	7	6	5	4	3	2	1
	Σ*t*=	>	*49*	*44*	*40*	*36*	*32*	*28*	*24*	*20*	*16*	*12*	*9*	*6*	*3*
Taxa	↓GI↓	50	45	41	37	33	29	25	21	17	13	10	7	4	1
Chloroperlidae *Perlidae* *Perlodidae* *Taeniopterygidae*	9	20	20	20	19	18	17	16	15	14	13	12	11	10	9
Capniidae *Brachycentridae* *Odontoceridae* *Philopotamidae*	8	20	20	19	18	17	16	15	14	13	12	11	10	9	8
Leuctridae *Gloososomatidae* *Beraeidae* *Goeridae* *Leptophlebiidae*	7	20	19	18	17	16	15	14	13	12	11	10	9	8	7
Nemouridae *Lepidostomatidae* *Sericostomatidae* *Ephemeridae*	6	19	18	17	16	15	14	13	121	11	10	9	8	7	6
Hydroptilidae *Heptageniidae* *Polymitarcidae* *Potamanthidae*	5	18	17	16	15	14	13	12	11	10	9	8	7	6	5
Leptoceridae *Polycentropodidae* *Psychomyidae* *Rhyacophilidae*	4	17	16	15	14	13	12	11	10	9	8	7	6	5	4
Limnephilidae[a] *Hydropsychidae* *Ephemerellidae*[a] *Aphelocheiridae*	3	16	15	14	13	12	11	10	9	8	7	6	5	4	3
Baetidae[a] *Caenidae*[a] *Elmidae*[a] *Gammaridae*[a] *Mollusques*	2	15	14	13	12	11	10	9	8	7	6	5	4	3	2
Chironomidae[a] *Asellidae* *Achetes* *Oligochetes*[a]	1	14	13	12	11	10	9	8	7	6	5	4	3	2	1

[a]Taxa represented by at least 10 individuals, the others by at least 3 individuals.

A representative sample consists of eight samplings. The sorting and identification of the sampled taxons are performed to determine the taxonomic variety of the sample and its fauna indicator group. These two criteria allow a statement on the biogenic quality of the station using an index established with the help of Table 5.1.

The habitats located in calm water (lentic facies) are prospected with the help of a shrimper, using traction over 50 cm, or, by default, by back-and-forth movement over an equivalent surface (the additional surface compared with that of the Surber compensates for the loss of a portion of the individuals).

A ring-shaped illumination lens is used for the sorting in a binocular lens (stereoscopic microscope $G \leq 50$) issued for the identification of the taxons.

5.3.1 SAMPLING

The IBGN is determined per station, which is defined as the segment of a water stream whose length is virtually equal to 10 times the width of the stream bed at the time of the sampling. The detection of disturbance is facilitated in extreme situations at the moment of low waters (minimum flow, maximum temperature) or during critical periods (discharges and seasonal human activities). The samples must be taken during a period of stabilized flow for at least 10 days.

For each station, the benthic fauna sample consists of eight samplings of 1/20 m^2 each (volume sampled for the loose substrates: 0.5–1 L) performed separately in eight different habitats selected from among the combinations defined for each station. The eight samples together must provide a representative picture of the mosaic of habitats of the station. Each habitat is characterized by a support-speed set.

In the absence of certain habitats, the samples can be obtained according to the strata. Each stratum, sampled separately, constitutes a complete sampling. For example, in the absence of a lentic habitat in a mountain stream, the surface of the grid is sampled; then, separately, the inside surface and the underlying substrate are sampled a second time.

5.3.2 SAMPLING PROTOCOL

The samples are taken with the help of the sampling devices. Each sample is immediately fixed on site by the addition of a 10 percent (v/v) formaldehyde solution, placed in a plastic bag, transported packed in ice, and then stored in the refrigerator. The surface speeds are estimated for each habitat.

The support categories (S) are studied in the order of the succession (from 9–0). This table layout recommends that the habitats most friendly to fauna should be prospected first. For each support category, the sampling is made in the speed class where the support is best represented. The speed classes (5 to 1) are listed in decreasing order.

When a monotonous station (straightened course, silted bed, or canal) does not include the eight different types of support, the number of samplings is extended to eight through samplings taken of the dominant support. The percentage of coverage of each habitat (SV set) can be estimated from the following:

% r =	>75%	50%	25%	10	>10
class =	5	4	3	2	1

5.3.3 BIOLOGICAL ANALYSIS

5.3.3.1 List of taxons

The selected taxonomic unit is the family with the exception of some fauna groups (branches or classes) with little representation or where the taxonomic analysis unveiled specializations. The repertory includes 138 taxons which may be included in the overall variety (Σt), of which 38 are indicator taxons that form the nine indicator fauna groups (GI in Table 5.1). The Mollusks and the Achetes are also listed.

The collected organisms are sorted and determined according to larval, nymph or adult stage, provided that this latter stage is aquatic. Empty sheaths or shells are not taken into account.

To facilitate the interpretation of the results, the samplings should not be mixed and the fauna list of the station should be prepared by indicating the distribution of the taxons in the eight habitats.

5.3.3.2 Determination of the global biological index (IBGN)

The IBGN is determined on the basis of the information in Table 5.1, which lists the nine indicator fauna groups (GI) and the 14 taxonomic variety classes. The following must be determined sequentially:

1. The taxonomic variety of the sample (Σt), equal to the total number of taxons collected even if they are represented only by a single individual. This number is compared to the classes included in Table 5.1.
2. Indicator fauna class (GI) considering only the indicator taxa represented in the samples by at least three individuals or 10 individuals depending on the taxons. GI is determined by prospecting the taxa listed in Table 5.1 from top to bottom (GI 9 to GI 1) and halting the examination at the first significant presence ($n > 3$ individuals or > 10 individuals) of a taxon in the list on the ordinate of Table 5.1. IBGN can be derived from the Σt and GI values. For example:

GI = 8, Σt = 33 >>> IBGN = 17
GI = 5 Σt = 30 >>> IBGN = 13
GI = 3, Σt = 14 >>> IBGN = 7

Because of the significant absence of indicator taxa (3 or 10 individuals), the IBGN score equals 0.

5.3.3.3 Test report

For each station, the test report must include the date; the exact geographic location (Lambert coordinates); the ecological type, if known; the distance from the source; the altitude; the length of the wet bed at the time of the sampling; the water temperature; the nature of the support and the flow rate pertaining to the eight samplings performed for the station (SV set) with an indication of the dominant habitat or, preferably, the approximate collected classes; the list of sampled taxons with their distribution over the eight habitats, with a possible indication of their relative abundance; the taxonomic variety of the sample (Σt); the indicator fauna group (sequence number of GI); and the standardized global biological index (IBGN).

For cartographic representation of the results, each segment of the stream can be assigned one of the following colours, depending on the value of IBGN:

IBGN	≥ 17	16–13	12–9	8–5	≤ 4
colour	blue	green	yellow	orange	red

The IBG variations throughout a segment or a water stream in its entirety can be plotted in a graph where the distance from the source is the abscissa and the index values are the ordinate.

5.3.4 EXAMPLE

An illustration has been prepared by the author of the Pont de Fleurey on the Dessoubre stream (affluent of the Doubs) at Jura Massif in France. The Dessoubre, a mesorhithron stream with the association Tadpole–Trout–Grayling–Minnow–Loach, presents a habitat diversity and a water quality corresponding to its ecology type. The start of a trend manifested by the most stenoecious fauna to leave the habitats of the lenitic facies should be noted. In 1981, this station was one of the stations used for the sampling of the range of index values in search of optimal values.

5.4 DETERMINATION OF THE GENERAL BIOLOGICAL QUALITY OF LAKES

5.4.1 BASIS

Although benthic macrofauna, because of its variety and abundance, constituted the material of chcice for the establishment of practical biological methods for the assessment of the general status of streams, at present no similar methods can be applied to still-water systems, although Limnology emerged with lacustrine investigations. This can be explained by several factors.

5.4.1.1 The nature of the organisms

Whereas benthos constitutes the core of the organism of streams, lakes are, however, characterized by microscopic planktonic organisms (phytoplankton and zooplankton) that have very brief developmental cycles, and present significant spatial and time-related variations. Thus, this material is difficult to use to determine the significance to the entire system. Therefore, employment of the zoomacrobenthos, whose integration power is much greater, has been projected. However, a portion of the species only colonizes in the littoral zones whose habitat mosaics prove to be very different from one lacustrine basin to the next. Brinkhurst (1974) shows the general phenomenon of a decrease in fauna (here, generic) variety with depth.

The main components of the macrobenthos capable of colonizing lacustrine sediments up to depths of 250–300 m belong to the "difficult groups," such as Mollusks, more specifically Pisidies, the Oligochetes (Brinkhurst, 1974), and especially the Chironomide dipters for which the analysis of a great number of species associations has offered for a long time the basis for lacustrine biotypology with the work of the great forerunners such as Thienemann (1920–1931) or Brundin in the late 1940s. The studies of comparative biocenotics, performed with this material, can be conducted only by true specialists, which unfortunately is increasingly less the case.

5.4.1.2 Interpretation ambiguities

While simplified methods are proposed based on the single phytoplankton or on the basis of the species of a single faunistic class, order, or family, the challenge is in uncovering the meaning of the analytical findings, especially when a global qualitative perspective requires considerable integration of widely diverse information. Therefore, the indices proposed by Lafont *et al.* (1991) by a simplified analysis of the Oligochete populations objectively express the biological quality of the water–sediment interface; Saether (1979) considers the communities of Chironomide dipters of the deep zone to be the indicators of the "quality of the waters," and links his results to a "trophic level" relative to the system, pollutions, and dysfunctions that are included and not differentiated.

The application of this method to the lakes of the Jura approximates "eutrophic" effects; the phytoplanktonic biomasses mark a varied range of partial primary productions, and physicochemical analyses of the sediment reveal a great variety of sedimentological types. Yet, equally apparent is the absence of relationships between the global sedimentary composition (% carbonates and MO), the primary production, and the depths of the basins.

Two main conclusions can be drawn from these comparisons:

1. The need to differentiate the trophic level from the nutritive substance content which express a potential and the "trophic status" of a system, by expressing a functioning or dysfunctioning mode whose sediments and fauna supply images for which interpretations must be found.
2. The usefulness to have available a practical biological method for the assessment of the general biogenic aptitude of a lake, which would offer sufficient synthetic significance.

Besides the recent proposals of Lafont *et al.* (1991) and Mouthon 1993) which propose, respectively, simple assessment methods for the biological quality of lacustine systems on the basis of Oligochetes and Mollusks, all other proposals tend to define different "trophic levels," but not the resulting biogenic aptitudes.

5.4.1.3 IBL: A method to evaluate the biogenic quality of lakes

An experimental classification of lacustrine systems based on a comparative analysis of the benthic fauna has been proposed. This method is called the Lacustrine Biogenic Index (IBL; Verneaux, 1993), and includes a comparative sampling protocol, original biologic descriptors, and a standard table that allows the definition of the biological type and the biogenic index of a lake.

Sampling

Only fine sediment over 5 cm is collected using a modified Ekman bucket with the addition of lateral ballast as well as a penetration limiter. Coarse substrates and hydrophytes are avoided, as are certain sites such as beaches, harbours or substrate enclosures. Two samplings are performed each station to form a station sample; and two depths, to which two isobaths correspond, are prospected (*Zo* at 2–2.5 m; and *Zf* at ¾ *Z* maximum relative depth). The number of stations per isobath is proportional to the length of each isobath, and should be determined using the following factors:

$$\text{at } Zo,\ no = 1.8\sqrt{10L} \tag{1}$$

$$\text{at } Zf,\ nf = 1.4\sqrt{10L} \tag{2}$$

where L is the length of the isobath in question, expressed in km.

The stations are distributed regularly (virtually in an equidistant manner) over each isobath. The samples are taken during a single sampling trip during an isothermia period which follows thawing and springtime circulation. Depending on the altitude, in the Jura lakes the expeditions took place in April or in May. Each sample (consisting of two samplings) is filtered through a conic net with a mesh width of 250 μm; then 5 percent formaldehyde solution is added and the sample is then placed in a plastic bag with the air removed. The samples are transported on ice, and stored in a refrigerator.

Sorting and determination of the taxa

The samples are analysed separately. The macroinvertebrates are identified but not counted out. The selected taxonomic unit is genus except for the Oligochetes, Nematodes, Hydracarians, and Ceratopogonidae. A listing shows the fauna data and the frequency of each taxon is expressed in percentage of occurrence in relation to the *no* and *nf* counts of the stations (or the samples per isobath).

Descriptors employed

The quality coefficient of the fauna of the fine sediment is determined where the found taxa are classified in the decreasing order of sensitivity to the physico-chemical quality of the water-sediment interface. Only taxon indicators are selected whose frequency is at least equal to 50 percent of the number of samples *no* taken at depth *Zo*. The descriptors include: vo = fauna variety (≃ generic) at depth *Zo*, qo = quality of the fauna at depth *Zo*, and df = bathymetric distribution coefficient at depth *Zf*

$$df = k\,\frac{vf}{vo} \tag{3}$$

where $k = 0.047\ vo + 1$, F = relative faunistic deficit index, and $F = ?df \cdot qo$. For example, if $vo = 38$ and $F = 0.77$ for type B_4, then biogenic index/20 = 15; and if $vo = 23$ and $F = 0.38$ for type B_3, then biogenic index/20 = 08.

Qualitative levels include eubiotic lake, eumerobiotic lake, merobiotic lake, merodysbiotic lake, and dysbiotic lake; and the quantitative levels include oligobiotic lake, oligomesobiotic lake, mesobiotic lake, mesopolybiotic lake, and polybiotic lake. The combinations of the two series of information is used to define the type of lake as either euoligobiotic or mesomerobiotic or dyspolybiotic lake.

Interpretation

The variety of endobenthic fauna sampled in the littoral zone beyond the river zone (*Zo* = -2, -2.5 m) constitutes a good indication of the biogenic potential of the system in relation to consumer organisms. The fauna distribution index (*F*) which takes into account the nature of the fauna (*qo*) and the significance of the corrected relative fauna deficit (*df*) expresses the operating mode of the system. The index (IBL) constitutes a mark of the resulting biogenic aptitude trend that can assume the same values in the cases of different figures. For example, an oligotrophic lake that is both eubiotic and oligobiotic (type A1, IBL = 10) with highly oxygenated water but with few minerals, and a hypertrophic lake (type D4, IBL = 10) of moderate depth and very rich in mineral salts and deoxygenated as of moderate depth have the same biogenic index of 10. The determination of the biological type of the lake allows an interpretation of the resulting biogenic index. The general quality of the water, corresponding to categories A, B, C, D, E, can be expressed by five colours ranging from blue to red to indicate decreasing quality.

5.5 REFERENCES

AFNOR (1992) *Détermination de l'indice biologique global normalisé (IBGN) [Determination of the Standardized Global Biological Index (IBGN)].* Report No. NF T 90350. AFNOR, Paris, 9 pp.

Amanieu, M., Gonzales, I.L., and Guelorget, O. (1981) Critères de choix d'un modèle de distribution d'abondance. Application à des communautés animales en Ecologie benthique [Selection criteria for an abundance distribution model. Application to the animal communities in benthic ecology]. *Oecol. Gen.* **2**, 265–286.

Armitage, R.D., Mass, D., Wright, J.F., and Furse, M.T. (1983) The performance of a new biological water quality score system based on macroinvertebrates over a wide range of unpolluted running-water sites. *Water Res.* **17**, 333–347.

Bartsch, A.F., and Ingram, W.M. (1966) Biological analysis of water pollution in North America. *Verh. Int. Verein. Limnol.* **16**, 786–800.

Benzecri, J.P., *et al.* (1973) *L'Analyse des Données. I: La Taxinomie. L'Analyse des Correspondances. [Data Analysis. I: Taxonomy. II: Analysis of Correlations]* Dunod, Paris, 615 pp and 619 pp.

Brinkhurst, R.O. (1974) Factors mediating interspecific aggregation of tubificid oligochaetes. *J. Fish Res. Bd. Can.* **31**, 460.

Chandler, J.R. (1970) A biological approach to water quality management. *Water Pollut. Control* **69**, 415–422.

Descy, J.P., and Coste, M. (1991) A test of methods for assessing water quality based on diatoms. *Verh. Int. Verein. Limnol.* **24**, 2112–2116.

Downing, J.A., and Rigler, H. (1984) *A Manual on Methods for the Assessment of Secondary Productivity in Fresh Waters.* Blackwell Scientific, Oxford, London, Edinburgh, Boston, Melbourne, 501 pp.

Hellawell, J.M. (1986) *Biological Indicators of Freshwater Pollution and Environmental Management.* Elsevier Applied Science, London, New York, 546 pp.

Hill, M.O. (1974) Correspondence analysis: a neglected multivariate method. *J. Roy Stat. Soc. Ser. C Appl. Stat.* **23**, 340–354.

Hurlbert, S.H. (1971) The nonconcept of species diversity: a critique and alternative parameters. *Ecology* **52**, 577–586.

Ilies, J., *et al.* (1978) *Limnofauna Europaea.* Fischer, Stuttgart, 53 pp.

Kothé, P. (1962) Der "Artenfehlbetrag," ein einfaches Gütekriterium und seine Anwendung bei biologischen Vortfluteruntersuchungen ["Species deficit," a simple quality criterion and its application to biological investigations of streams]. *Dt. Gewässerkundl. Mittl.* **6**, 60–65.

Lafont, M., Juget, J., and Rofes, G. (1991) Un indice biologique lacustre basé sur l'examen des Oligochètes [A lacustrine biological index based on the investigation of Oligochetes]. *Rev. Sci. Eau* **4**, 253–268.

Legendre, L., and Legendre, P. (1979) *Ecologie Numérique [Numerical Ecology]*, Vols. 1 and 2. Masson, Paris, 197 pp. and 247 pp.

Lloyd, M., and Gehlardi, R.J. (1964) A table for calculating the "equitability" component of species diversity. *J. Anim. Ecol.* **38**, 317–325.

Margalef, R. (1974) *Ecologie.* Omega, Barcelona, 951 pp.

Mouthon, J. (1993) Un indice biologique lacustre basé sur l'examen des peuplements de Mollusques. [A lacustrine biological index based on the investigation of the Mollusk populations.] *Rev. Ecol. Appl.*

Mouthon, J., and Faessel, B. (1978) *Méthodes d'Estimation de la Qualité Biologique des Eaux Courantes. Etude des Torrents de Parme, Stirone, Chiara et d'un Tronçon du Pô [Methods for Assessing the Biological Quality of Streams. Study of the Flow of the Parma, Stirone, Chiara and a Segment of the Po].* Technical Paper CEE, 20 pp.

Mouvet, C. (1986) *Métaux Lourds et Mousses Aquatiques. Synthèse Méthodologique [Heavy Metals in Aquatic Foams. Methodological Summary].* Agence de l'Eau R.M.C. et Lab. Ecol. Univ. Metz, 110 pp.

Orlóci, L. (1975) *Multivariate Analysis in Vegetation Research.* Dr. W. Junk B.V., The Hague, 276 pp.

Patrick, R., Kohn, M.H., and Wallace, J.M. (1954) A new method for determining the pattern of the diatom flora. *Notul. Nat. Acad. Sci. Philadelphia* **259**, 12 p.

Patten, B.C. (1962) Species diversity in net phytoplankton of Raritan Bay. *J. Mar. Res.* **20**, 57–75.

Pielou, E.C. (1975) *Ecological Diversity.* J. Wiley & Sons, New York, 165 pp.

Preston, F.W. (1948) The commonness and rarity of species. *Ecology* **29**, 254–283.

Shannon, C.E. (1948) A mathematical theory of communication. *Bell System Tech. J.* **27**, 379–423, 623–656.

Verneaux, J. (1973) *Cours d'Eau de Franche-Comté (Massif du Jura). Essai de Biotypologie [Streams in Franch-Comté (Jura Massif). Biotypology Assay].* Mém. thèse d'Etat, Univ. Fr. Comté, Besançon, France, 260 pp.

Verneaux, J. (1976) Biotypologie de l'écosystème "eau courante." La structure biotypologique [Biotypology of the ecosystem "stream." The biotypological structure]. *C.R. Acad. (Paris)* **283**, 1665–1666.

Verneaux, J. (1981) Les poissons et la qualité des cours d'eau [Fish and the quality of streams]. *Ann. Sci. Univ. Fr. Comté Biol. Anim.* **4**(2), 33–41.

Verneaux, J., and Tuffery, G. (1967) Une méthode zoologique pratique de détermination de la qualité biologique des eaux courantes. Indices biotiques [A practical zoological method for determination of the biological quality in streams. Biotic indices]. *Ann. Sci. Univ. Fr. Comté Zool.* **3**, 79–90.

Verneaux, J., Faessel, B., and Malesieux, G. (1976) *Note Préliminaire à la Proposition de Nouvelles Méthodes de Détermination de la Qualité des Eaux Courantes [Introductory Statement of the Proposal for New Methods for Determining the Quality of Streams].* Trav. Lab. Hydrobiol. Univ. Fr. Comté et CEMAGREF, Besançon and Paris, 148 pp.

Verneaux, J., *et al.* (1981) *Expression Biologique Pratique de l'Aptitude des Course d'Eau au Développement de la Faune Benthique. Un Coefficient d'Aptitude Biogène: le Cb2. Protocole Expérimental [Practical Biological Expression of the Capacity of Streams in Relation to Benthic Fauna. A Biogenic Capacity Coeffecicient: Cb2. Experimental Protocol].* Trav. Lab. Hydrobiol. Hydroécol. Univ. Fr. Comté, Besançon, France, 19 pp.

Verneaux, J., *et al.* (1982) Une nouvelle méthode pratique d'évaluation de la qualité des eaux courantes. Un indice de qualité biologique générale (IBG) [A new practical method for assessment of the quality of streams. A general biological quality index (IBG)]. *Ann. Sci. Univ. Fr. Comté Biol. Anim.* **4**(3), 11–21.

Woodiwiss, F.S. (1964) The biological system of stream classification used by the Trent River Board. *Chem. Ind.* **11**, 443–447.

6 Methods to Assess the Effects of Chemicals on Fresh Waters

Peter Calow
University of Sheffield, United Kingdom

6.1 INTRODUCTION

The sources and types of chemical pollution to which freshwater systems can be exposed are many and varied. The origins of pollutants entering these systems range from those that are predominantly point sources (i.e., from sewage treatment works, fish farms, and industry) to those that are mainly diffuse, such as agricultural run-off and acid deposition. Some of these are effectively continuous, whereas others are intermittent or episodic.

Pollutants can also be characterized by their quality: general organic loading, specific organic toxicants, inorganic toxicants, and acids. (Holdgate, 1979). While most of the sources listed above produce mixtures of these characteristics, particular components are enhanced. Thus, sewage treatment works, fish farms, and farmyards produce general organic loading, but may also be the source of pesticides, and organic and inorganic toxicants. Industrial effluents generally are the source of specific toxicants, but might also lead to general organic loading.

Some consideration should also be given to the types of system exposed to these pollutants. Most freshwater systems flow, but some more rapidly than others; that is, lakes and ponds have a slow throughput, whereas that for rivers and streams is rapid. Because they flow rapidly, the so-called lotic systems have been used for the transport of materials, such as removing industrial pollutants from factories and organic effluents from sewage works (Hellawell, 1986).

A view exists that lotic systems that depend substantially on organic inputs from terrestrial ecosystems as a basis for their economy should have the capacity for self-cleansing of at least organic pollutants (Calow *et al.*, 1990). However, this perception has been challenged recently (Royal Commission on Environmental Pollution, 1992). Of course, "still" or so-called lentic waters may also be important repositories of pollutants, and are certainly exposed to diffuse inputs from agricultural land leading to eutrophication (Harper, 1992).

Thus, the challenge for aquatic ecotoxicology is to develop methods that can not only assess but also estimate the ecological impact of chemicals on a variety of complex ecosystems under diverse and complicated circumstances (e.g., periodic

Methods to Assess the Effects of Chemicals on Ecosystems
Edited by R. A. Linthurst, P. Bourdeau, and R. G. Tardiff

perturbations or steady trickles, and mixtures of varying complexity). The predictive approach is important in risk assessment used in the regulation of chemical pollutants. Assessment, at times referred to as a "retrospective approach," is used to determine whether particular contaminants have an impact on specified natural ecosystems. This chapter reviews current methods, first predictive and then retrospective approaches, to address such problems. Initially, one must define what is to be measured, for it has relevance to both approaches.

6.2 WHAT TO MEASURE

Effects of contaminants can be identified across the entire biological hierarchy from subcellular systems to ecosystems. Moving down in scale, effects are usually measured more quickly and with more experimental rigour and control. Moreover, in this direction, systems are more general; most organisms contain DNA, but most ecosystems differ considerably in species composition. Since ecotoxicology is concerned with protecting ecological systems, this generally means consideration of populations and communities that compose ecosystems.

Considerable debate continues about the ecological relevance of observations in individual organisms and of the processes functioning within them (Calow, 1989). One view is that ecological processes are driven "bottom up," so that specific functional links can be made between effects observed at the organismic and suborganismic levels and those at the population–community levels. Thus, any impact on the physiological materials and energy flows in organisms can be linked with consequences for developmental rates, survival chances, and reproductive output; these then can be linked to consequences in population dynamics and, alternatively, to the role of a species or population within the community (Calow and Sibly, 1990). By contrast, the impact of a pollutant on keystone predators (Paine, 1988) is likely to alter the interaction between prey species in a way that could not have been predicted from ecotoxicological observations on the species individually.

The relative values of "bottom-up" and "top-down" regulation may vary from one ecosystem to another, and may depend upon the extent to which particular ecosystems are structured. Thus flowing-water systems that may be relatively unstructured, because of their temporal dynamism (Peckarsky, 1984; Hildrew and Townsend, 1987), might be subject predominantly to "bottom-up" control. This outcome could be a justification for use of organismic-suborganismic test systems in defining the hazards associated with chemicals to which the organisms are likely to be exposed. More research is needed to clarify these relationships.

Even for ecosystems difficult questions have to be considered to decide which attributes should be measured. Unlike for organisms (Calow, 1992), ecosystem "goal states" are unlikely, although states related to ecosystem stability and resilience are known to exist (Westman, 1978). Traditionally in ecotoxicology, measurements of effects have been made of physicochemical variability, species composition and diversity (i.e. Crossland and Wolff, 1985), and, less commonly, functional features (Odum, 1985; Cairns and Pratt, 1986). Yet food-web structure might be as important (Hildrew, 1992), as might be the analysis of functional components and their ability to maintain balance of energy flow and of material cycles within systems. Protecting species deemed important for humans might also be important.

6.3 PREDICTIVE TESTS

These theoretical considerations notwithstanding, single-species tests have dominated studies in aquatic ecotoxicology for the past 20 years (Maltby and Calow, 1989). Moreover, certain taxa, particularly the daphnids, have dominated these single-species studies (Maltby and Calow, 1989), because single species tests can be accomplished more quickly and easily and, in principle, daphnids are easily maintained in laboratory culture for such tests.

6.3.1 SINGLE-SPECIES TESTS

These tests can be classified according to either (1) the form of the response measured: discontinuous or quantal (i.e., death or cessation of a specified behaviour) or continuous (e.g., a reduction in growth or reproduction) or (2) the relationship between the time-course of the response and the generation time (GT) of the test organism, may be short or long relative to GT. Tests are usually referred to as acute (quantal and short term) or chronic (continuous or long term).

Most tests in aquatic ecotoxicology have been performed with a single species and of acute duration. Both acute and chronic test systems have been incorporated into legislation concerned with the assessment of ecological hazards of both new and existing chemicals and detailed descriptions of these standard test systems are provided by Persoone and Janssen (1993) and Solbe (1993). Their use has arisen largely from the convenience of some *post hoc* justification of relevance. Thus, the use of daphnids as test organisms can be justified by (1) the key roles that they and their taxonomic associates play in freshwater ecosystems such as zooplankton (interestingly, not often the case for lotic systems) and (2) their proven sensitivity relative to other freshwater species (Baudo, 1987).

Convenience has been a criterion in the choice of test systems, because it enables tests to be executed on a routine basis and to be standardized. The latter is paramount, since without the ability to get the same result from the same test system in different laboratories, something not always possible even for daphnids (Baird *et al.*, 1989), the whole legal and scientific basis of the test is undermined (Calow, 1993).

For the following reasons, attention must also be paid to the ecological relevance of the test. First, demonstrating that the test species is pivotal to estimating broad environmental consequences, its ecological legitimacy would be strengthened. However, often too little understanding exists of the structure of freshwater ecosystems to know if keystone species exist at all (Hildrew, 1992). Second, ecological relevance might be claimed if the test species were known to be more sensitive than most other species. The daphnids appear to be quite sensitive, but much variation in sensitivity to numerous toxicants is apparent even between genotypes of *Daphnia magna* (Baird *et al.*, 1991). The concept of the more sensitive indicator species seems to have been undermined by Cairns (1986). Much more needs to be known about the distribution of sensitivities to toxicants both within and between species in communities (Kooijman, 1987; Van Straalan and Denneman, 1989).

Moreover, the emphasis on pelagic organisms also needs careful consideration. Epibenthic and sediment systems may be as important, if not more so, in freshwaters as are pelagic ones; yet, because of difficulties in manipulation, sediments have been relatively neglected. Sediment ecotoxicity is nevertheless developing, as indicated by useful information provided by Reynoldson and Day (1993).

Given the predominance of acute tests and the interest in using them to set priorities and environmental standards through the application of safety factors, the relationships between acute and chronic responses must be well understood. Pioneering work on this topic reported by Sloof *et al.* (1986) and carefully performed studies reported by ECETOC (1993) support the general view that chronic effects are approximately equal to LC_{50}/LC_{40}. Since this finding is based on a correlation, no guarantee exists that this relationship is likely to apply to any particular chemical, particularly given differences in activation of toxicants (Maltby and Calow, 1989). To increase scientific confidence, the mechanistic basis of the relationship between acute short-term and chronic longer-term responses needs to be supported by additional data.

6.3.2 MULTI-SPECIES TESTS

Two reasons that multi-species systems might bring more ecological relevance to ecotoxicity tests include:

1. The test results may expose crucial emergent effects that arise out of species interactions.
2. The test findings from testing in complex milieu may provide insight into the distribution (fate) of toxicants between various compartments of the systems (i.e., sediment and water column) and hence on bioavailability.

If a divergence occurs between single- and multi-species responses, knowing whether it arises from basis 1 or 2 above would prove useful.

Multi-species tests can certainly be conducted for a variety of scales and using a number of alternative microcosms and macrocosms. However, no generally accepted distinction exists among systems based solely on size (SETAC Europe, 1991; SETAC, 1992). They can be based indoors or outdoors, and they can be closed or open. Closed systems are observed more frequently than are open ones, largely because of the complexity of the "plumbing" needed for open systems; however, both indoor and outdoor artificial streams have been constructed and used (Cairns and Cherry, 1993).

Mesocosms have been used in a legislative context: for pesticide registration in the US where the USEPA can require a mesocosm analysis if information from other tests is insufficient to "negate a presumption of risk" or indicates that significant biological populations will be at risk. The best practice in this context has been evaluated in Workshops organized by the Society of Environmental Toxicology and Chemistry for the use of both microcosms (SETAC, 1992) and mesocosms (SETAC Europe, 1991). Standard multi-species ecotoxicological test systems have been described, and reviewed by Cairns and Cherry (1993).

For the effort and expense involved in establishing and managing these systems, careful consideration needs to be given to their usefulness and relevance. Three major problems are apparent: (1) because of the size and complexity of these systems, data obtained from them are difficult to replicate, so rigorous statistical analysis is not always possible; (2) methods of construction differ sufficiently to increase the difficulty in reproducing results; and (3) uncertainty exists over exactly which properties should be measured, as it does for natural ecosystems.

Multi-species systems can either be set up by (1) sampling from natural systems (e.g. enclosing a sample of plankton in a closed flask; Leffler, 1981) or (2) colonization (e.g., leaving the pond or stream open for colonisation; Eaton *et al.*, 1985; Crossland and Wolff, 1985); or (3) by deliberate construction (Taub and Read, 1982). Techniques (1) and (2) cannot guarantee reproducibility, and the enclosing effects in (1) and (3) could lead to peculiar dynamics that interact with or potentially mask other perturbations. These effects need to be thoroughly investigated in control systems. These features mean that standardization is probably difficult to achieve, and ecological relevance may be illusory.

6.4 APPLICATION OF TOXICANTS

Most of the single- and multi-species test systems apply the toxicant continuously; however, episodic events are also common in freshwater systems. A few studies have been conducted in which the toxicant was applied as a pulse. Thurston *et al.* (1981) studied the effects of fluctuating ammonium on trout, and concluded that, over a 96-hour period, they could withstand a fixed concentration of ammonia better than fluctuating concentrations with the same mean. By contrast, Brown *et al.* (1969) found that, with either zinc or mixtures of zinc and ammonia, no significant difference in survival was observed for trout exposed either continuously or alternately to low and high concentrations.

To distinguish between standard toxicity tests and those including post-exposure mortality, Abel (1980) proposed the use of median lethal exposure time (i.e., time required to kill half the population within a predetermined post-exposure period). A similar index, median-post-exposure-lethal time (i.e., time from end of exposure period by which 50 percent of the animals are dead), has been proposed by Pascoe and Shazili (1986).

Within the context of episodic events, more careful consideration needs to be given to the effects of amplitude, duration, and frequency of pulse of the dose on populations and communities, not only in terms of survivorship but also of developmental rates and reproductive output.

A further complication to be addressed is that test substances are most often applied singly to test systems, because complex mixtures are more difficult to handle and because regulations focus predominantly on single substances. All the most appropriate mixtures and conditions in the field for carrying out tests would be impossible to imagine, let alone construct experimentally. Yet interest in the effects of complex effluents on natural systems has directed attention to this problem (USEPA, 1986; de March, 1987a, 1987b; McCarty *et al.*, 1992). Many mixtures of organic chemicals appear to act primarily by simple, sometimes joint, actions so that the contribution of the components to the total mixture are additive (Broderius and Kahl, 1985; Deneer *et al.*, 1988); however, not all toxicants produce this effect (Marking, 1985).

Understanding of basic mechanisms of ecotoxicological action needs to be extended to ascertain how toxicants exert their primary effects and how toxicants might interfere with or enhance each other's actions.

6.5 RETROSPECTIVE ASSESSMENT

Assessments can be carried out on different scales: (1) surveying the general state of fresh waters regionally, nationally (National Rivers Authority, 1991) and, locally; (2) assessing impacts of contamination at a particular site; and (3) monitoring effluent on a site on a continuous basis to assess quality. In principle, all these activities can be carried out using effects at all biological levels described previously; often, however, organismic and suborganismic responses are convenient to use in limited studies, whereas community and ecosystem properties are more convenient in the more extensive ones. In all, ecological relevance remains a critical determinant.

Chemical analyses on their own are rarely adequate in assessments, because the dynamic complexity of inputs means that samples taken at specific times (the norm for chemical analysis) may be misleading. Also, the qualitative complexity of inputs implies extreme difficulty in assessing ecological impact from a database comprised of only those substances present. Biological systems, however, integrate effects over time of complex mixtures of contaminants.

Two key issues in designing and interpreting retrospective biological assessments are: (1) what to measure and (2) how to identify cause from correlation (related to problems of design of sampling programmes).

6.6 THE MEASUREMENT PROBLEM REVISITED

Are there characteristics of healthy ecosystems that can be defined and against which divergences can be gauged? As noted previously, while "goal states" are not appropriately applied to ecosystems, steady or stable states might be. One approach based on an attempt to define a characteristic assemblage of benthic invertebrates for a particular river habitat is the River Invertebrate Predictive And Classification System (RIVPACS) (Armitage *et al.*, 1987). This system defines relationships between physicochemical variables and species assemblages on the basis of an initial survey of clean sites, and then uses the model based on the results to estimate expected assemblages at particular sites from observation of the physico-chemical variables. The deviation of observed from predicted findings is considered an index of disturbance that can be used for both surveys and impact assessments. However, this and similar approaches developed for other taxa (i.e., fish) are based on correlations between what is presumed to be dependent and independent variables (Bayley and Li, 1992); thus, deviations from expectation that correlate with an extra variable (i.e., presence of a contaminant) need not be causally related.

Another approach focuses not so much on the details of species occurrence as on the general properties of ecological diversity (Wilhm, 1972). Disputes on how this property should be measured remain (Magurran, 1988), and certainly no *a priori* theory that predicts what levels of diversity might be expected in particular habitats exists. How diversity should respond—either positively or negatively—to stress is simply unclear; that is, diversity of plankton appears to diminish continuously with organic enrichment, and to increase and then diminish for benthic invertebrates (Calow, 1984).

The expectations for the functional organization of ecosystems and how it might change under stress are defined relatively easily: for any stable ecosystem, the input of energy (either autochthonous or allochthonous) should balance output. Organic loading can destabilize the balance by causing input to exceed output; alternatively, toxic stress can destabilize this process by causing output to exceed input, resulting from the increased metabolic work needed to counteract this kind of pollution. This situation leads to changes in ecosystem functional ratios involving production, biomass, and respiration (Odum, 1985). Such changes will also be closely linked to the cycling and spiralling of specific nutrients, a topic beginning increasingly to attract attention (Newbold, 1992).

The extent to which structure and function are coupled in ecosystems is problematic (Cairns and Pratt, 1986), so that the one probably cannot be used as a surrogate for the other. This situation means that a choice has to be made either to measure either structural or functional responses, but not both, or to combine structural and functional indices as a basis for inferences about divergences from the norm (Karr, 1991). For the latter approach, care has to be taken to ensure that the different indices are not autocorrelated (Barbour *et al.*, 1992).

A close coupling exists between functioning and trophic structure embodied in food webs. The properties of these are coming under increased scrutiny. Structural-functional attributes are likely to be identified for use in defining relative stability as well as resilience to disturbance in ecosystems (De Angelis, 1992). Benthic invertebrates have been classified into trophic functional groups (Cummins and Klug, 1979), and predictable shifts in the relative abundance of these along rivers (from shredders to collectors) have been claimed as part of the River Continuum Concept (Cummins, 1992). Thus, deviations from these could signal disturbances. Two caveats persist: (1) predictions are not yet precisely defined, and the RCC theory is not universally accepted (Townsend, 1989); and (2) causation should not be mistaken for correlation.

If ecosystem properties and processes are difficult to measure, an alternative is to use alterations in individual species populations as indicators of change. Searching for key species may be tempting; however, "key" species may have several definitions: sensitivity, functional role, or impact on the dynamics of other species. Approaches relying on indicator species (involving presence or absence of a single species; ratios of sensitive to tolerant species; and biotic scores) are based largely on the construct of "sensitive" species, a philosophy whose difficulties have already been noted (Metcalfe, 1989). Moreover, separated populations within species can display considerable variation in response to pollutants based possibly on differential local conditions and selection pressures (Maltby *et al.*, 1987). Shifts from sensitive to tolerant forms of organisms (both inter- and intra-species) have been used independently as indicators of stress as is the case for the Pollution Induced Community Tolerance (PICT) system (Blanck *et al.*, 1988).

6.7 CAUSE OR ONLY CORRELATION?

In complex field situations, defining confidently causal links from direct observations is always difficult. Usually a change is observed to be correlated with a variable of interest (e.g., a reduction in diversity or density downstream of an effluent), but to say that one caused the other is unjustified without further research, because important contributing variables may be present but hidden from view.

However, without some *a priori* yardstick against which to compare observed systems, divergences from normal have to be judged comparatively by reference to the state of the system before the presumed disturbances or with similar systems in habitats not exposed to the pollutants of interest (e.g., upstream and downstream comparisons). Some profound problems are associated, however, with being able to distinguish between natural variation through space and time and that caused by the putative disturbance—the so-called sampling problem of pseudodesign (Norris *et al.*, 1992).

The only way to identify causal links with more certainty is to associate the field observations with an experimental programme. Testing ecological hypotheses derived from field observations in more rigorously controlled experimental circumstances can be used widely in ecology, and ought to be used more in ecotoxicology. For example, conclusions about putative interactions between contaminants and organisms in field situations could be tested using classical dose-response experiments. The reverse is sometimes undertaken; that is, quantitative dose-response comparisons can result from simple laboratory tests using the more complex microcosms, mesocosms, and with true ecosystems (Crossland *et al.*, 1992).

6.7.1 PLANTED SYSTEMS

Planted systems are often used to monitor effluents or even whole rivers in which physiological or behavioral features are used to signal change in systems that range from caged fish to microbes (Pascoe and Edwards, 1984).

Planted systems can also be used in upstream-downstream, before–after studies as *in situ* bioassays. They avoid problems of pseudodesign, because in principle all replicates start in the same condition. Of course, these systems raise questions of relevance. Careful choice of a system and of effects may help to overcome this difficulty. The *Gammarus* Scope for Growth Bioassay (Naylor *et al.*, 1989) serves to illustrate this point: the organism, *G. pulex* was chosen as the test animal because of its relative sensitivity to a wide range of toxicants and because of its important role in the decomposer food chain is of considerable importance in the economy of flowing-water systems. Furthermore, a physiological response involving the net energy available for somatic and reproductive products was chosen, because it is sensitive to contaminants, can be linked causally to population dynamics (Calow and Sibly, 1990), and is obviously related to important ecosystem processes in the decomposer food chain.

6.8 REFERENCES

Abel, P.D. (1980) A new method for assessing the lethal impact of short term high level discharges of pollutants on aquatic animals. *Progr. Water Technol.* **13**, 347–352.

Armitage, P.D., Gunn, R.J.M., Furse, M.J., Wright, J.F., and Moss, D. (1987) The use of prediction to assess macroinvertebrate response to river regulation. *Hydrobiologia* **144**, 25–32.

Baird, D.J., Barber, I., Bradley, M., Calow, P., and Soares, A.M.V.M. (1989) The *Daphnia* bioassay: A critique. *Hydrobiologia* **188/189**, 403–406.

Baird, D.J., Barber, I., Bradley, M., Soares, A.M.V.M., and Calow, P. (1991) a comparative study of genotype sensitivity to acute toxic stress using clones of *Daphnia magna* Straus. *Ecotoxicol. Environ. Safety* **21**, 257–265.

Barbour, M.T., Plafkin, J.L., Bradley, B.P., Graves, C., and Wisseman, R.W. (1992) Evaluation of EPA's rapid bioassessment benthic metrics: metric redundancy and variability among reference sites. *Environ. Toxicol. Chem.* **11**, 437–449.

Baudo, R. (1987) Ecotoxicological testing with *Daphnia.* In Peters, R.H., and de Bernardi, R. (Eds.) *Daphnia*, Vol. 45, pp. 461–482. Mem. Ist. Ital. Idrobiol.

Bayley, P.B., and Li, H.W. (1992) Riverine fishes. In: Calow, P., and Petts, G. (Eds.) *Rivers Handbook*, Vol. 1, pp. 251–281. Blackwell Scientific, Oxford.

Blanck, H., Wängberg, S.A., and Molander, S. (1988) Pollution-induced community tolerance—a new ecotoxicological tool. In: Cairns, J., and Pratt, J.R. (Eds.) *Functional Testing of Aquatic Biota for Estimating Hazard of Chemicals*, pp. 219–230. AST STP 988. American Society for Testing and Materials, Philadelphia.

Broderius, S., and Kahl, M. (1985) Acute toxicity of organic chemical mixtures to the fathead minnow. *Aquat. Toxicol.* **6**, 307–322.

Brown, V.M., Jordan, D.H.M., and Tiller, B.A. (1969) The acute toxicity of rainbow trout in fluctuating concentrations of mixtures of ammonia, phenol and zinc. *J. Fish Biol.* **1**, 1–9.

Cairns, J. (1986) The myth of the most sensitive species. *Bioscience* **36**, 670–672.

Cairns, J., and Cherry, D.S. (1993) Freshwater multispecies test systems. In: Calow, P. (Ed.) *Handbook of Ecotoxicology*, Vol. 1. Blackwell Scientific, Oxford.

Cairns, J., and Pratt, J.R. (1986) On the relation between structural and functional analysis of ecosystems. *Environ. Toxicol. Chem.* **5**, 785–786.

Calow, P. (1984) Factors affecting the diversity of littoral invertebrates: A hypothesis. *Ing. Sanit.* **30**, 41–46.

Calow, P. (1989) The choice and implementation of environmental bioassays. *Hydrobiologia* **188/189**, 61–64.

Calow, P. (1992) Can ecosystems be healthy? Critical consideration of concepts. *J. Aquatic Ecosys. Health* **1**, 1–5.

Calow, P. (1993) Seeking standardization in ecotoxicology. In: Soares, A.M.V.M., and Calow, P. (Eds.) *Progress in Standardization of Aquatic Toxicity Tests.* Lewis, Boca Raton, Florida.

Calow, P., and Sibly, R.M. (1990) A physiological basis of population processes: ecotoxicological implications. *Funct. Ecol.* **4**, 283–285.

Calow, P., Armitage, P., Boon, P., Chave, P., Cox, E., Hildrew, A., Learner, M., Maltby, L., Morris, G., Seager, J., and Whitton, B. (1990) *River Water Quality*, British Ecological Society Ecological Issues No. 1. Field Studies Council, Montford Bridge, Shrewsbury.

Crossland, N.O., and Wolff, C.J.M. (1985) Fate and biological effects of pentachlorophenol in outdoor ponds. *Environ. Contam. Toxicol.* **4**, 73–86.

Crossland, N.O., Mitchell, G.C., and Dorn, P.B. (1992) Use of outdoor artificial streams to determine threshold toxicity concentrations for a petrochemical effluent. *Environ. Toxicol. Chem.* **11**, 49–59.

Cummins, K.W. (1992) Invertebrates. In: Calow, P., and Petts, G. (Eds.) *Rivers Handbook*, Vol. 1, pp. 234-250. Blackwell Scientific, Oxford.

Cummins, K.W., and Klug, M.J. (1979) Feeding ecology of stream invertebrates. *Ann. Rev. Ecol. Sys.* **10**, 147–172.

De Angelis, D.L. (1992) *Dynamics of Nutrient Cycling and Food Webs.* Chapman and Hall, London.

de March, B.G.E. (1987a) Mixture toxicity indices in acute lethal toxicity tests. *Arch. Environ. Contam. Toxicol.* **16**, 33–37.

de March, B.G.E. (1987b) Simple similiar action and independent joint action—two similar models for the joint effects of toxicants applied as mixtures. *Aquat. Toxicol.* **9**, 291–304.

Deneer, J.W., Seinen, W., and Hermens, J.L.M. (1988) Growth of *Daphnia magna* exposed to mixtures of chemicals with diverse modes of action. *Ecotoxicol. Environ. Safety* **15**, 72–77.

Eaton, J., Arthur, J., Hermanutz, R., Kiefer, R., Mueller, L., Anderson, R., Erickson, R., Nordling, B., Rogers, J., and Pritchard, H. (1985). Biological effects of continuous and intermittent dosing of outdoor experimental streams with chlorphyrifos. In: *Aquatic Toxicology and Hazard Assessment*. Eighth Symp. ASTM STP 891. American Society for Testing and Materials, Philadelphia.

ECETOC (1993) *Environmental Hazard Assessment of Substances*. Technical Report No. 51. ECETOC, Brussels.

Harper, D. (1992) *Eutrophication of Freshwaters*. Chapman and Hall, London.

Hellawell, J.M. (1986) *Biological Indicators of Freshwater Pollution and Environmental Management*. Elsevier, London.

Hildrew, A.G. (1992) Food webs and species interaction. In: Calow, P., and Petts, G. (Eds.) *Rivers Handbook*, Vol. 1. Blackwell Scientific, Oxford.

Hildrew, A.G., and Townsend, C.R. (1987) Organization in freshwater benthic communities. In: Gee, J.H.R., and Giller, P.S. (Eds.) *Organisation of Communities: Past and Present*, pp. 347–371. 27th Symposium of the British Ecological Society. Blackwell Scientific, Oxford and London.

Holdgate, M.W. (1979) *A Perspective of Environmental Pollution*. Cambridge University Press, Cambridge.

Karr, J.R. (1991) Biological integrity: a long neglected aspect of water resource management. *Ecol. Appl.* **1**, 26–35.

Kooijman, S.A.L. (1987) A safety factor for LC_{50} values allowing for differences among species. *Water Res.* **21**, 269–276.

Leffler, J.W. (1981) *Aquatic Microcosms and Stress Criteria for Assessing Environmental Impact of Organic Chemicals*. Office of Pesticides and Toxic Substances, US Environmental Protection Agency, Washington, D.C., pp. 50.

McCarty, L.S., Ozburn, G.W., Smith, A.D., and Dixon, D.G. (1992) Toxicokinetic modelling of mixtures of organic chemicals. *Environ. Toxicol. Chem.* **11**, 1037–1047.

Magurran, A.E. (1988) *Ecological Diversity and Its Measurement*. Croom Helm, London.

Maltby, L., Calow, P., Cosgrove, M., and Pindar, L. (1987) Adaptation to acidification in aquatic invertebrates: Speculation and preliminary observations. *Ann. Soc. Roy. Zool. Belgique* **117**, 105–115.

Maltby, L., and Calow, P. (1989) The application of bioassays in the resolution of environmental problems; past, present and future. *Hydrobiologia* **188/189**, 65–76.

Marking, L.L. (1985) Toxicity of chemical mixtures. In: Rand, G.M., and Petrocelli, S.R. (Eds.) *Fundamentals in Aquatic Toxicology*, pp. 164–176. Hemisphere, Washington, D.C.

Metcalfe, J.L. (1989) Biological water quality assessment of running waters based on macroinvertebrate communities: history and present status in Europe. *Environ. Pollut.* **60**, 101–139.

National Rivers Authoity (1991) *The Quality of Rivers, Canals and Estuaries in England and Wales. Report of the 1990 Survey*. National Rivers Authority, Cambridge.

Naylor, C., Maltby, L., and Calow, P. (1989) Scope for growth in *Gammarus pulex*, a freshwater benthic detritivore. *Hydrobiologia* **188/189**, 517–523.

Newbold, J.D. (1992) Cycles and spirals of nutrients. In: Calow, P., and Petts, G. (Eds.) *Rivers Handbook*, Vol. 1, pp. 379–408. Blackwell Scientific, Oxford.

Norris, R.H., McElravy, E.P., and Resh, V.H. (1992) The sampling problem. In: Calow, P., and Petts, G. (Eds.) *Rivers Handbook*, Vol. 1, pp. 282–306. Blackwell Scientific, Oxford.

Odum, E.P. (1985) Trends expressed in stressed ecosystems. *Bioscience* **35**, 419–422.

Paine, R.T. (1988) On food webs: road maps on interaction or grist for theoretical development? *Ecology* **69**, 1648–1654.

Pascoe, D., and Edwards, R.W. (Eds.) (1984) *Freshwater Biological Monitoring*. Pergamon Press, Oxford.

Pascoe, D., and Shazili, N.A.M. (1986) Episodic pollution—a comparison of brief and continuous exposure of rainbow trout to cadmium. *Ecotoxicol. Environ. Safety* **12**, 189–198.

Peckarsky, B.L. (1984) Predator–prey interactions among aquatic insects. In: Resh, V., and Rosenberg, D.M. (Eds.) *The Ecology of Aquatic Insects*., pp. 196–254. Praeger, New York.

Persoone, G., and Janssen, C.R. (1993) Freshwater invertebrate toxicity tests. In: Calow, P. (Ed.) *Handbook of Ecotoxicology*, Vol. 1. Blackwell Scientific, Oxford.

Reynoldson, T.B., and Day, K.E. (1993) Freshwater sediments. In: Calow, P. (Ed.) *Handbook of Ecotoxicology*, Vol. 1. Blackwell Scientific, Oxford.

Royal Commission on Environmental Pollution (1992). *Freshwater Quality*. Sixteenth Report. HMSO, London.

SETAC (Society of Environmental Toxicology and Chemistry) Europe (1991) Guidance document on testing procedures for pesticides in freshwater mesocosms. In: *Proceedings of a Workshop. A Meeting of Experts on Guidelines for Static Field Mesocosm Tests.* SETAC Europe.

SETAC (1992) *Workshop Report on Aquatic Microcosms for Ecological Assessment of Pesticides.* Society of Environmental Toxicology and Chemistry, Arlington, Virginia.

Slooff, W., Van Oers, J.A.M., and De Zwart, D. (1986) Margins of uncertainty in ecotoxicological hazard assessment. *Environ. Toxicol. Chem.* **5**, 841–852.

Solbe, J.F. de L.G. (1993) Freshwater fish. In: Calow, P. (Ed.) *Handbook of Ecotoxicology*, Vol. 1. Blackwell Scientific, Oxford.

Taub, F.B., and Read, P.L. (1982) *Final Report and Protocol, Model Ecosystems; Design, Development, Construction and Testing*. Contract No. 223-80-2352. US Department of Health and Human Services, Food and Drug Administration, Washington, D.C.

Thurston, R.V., Chakoumakos, C., and Russo, R.C. (1981) Effect of fluctuating exposures on the acute toxicity of ammonia to rainbow trout (*Salmo gairdneri*) and cutthroat trout (*S. clarki*). *Water Res.* **15**, 911–917.

Townsend, C.R. (1989) The patch dynamics of stream community ecology. *J. N. Am. Benthol. Soc.* **8**, 36–50.

USEPA (US Environmental Protection Agency) (1986) Guidelines for the health risk assessment of chemical mixtures. *Fed. Reg.* **51**, 34014–34025, 24 September.

Van Straalen, W.M., and Denneman, C.A.J. (1989) Ecotoxicological evaluation of soil quality criteria. *Ecotoxicol. Environ. Safety* **18**, 241–251.

Westman, W.E. (1978) Measuring the inertia and resilience of ecosystems. *Bioscience* **28**, 705–710.

Wilhm, J. (1972) Graphic and mathematical analysis of biotic communiies in polluted streams. *Ann. Rev. Entomol.* **17**, 223–252.

7 Methods to Assess Effects on Brackish, Estuarine, and Near-Coastal Water Organisms

M. H. Depledge
Odense University, Denmark

S. P. Hopkin
University of Reading, United Kingdom

7.1 INTRODUCTION

In 1956, 2000 cases of alkyl-mercury poisoning resulting from ingestion of contaminated fish and shellfish were reported among fishermen and their families at Minimata Bay, Japan (Clark, 1986). This and other similar poisoning events stimulated marine pollution studies throughout the 1960s and 1970s. Research focused primarily on detecting the dangers posed to human health from contaminants in marine food products.

During the past decade, at least two major changes in emphasis have occurred. First, marine ecosystems are worth preserving in their own right and not simply to limit indirect threats to human health. Second, the science of ecotoxicology has revealed that the effects of pollutants in natural ecosystems are diverse, complex, and often unpredictable. Dissatisfaction is increasing with the lack of ecological relevance of many standard ecotoxicological tests and a realisation that extrapolation from laboratory findings to real-world situations is often impractical (Giddings, 1986). This situation has spurred efforts to find more ecologically relevant methods to assess the effects of pollutants in marine, estuarine and brackish waters (Hopkin, 1993). An objective of this chapter is to review these methods.

7.2 POLLUTANTS, CONTAMINANTS, AND EFFECTS

A key problem in ecotoxicology is that many of the terms used are ambiguous. Therefore, the terminology used here will be defined.

The Group of Experts on the Scientific Aspects of Marine Pollution (GESAMP)

Methods to Assess the Effects of Chemicals on Ecosystems
Edited by R. A. Linthurst, P. Bourdeau, and R. G. Tardiff

has defined marine pollution as "the introduction by man, directly or indirectly, of substances or energy into the marine environment (including estuaries) that result in such deleterious effects as harm to living resources, hindrance to marine activities including fishing, impairment of quality for the use of seawater and reduction of amenities" (GESAMP, 1990). In this chapter, brackish water bodies, ranging in size from coastal lagoons to the Baltic Sea, are also considered subject to marine pollution.

A "contaminant" is a substance that can be detected in an ecosystem above its background concentration, but which has not been demonstrated to give rise to one or more of the adverse effects mentioned above.

When discussing adverse effects of pollutants, some consider the accumulation of pollutant residues in the tissues of organisms to be adverse. Others consider an effect injurious only if changes occur in physiological processes in organisms, such as alterations in cellular morphology, metabolic activity, or physiological rates. Ecologists might restrict this definition still further to only those pollutant-induced effects that give rise to ecologically significant changes, i.e., those at the population level. These latter effects are of prime interest, and they will receive the most attention in this chapter. The other effects mentioned are also important, since they represent stages in the progression of ecological adaptation.

7.3 GENERAL STRATEGIES

Effects of pollutants can be detected at several different levels of biological organization, ranging from the level of the whole ecosystem to that of the subcellular and molecular. Before selecting methods to detect specific changes, the level of organization (individuals, a population, a community, or an ecosystem) for study must be clearly defined. For instance, detecting changes at the molecular level in one tissue may have little significance for the health and survival of the entire individual. Likewise, alterations at the level of individuals may not be evident at population or community levels (Moriarty, 1983). Assessments may be conducted at any location and at any moment in time, and may be repeated periodically to detect insidious changes apparent only as trends over long time periods. Such an approach may help to estimate the future changes from proposed introductions of new chemicals into the environment. While a desirable goal may be to visit a site only once, make non-destructive measurements, and determine the consequences of specific concentrations of one or more substances on the biota at that site, attainment of this goal is not possible because most methods are empirical, and require temporal or spacial comparisons between control and test sites.

A most difficult issue to address is that when an effect can be characterised as truly adverse. This matter will be addressed only superficially herein; rather the focus is on the detection and quantification of biological changes induced by pollutant exposure. The reader is referred to Forbes and Forbes (1993) for a fuller discussion of the scientific, managerial, and political issues involved in deciding when effects caused by pollutants are of sufficient impact to warrant remedial action.

7.4 MONITORING AT DIFFERENT LEVELS OF BIOLOGICAL ORGANIZATION

The most ecologically relevant measurements to assess ecotoxicity are those that describe changes in ecosystem structure and function (Kelly and Harwell, 1989). However, such measurements are often difficult and time consuming to make, and are seldom predictive. Thus, by the time that a significant change can be measured, for example, in nutrient cycling, the ecosystem may have been severely damaged already. Another limitation of this type of measurement is that relating the degree of ecosystem change to a particular level of environmental contamination (even major pollutants) may often be impossible.

Descending through the biological hierarchy, a range of measurements can be made at the level of communities, such as those of algae or encrusting species such as bryozoans, hydroids, or barnacle assemblages (Blanck *et al.*, 1988; Bayne *et al.*, 1985). However, similar measurements are much less practical when dealing with larger, free-moving organisms, such as macroinvertebrates, fish, marine mammals, and birds that have longer generation times.

Holwerda and Opperhuizen (1991) presented a contemporary overview of physiological and biochemical approaches to the toxicological assessment of environmental pollution. However, much of the work presented does not define the relevance of biochemical and physiological effects to consequences in populations or communities. Blanck *et al.* (1988) postulated that pollutants that do not exert a selection pressure can cause no significant biological effects to an ecosystem, since these substances are unable either to restructure communities or to change the genotypic distribution in the populations. Thus, inter-individual variability in responses to pollutants becomes of major importance, since it is the key to understanding the mechanisms of selection underlying pollution-induced (as well as naturally occurring) ecological change (May, 1986; Depledge, 1990a).

At the cellular and molecular levels, numerous pathological changes and biochemical markers have been identified that signal exposure to pollutants (McCarthy and Shugart, 1990; Huggert *et al.*, 1992). Relationships between particular pollutants at known concentrations and pathological or biochemical responses to which they give rise have been established readily; however, as with physiological effects, relating effects in individuals to alterations in populations and communities has proved to be difficult at best.

7.5 MARINE, BRACKISH WATER, AND ESTUARINE POLLUTION

When comparing data obtained from monitoring of brackish and fully marine habitats, certain factors should be kept in mind. Species diversity is often greater in fully saline coastal waters than in estuaries and brackish waters (Remane and Schlieper, 1971); thus, pollution may result in the early loss of sensitive species. Many estuarine and brackish water species can tolerate the fluctuating conditions of such environments because they are able to maintain life in such environments (Depledge, 1990b). Hence, these tolerant species may be preadapted to tolerate greater stress such as that arising from pollution (Gray, 1974; Howell, 1984; Depledge, 1990b). Alternatively, the biota of estuarine and brackish waters may already be living near their tolerance limits due to continuous exposure to natural stressors. Some researchers, therefore, have concluded that estuarine and brackish water species are more vulnerable than their marine counterparts to additional pollutant-induced stress. This matter has yet to be resolved, and is discussed more extensively by McClusky *et al.*, 1986.

Salinity differences may also strongly alter the bioavailability of pollutants to organisms, and hence may impact toxicity. With cadmium in water, for instance, reducing salinity also reduces chloride complexation, thereby increasing the bioavailability of cadmium to organisms (Mantoura *et al.*, 1978). However, this situation may not pertain if interactions occur with calcium in the water (Bjerregaard and Depledge, 1992; Depledge, 1990b). Differences in pollutant bioavailability in waters of different salinity emphasize the utility of bioindicator studies in which only the pollutant load accumulating in the organism is assessed.

7.6 BIOINDICATORS OF MARINE, BRACKISH WATER, AND ESTUARINE POLLUTION

In the early 1970s, measurements of pollutant residues in aquatic organisms were recognized to be a valuable addition to analyses of water and sediments (Phillips, 1980). The idea behind such biomonitoring approaches is to use an organism's pollutant load as an index of exposure. According to Phillips (1980), the advantages of measuring pollutant residues in aquatic organisms are:

1. the concentration of many pollutants is higher than in the surrounding water;
2. only the fraction of the pollutant that is biologically available is measured; and
3. a time-averaged index of pollution can be obtained, provided the rates of uptake and excretion are known.

An important reason to measure the concentrations of contaminants in marine organisms is that the analysis of seawater and sediments is difficult and expensive. Usually, contaminant concentrations are much lower in water and sediment samples than in biota (often below analytical detection limits). Furthermore, water current may disperse pollutants, even though organisms remaining at a locality have been contaminated. This situation makes difficult the prediction of the extent of accumulation in marine organisms from the results of analysis of abiotic samples.

Ratios of organism to water concentration usually span several orders of magnitude, even at the planktonic level (Skwarzec and Bojanowski, 1988). For example, tissues of the sea skater, *Halobates micans*, contain remarkably high concentrations of cadmium, despite the presence of much lower concentrations of cadmium in the water that the organism inhabits (Schulz-Baldes, 1989). The metal appears to be concentrated in the surface microlayer of the seas, and is accumulated by *Halobates*. The insect clearly provides a far more significant route for cadmium transfer to its predators than would have been predicted from analysis of seawater alone.

Likewise, some pyrethroid insecticides cause substantial acute mortality in estuarine populations at concentrations well below the detection limits in seawater (Schimmel *et al.*, 1983; Clark *et al.*, 1989). Monitoring of water, sediment, and biota from estuaries in South Carolina (US) indicated little potential impact on the fauna resulting from insecticide exposure (Trim and Marcus, 1990). Nonetheless, 30 insecticide-related fish kills representing 11.5 percent of all estuarine kills occurred during the monitoring period. Thus, monitoring data allowed no early detection of either pesticide impact or identification of pollutant sources (Trim and Marcus, 1990).

Quantifying the relative proportions of pollutants taken up from food, seawater, or sediments is important to assess the ecotoxicological significance of particular contaminant loads (Depledge and Rainbow, 1990). The influence of temperature and reduced salinity on bioavailability should also be considered (Newman and McIntosh, 1989), as in the case of hydrocarbon residues from the Amoco Cadiz oil spill which are locked up in anaerobic sediments from which they are being released slowly to contaminate oyster tissues (Berthou *et al.*, 1987). Because this process is so complex, accurate estimates of the rates at which the residues will accumulate in oysters can be made only with much uncertainty.

Both biotic and abiotic variables are known to influence pollutant kinetics, and must be taken into account when interpreting biomonitoring data (Phillips, 1980). Although some strong correlations between environmental pollution and pollutant content in tissues of some species have been found (Phillips, 1980), the interpretation of pollutant body burdens may not be as simple as originally anticipated. Depledge and Rainbow (1990) emphasized that the systemic mechanisms of handling metals and an organism's physiological condition can determine the significance of the body burden of a specific metal. Thus, differences in the partitioning of metals among different tissues may largely influence toxicity, but may be masked when measurements are restricted to those of whole body concentrations (Depledge and Rainbow, 1990). Until a greater understanding of trace metal handling by organisms emerges, monitoring metal loads in biota may not always be a suitable approach to mapping the influence of metal pollution in the environment.

Very few species meet all, or even most, of the criteria for an ideal bioindicator species. Bivalve molluscs appear to be one of the most suitable groups; by far the greatest research effort has been directed towards them (Phillips, 1980). Bivalves are widespread and commonplace, are easily collected in large numbers, and are sedentary. Because bivalves are filter feeders, they pass large volumes of water over their body surfaces, and accumulate pollutants almost continuously. Thus, they act as integrators of exposure over long time periods.

Mytilus edulis has been analyzed most frequently, since it is common, and has an almost global distribution. However, caution should be exercised in using *Mytilus edulis* as the identifier, because it is now known to be a complex of three species (Lobel *et al.*, 1990). *Mytilus* has become a part of global Mussel Watch programmes (Bayne, 1989; Cossa, 1988; Cossa, 1989), and has demonstrated interesting trends in pollutant concentrations. For example, Fischer (1989) demonstrated that concentrations of cadmium in *Mytilus edulis* in Kieler Bight in the western Baltic in 1984 had declined to about 30 percent of the levels in 1975. Reductions in cadmium and lead concentrations in mussels and oysters in the US over a similar time period were found by Lauenstein *et al.* (1990). The long durations of many of the programmes will ensure that long-term trends in bioavailability of inorganic and organic pollutants to bivalves and their predators can be separated reliably from natural fluctuations (Sericano *et al.*, 1990).

Crustacea have also been used extensively in biomonitoring. Rainbow *et al.* (1989) identified sites in Scottish coastal waters where concentrations of copper in amphipods were higher than findings in abiotic samples would have predicted, suggesting that copper may have been discharged into the sea from whisky distilleries. This work has continued with studies on copper and zinc in *Orchestia gammarellus* on North Sea coasts, where several pollution hotspots have been identified (Moore *et al.*, 1991).

Concentrations of copper and zinc in barnacles from contaminated sites are among the highest of any marine organism (Chan *et al.*, 1986; Powell and White, 1990). Transplant experiments have shown that the barnacles reflect environmental levels by rapidly accumulating zinc and copper present at contaminated sites (Al-Thaqafi and White, 1991). They exhibit an uptake-storage mechanism of metal detoxification unlike most other marine Crustacea that exhibit some degree of metal regulation resulting in maintenance of a more or less constant body concentration. Sedentary, adult barnacles are potentially ideal long-term *in situ* monitors of pollution, particularly of copper and zinc.

Depledge (1990a) emphasized the importance of considering inter-individual differences in pollutant load among the representatives of a population of organisms. Thus, knowing the mean value of a pollutant concentration for a population sample is often insufficient. When sampling organisms or tissues to assess the bioavailability of pollutants at a particular locality, the distribution of pollutant loads in the population must be ascertained for predictions of ecological significance. Depledge and Bjerregaard (1989) and Lobel *et al.* (1989) pointed out that the frequency distributions of particular trace metal concentrations may not be normal, a fact almost invariably ignored in biomonitoring surveys. The problems that arise by assuming a normal distribution when actually the data are skewed has been clearly demonstrated by Lobel *et al.* (1982), and Blackwood (1992) has provided a useful discussion of the legitimacy of using the log-normal distribution for describing environmental data.

For zinc in *Mytilus edulis* collected from the Tyne estuary (UK), the mean zinc concentration was only 75 percent of the mid-range value. In a comparison of three sites contaminated with zinc to varying extents, the lowest tissue concentrations recorded at each site were similar (0.83, 1.5, and 1.11 μmol Zn/gm), whereas the highest concentrations were markedly different (3.32, 10.0, and 20.5 μmol Zn/gm). The distributions of tissue concentrations from animals at each site were skewed positively. Statistical techniques are already available to quantify and compare the residual variability of trace metal concentrations in biological tissues (Lobel *et al.*, 1989).

7.7 ASSESSING POLLUTANT EFFECTS AT THE ECOSYSTEM LEVEL

When measuring effects of chemicals on ecosystems, the uniformity, intensity, frequency and duration of exposure, and uniqueness of the chemical (i.e., whether found normally in ecosystems but in lower concentrations or whether it is man-made) are all important to consider. These and other factors have been discussed in detail by Kelly and Harwell (1989). An extensive debate has been ongoing as to whether changes in ecosystem structure precede changes in function or vice versa; however, this issue remains unresolved (Ford, 1989). The effects of chemical stressors are manifest initially as a loss of sensitive species and changes in the relative abundance of rapidly reproducing taxa. However, changes in ecosystem processes do not always occur as pollution increases, nor is the rate of ecosystem change constant when it does occur (Levine, 1989). Since some ecosystem processes (e.g., nitrogen and phosphorus cycling) can be taken over by other species as more vulnerable species are eliminated, structural changes still offer probably the best prospect of signalling pollutant-induced change in most aquatic ecosystems.

Numerous authors have attempted to find indices of overall ecosystem health, such as the rate of photosynthesis, the photosynthesis–respiration ratio, and the activity of electron transport systems (Woodwell, 1962; Ivanovici and Weibe, 1981; Levine, 1989). Critical reviews of the concept of ecosystem health have recently been issued by Calow (1992) and Rapport (1992). A key problem with this approach is that little attention has been paid to relating changes in ecosystem health to particular degrees or types of pollution. Babich *et al.* (1983) attempted to overcome this limitation by suggesting establishment of dose-response relationships for ecosystems in which a pollutant load which causes a 50 percent or a 10 percent reduction in an ecosystem process is determined (EcD_{50} or EcD_{10}, respectively). So far, these indices have proved to be of no practical utility.

Critical concentrations for pollutants in soils and sediments have been proposed to be set at levels that protect 95 percent of species from poisoning; however, this concept has limited acceptance for the marine environment (Hopkin, 1993).

7.8 COMMUNITY-LEVEL EFFECTS

For marine benthic communities, several studies have revealed important changes in species composition following exposure to organic pollution from pulp mills and oil spills (Pearson, 1970, 1971, 1975; Rosenberg 1972; Rosenberg, 1973; Sanders *et al.*, 1991). In these and other instances, the affected communities are dominated by a very few species, usually of annelid worms. Such pollution indicator organisms are potentially useful to assess the extent to which benthic communities are affected. Ecological succession occurring in response to a pollution gradient is complex, and may be influenced by a great many factors other than pollutant exposure (Pearson and Rosenberg, 1978), so the approach should be used cautiously. Nevertheless, changes in community composition appear to be reasonably consistent when addressed using the above approach.

Reductions in species diversity, retrogression to opportunistic species, and shifts to smaller sized species are all well-documented responses of communities to stress (Forbes and Forbes, 1993). Diversity indices usually integrate the relative abundances and numbers of species in a community, and provide an estimate of community complexity. A fall in diversity usually indicates significant pollutant-induced change (Gray, 1980), but predation, competition, spatial heterogeneity, and successional change may also influence diversity indices, and have led to valid criticisms by Gray (1980). Thus, Ford (1989) concluded that general agreement exists that diversity indices do not reliably reflect pollutant disturbances over time or space.

Practical difficulties associated with the use of diversity indices include problems with the accurate identification of organisms to the species level. While this characterization may be accomplished in well-studied, temperate, brackish water, and estuarine ecosystems, greater problems may be encountered in coastal regions. Furthermore, practising this approach in the subtropics and tropics (where species diversity increases enormously and where many species have yet to be classified) may be impossible at present.

One of the simplest methods to detect pollution-induced changes in communities of marine benthic organisms is to analyse the log-normal distribution of individuals per species in sediment samples. This approach assumes that a community equilibrium exists (Connell and Sousa, 1983; Williamson, 1987). For example, temporary departures from a log-normal distribution may occur during seasonal recruitment of juveniles. Gray (1980), therefore, recommended that monitoring year-to-year variations in species abundance could be conducted most effectively during the winter in temperate habitats or at those times of the year when populations are not actively recruiting. A detailed study has recently been conducted by Ferraro and Cole (1992) on the taxonomic level that needs to be adopted to assess the impact of pollutants on macrobenthic communities.

In many samples of benthic communities, the most abundant class is not that represented by one individual per species, but often lies between classes with either three or six individuals per species. Thus, the curve relating the number of individuals per species (*x* axis) to the number of species (*y* axis) is often strongly skewed. This curve can be converted to a normal one by plotting the number of individuals per species on a geometric scale (usually 2×). Plotting the geometric classes on the *x* axis (Class I = 1, Class II = 2 to 3, Class III = 4 to 7, Class IV = 8 to 15, and so on) against cumulative percent of species on the *y* axis invariably gives a straight line. At polluted sites, a break in the line often exists, indicating departures from an equilibrium community. A persistence of this break over several sampling occasions is indicative of pollution-induced disturbance. The log-normal distribution procedure has been described by Gray (1981).

Pollution-induced community tolerance (PICT) can be used in the assessment of pollution effects. Blanck and Wangberg (1988) investigated PICT in periphyton communities established on 1.5 cm^2 glass discs. When exposed to severe arsenate stress, the communities developed 17000-fold tolerance to the pollutant, as measured by photosynthetic activity. Arsenate exerted a selection pressure leading to replacement of sensitive species by tolerant ones. This effect caused overall arsenate tolerance of the community to increase. The stress also affected the rate of increase of biomass. An important consideration when using this approach is that exposure to one pollutant may confer tolerance to another (i.e., co-tolerance). Although this effect was reported to be insignificant for arsenic (Blanck and Wangberg, 1988), situations can be envisaged in which tolerance to a particular trace metal or organic pollutant may be due to exposure to a quite different trace metal or organic compound having the same or similar mechanism of action, leading to the incorrect conclusion that a community has been exposed to a chemical which actually it has never encountered.

Warwick *et al.* (1990) studied the structure of benthic communities in relation to pollution in Hamilton Harbour, Bermuda. Multivariate analysis of the fauna detected differences in community composition that could be related to the pollution gradient in the Harbour. For the statistical analysis of the macrobenthos, aggregation of species data to family level was acceptable. However, for nematodes, aggregation from genus to family level resulted in a significant loss of information.

One of the most detailed studies on the responses of marine organisms to a putative pollution gradient was conducted on the sublittoral fauna of the Frierfjord-Langesundfjord (Norway) by Gray *et al.* (1988). This valuable research includes multivariate statistical analyses, which discriminate among sites on their faunistic attributes, and univariate measures of community stress, and provides a very comprehensive, comparative account of methods for assessing community responses. The effects observed were actually attributed to seasonal anoxia rather than pollution.

7.9 POPULATION-LEVEL EFFECTS OF POLLUTANTS

Gray (1979) commented on the surprising paucity of information on the effects of pollutants on populations. Since then, some notable studies have been added to the scientific literature. Nonetheless, population-level effects of pollutants are perhaps the most neglected area of ecotoxicological research.

An excellent analysis of stress (including pollutant-induced stress) on natural populations was carried out by Underwood (1989). He noted that experimental studies on populations are difficult to conduct because of inertia (a lack of response in a population following exposure to pollution), resilience (the capability of the perturbation from which a population can recover), and stability (the rate at which a population recovers from a stress). The extent to which a population has the above attributes determines its range of responsiveness to intermittent, temporary, and acute and chronic pollutant exposure.

The timing of exposure to stress relative to the life cycle of individuals within the population is also critical. For instance, experimental deletion of patches of seaweed from a sublittoral kelp bed at certain times of the year leaves the kelp population depleted until after the following reproductive season. However, if kelp patches are removed when the plants are reproducing, little effect can be observed (Kennelly, 1987).

Effects of pollutants on populations may involve loss of individuals due to smothering with oil, poisoning, or incapacitation, which separately render some organisms more susceptible to predation or to mortality caused by other environmental stressors. Alternatively, population-level effects may emerge due to alterations in individual growth rates, reduced fecundity and longevity, and disturbance of endogenous biological rhythms (Gray, 1979; Moriarty, 1983; Depledge, 1984).

A most effective means of detecting changes in populations over time is in the use of life tables (Morris, 1959; Varley and Gradwell, 1960; Varley *et al.*, 1975). Life tables permit analysis of the role and importance of different mortality factors in relation to the overall mortality rate of individual populations (Begon and Mortimer, 1986). However, distinguishing between pollution-induced effects and those related to fluctuations in natural environmental factors may be difficult.

Luoma (1977) commented that the existence of one population of a species that is more resistant to a toxicant than another is direct evidence that the concentration of the toxicant in the environment of the resistant population is sufficient to elicit adverse biological effects. The presence of a toxicant-resistant population of one species in an ecosystem further suggests that other species may have been affected by the resistance-eliciting substance. Luoma (1977) stated that populations resistant to a particular pollutant are found only in areas of known contamination. An illustration is populations of the polychaete, *Nereis diversicolor*, found in English estuaries contaminated with copper, cadmium, lead, and zinc that are more tolerant to high concentrations of copper in seawater than are *N. diversicolor* found in less contaminated areas (Bryan and Hummerstone, 1971, 1973; Bryan, 1974). Thus, the implied selection of resistant genotypes within a population exposed to a contaminant constitutes the basis of the PICT assessment method. However, for *N. diversicolor*, enhanced genetic resistance to cadmium or lead has not been detected.

Futyuma (1986) reviewed data from several fields of research showing that genetic mutations conferring resistance occur independently of exposure, and, therefore, are not induced by pollutants. In support of this conclusion, Weeks and Depledge (1992) recently reported marked differences in mercury tolerance among three populations of the amphipod, *Platorchestia platensis*, living in similar habitats within 20 km of one another. None of the populations has been exposed to significant trace metal contamination *in situ*.

Gray (1979) stated that, in habitats under severe pollution stress, the species that dominate are those that have flexible life histories. Species having less flexible life histories increase in abundance under conditions of slight pollution. Thus, Gray (1979) concluded that the presence of a particular species in a polluted area may be more closely related to its life history strategy than to tolerance to adverse conditions. Nonetheless, when an intense selection of genotypes takes place, the population that survives at a polluted locality would be expected to have a different genetic composition when compared with populations of the same species in clean conditions. Modern genetic techniques have the potential to detect such phenomena (Bickham *et al.*, 1986).

The expression of particular genes confers the ability to survive in polluted conditions. Thus, phenotypic differences between populations of a species occupying clean and polluted conditions might be expect to occur. Such differences might be manifested as alterations in morphology, behaviour, and biochemical–physiological characteristics. Depledge (1992) has argued that of these factors, particular biochemical-physiological traits are most likely to confer resistance to chemical toxicity. He proposed that, by looking for changes in the relative proportions of resistant versus sensitive phenotypes (or "physiotypes" as they were called to emphasize the importance of biochemical-physiological characteristics), detection of changes in the tolerance distribution of individuals within populations exposed to marine pollutants may be possible. Similar studies have been conducted for many years to assess the development of resistance to pesticides among insects (Wood and Bishop, 1981) and by evolutionary biologists (Via and Lande, 1985; Schluter, 1988; Anholdt, 1991; Forbes and Forbes, 1993).

Invertebrates that have been shown by breeding experiments to be genetically resistant to high concentrations of metals include marine polychaetes (Grant *et al.*, 1989). These authors have suggested that genetically based tolerance to metals in polychaetes could be mapped, and used as an *in situ* monitor of the ecological impact of pollutants.

Several biochemical techniques are now available to detect genetic changes in organisms exposed to pollutants (Shugart *et al.*, 1992). For example, the DNA alkaline unwinding assay can be used to detect DNA-damaging substances in marine animals exposed to environmental pollutants. In gills of blue mussels caged at the New Bedford Harbour Superfund Site (Massachusetts, US) which is highly contaminated with many organic and inorganic substances, a significant increase in DNA strand breaks was detected after three days of exposure (Nacci and Jackim, 1989). In fish, DNA-xenobiotic adducts persisted in the liver for much longer than aromatic hydrocarbon (AH) metabolites that induced them (Varanasi *et al.*, in press). Thus, these effects on DNA may have a practical use in integrating cumulative exposure of fish to AHs.

7.10 TRANSPLANT TECHNIQUES AND MESOCOSM TESTING

Phillips (1980) described the use of organisms transplanted from clean sites to polluted sites or vice versa to examine the accumulation or depuration of pollutants. The value of such an approach was demonstrated in studies in Hong Kong on the uptake and release of PCB isomers in transplanted mussels (*Perna viridis*) (Tanabe *et al.*, 1987). Similar experiments have been attempted in which parameters of the well-being of organisms have been measured after transfer from one site to another (e.g., scope for growth in transplanted mussels) (Tedengren *et al.*, 1990).

The responses (in terms of changes in species composition) of fouling communities to pollution stress can be monitored *in situ* by reciprocal transplants. Climax communities are allowed to develop on submerged surfaces in a clean and a polluted site, and are then moved between sites. An experiment in Australia at Woolongong Harbour (uncontaminated) and Port Kemblar Harbour (polluted by discharges from heavy industry) demonstrated rapid changes in community structure in response to pollution (Moran and Grant, 1991). Within two months after transfer, communities moved from Woolongong to Port Kemblar were similar in structure to those that had developed wholly at Port Kemblar. Most changes occurred in the short term when sensitive species were killed by periodic discharges—an effect difficult to predict by measuring levels of pollutant in water. Space previously occupied by these species was quickly colonized by opportunists tolerant of pollutants, giving rise to changes in community structure.

Changes at the ultrastructural level can also be examined in organisms transplanted between clean and contaminated sites. Thomas and Ritz (1986) showed that much of the zinc accumulated by the barnacle *Elminius modestus* transplanted to a zinc-contaminated site was stored with phosphate in intracellular granules. These granules increased in size in proportion to the length of exposure. Few granules were lost from contaminated organisms when they were transplanted to a clean site. The full potential of transplant experiments in ecotoxicology has yet to be explored.

Caging organisms in mesocosms provides a means of bridging the gap between the laboratory and the field (Kuiper and Gamble, 1988). *Mytilus edulis* caged in the effluent from a factory producing titanium dioxide, accumulated titanium in excretory granules (Ballan-Dufrancais *et al.*, 1990). Mitochondria in the bivalves contained lesions and reduced cristae, indicating a serious effect of the effluent on the respiratory metabolism of the organisms (Coulon *et al.*, 1987).

7.11 BIOMARKERS FOR THE DETECTION OF POLLUTANT TOXICITY

The most powerful tools for the investigation of pollutants *in situ* are biomarkers. Depledge *et al.* (1992) and Depledge (1993) have discussed the rational basis of the biomarker approach. In the past, a variety of cellular biochemicals have been measured in tissue samples from aquatic animals (less so plants), and their concentrations have been related to exposures to specific pollutants or to various classes of pollutants. This approach has achieved very little, because biomarker responses are of unknown ecological significance. For example, a rise in mixed function oxidase (MFO) in the livers of fish taken from a polluted area may signify pollutant exposure; yet, the fish may continue to grow and reproduce normally, and the MFO response may be viewed as part of an acclimatization process to altered environmental conditions rather than a manifestation of an injury.

Much can be gained by relating biomarker responses to changes in Darwinian fitness parameters (Depledge, 1993). As an illustration, Sanders *et al.* (1991) measured stress protein responses and scope for growth (SFG) in *Mytilus edulis* exposed to various concentrations of copper in the laboratory. By relating the two, the investigators showed that stress protein responses occurred prior to reductions in SFG. As copper exposure increased further, SFG decreased and stress protein responses were even more marked. Thus, the stress protein biomarkers could be used to signify a change in Darwinian fitness that may have consequences for the entire population.

Recently the biomarker concept has been extended from purely biochemical measurements to include those of cellular pathology, physiological processes, and even behaviour of organisms exposed to varying concentrations of pollutants (Depledge *et al.*, 1992; Sanders *et al.*, 1991). This enhancement creates the possibility of using a hierarchy of biomarker measurements (Depledge, 1993). Initially, effects of pollutants might be detected by relatively non-specific biomarkers, usually high in the hierarchy (e.g., behavioural and physiological biomarkers). Detection of abnormalities with these non-specific biomarkers at a site at risk from pollution might then justify the measurements of more costly, lower hierarchy, specific biochemical and cellular biomarkers (e.g., MFO activity, metallothioneins, intracellular granules, and tissue lesions) to seek to identify the class of pollutant responsible for the exposure. The magnitude of biomarker responses together with determination of tissue residue concentrations of pollutants would contribute to the overall assessment of pollutant impact.

Biomarkers may also have an important role in unravelling the interactions between natural environmental stressors (e.g., hypoxia, thermal and salinity stress) and pollutants, and the effects of mixtures of contaminants in areas receiving several pollutants (Livingstone *et al.*, 1988). Where more than one stressor (natural factors plus one or more pollutants, or complex mixtures of pollutants alone) is present, the substances may act synergistically so that the combined effect is greater than the sum of the effects of the individual chemicals (Depledge, 1987; Walker and Johnston, 1989). However, in some circumstances, stressors (including pollutants) can act antagonistically. TBT and hydrocarbons, when present together, caused a much lower reduction in clearance rate in *Mytilus edulis* than had been predicted on the basis of proportional additivity (Widdows and Donkin, 1991). In their excellent summary of this problem, Widdows and Donkin (1991) described how reductions in SFG in *Mytilus edulis* in contaminated sites can be apportioned between specific pollutants. In a study conducted in Bermuda, Widdows *et al.* (1990) showed that the overall reduction in SFG of *Mytilus edulis* could be proportional such that, at the most contaminated sites, TBT accounted for 21 percent and hydrocarbons 74 percent of the observed effects.

When measuring biomarker responses in organisms sampled from the field (or when measuring pollutant residue levels in biota), only those organisms that have survived at the locality are considered. In moderately or severely polluted environments, situations can be envisaged in which the mean biomarker response is high initially and then declines, a change of response that might be due either to the return of clean conditions or to the dying out of the most severely affected organisms, so that only resistant or less severely stressed organisms remain. This complex matter deserves further study.

7.12 BIOCHEMICAL MARKERS

Biochemical markers (also called "biomarkers") of pollutant exposure and effects have been reviewed by McCarthy and Shugart (1990), Stegeman *et al.* (1992), and Fossi and Leonzio (1993). These works focus on the use of biomarkers to signal pollutant exposures rather than to detect injury to ecological systems. This limitation may reflect the complexities of validating the latter approach. Table 7.1 gives examples of biomarkers that have been used most often and pollutants known to initiate responses (Stegeman *et al.*, 1992).

As an example, microsomal cytochrome P_{450} enzymes normally function at low rates in the liver of fish, but are induced by chemicals with relatively flat molecular structures such as aromatic hydrocarbons (AH), dioxins, and co-planar PCBs. Thus, P_{450}induction has been advocated as a reliable indicator of organic chemical contamination in marine systems (Stegeman *et al.*, 1990). In the liver of dab and other fish, the induction of cytochrome P_{450} (IA family = P_{450IA}) is highly correlated with activity of the enzyme 7-ethoxyresorufin *O*-deethylase (EROD) (Goksyor *et al.*, 1989). Although the EROD assay is easier to perform than Western blotting or ELISA (enzyme linked immunosorbent assay) for P_{450IA}, circumstances exist whereby pollutants (e.g., PCBs) can inhibit EROD measurement (Goksyor *et al.*, 1991).

Table 7.1. Examples of biochemical markers of pollutant exposure

Biomarker	Pollutants initiating response
P_{450}	Aromatic hydrocarbons, dioxins
Metallothioneins	Cd, Cu, Zn, Hg, Co, Ni, Bi, Ag
Stress proteins	Thermal pollution, TBT, Cu and other metals, PAH, UV radiation
Glutathione transferases	PAH, PCB, BNF
Lipid peroxidation	Cd, PCB
Heme and porphyrins	Pb, As, Hg, PCB, dioxin, HCB

7.13 THE SPECIAL CASE FOR MORPHOLOGICAL BIOMARKERS

A very useful approach to assess dose-dependent pollutant toxicity in the marine environment is to observe and measure chemical-specific (e.g., tributyl tin, TBT) morphological changes in a widely distributed organism (molluscs, such as, the dogwhelk, *Nucella lapillus*). The extent of the changes reflects the degree of environmental exposure.

Organotin exposure in the marine environment occurs primarily as a result of TBT leaching from antifouling paints applied to the hulls of boats to inhibit settlement of marine invertebrates, especially barnacle and mollusc larvae. Other sources of organotin include PVC manufacture, wood preservatives, and general biocides (Fent *et al.*, 1988).

A combination of traditional laboratory toxicity studies, transplant experiments, and field observations have shown that TBT is probably the most toxic man-made substance ever to have been deliberately introduced into the marine environment (Bryan *et al.*, 1986). TBT is toxic to many marine organisms, and has had a severe economic impact on oyster fisheries and farms (Waldock *et al.*, 1983). In Australia, shell curling in oysters was induced in a hitherto pristine lake by mooring just two TBT-coated boats for only one month (Scammell *et al.*, 1991). In some European estuaries and coastal waters, TBT has had a dramatic effect on populations of deposit-feeding bivalves and the dogwhelk, *Nucella lapillus,* that are extremely sensitive to TBT (Langston *et al.*, 1990). Dogwhelks are now absent from many seashores in the UK, where they were common before the introduction of TBT paints (Bryan *et al.*, 1986).

TBT causes female dogwhelks to grow a vas deferens and a penis. These appendages block the opening of the female genital duct so that egg capsules cannot be released (Gibbs and Bryan, 1986). TBT appears to affect the hormone system that determines the sex of prosobranch molluscs (Gibbs *et al.*, 1988). Similar findings have been obtained with other *Nucella* species and related genera elsewhere in the World (Bright and Ellis, 1990; Stickle *et al.*, 1990; Smith, 1980).

The phenomenon in which male sex characters are imposed on the female snail is called "imposex," and is widespread in stenoglossan gastropods (Smith, 1980). The mean size of the female penis relative to males in a population of dogwhelks provides a sensitive indicator of the degree of imposex and hence of exposure of dogwhelks to TBT. Thus, determining the level of imposex provides a much easier and cheaper method to assess the bioavailability of TBT than direct measurement of the chemical in seawater (Langston *et al.*, 1990; Spence *et al.*, 1990).

7.14 PHYSIOLOGICAL MONITORING OF POLLUTANT EFFECTS

A description of the major physiological responses of marine organisms to chemicals is presented by Vernberg and Vernberg (1974), Vernberg *et al.* (1977, 1982), Dorigan and Harrison (1987), and Holwerda and Opperhuizen (1991). The rationale underlying extensive physiological monitoring, revealing disturbances of almost all physiological systems, has been fully discussed by Depledge (1989), providing encouragement that a strategy for changes in physiology to be related to alterations in the Darwinian fitness of individuals.

Acquisition of physiological data, whether in the laboratory or *in situ*, has been severely hampered by the lack of suitable transducers capable of recording data over long periods, from several test organisms, simultaneously, without imposing undue stress. However, the introduction of computer-aided monitoring systems and non-invasive transducer techniques is beginning to alleviate this situation (Depledge and Andersen, 1990; Aagaard *et al.*, 1990).

Mayer *et al.* (1992) recently reviewed methods to assess chemical-induced changes in whole animal physiology. Scope-for-growth has proved to be the most useful *in situ* assay procedure. However, much work remains before physiological monitoring can be performed routinely. As with population- and community-level approaches, responses detected at the physiological level are often difficult to relate to specific substances at known concentrations. Nonetheless, an integrated approach involving biochemical, physiological, and behavioural biomarkers may eventually prove valuable.

7.15 PATHOLOGY AND DISEASE AS INDICATORS OF POLLUTION

Detailed accounts concerning the relationships between tissue pathology and diseases in marine organisms associated with pollution have been provided by Bucke and Watermann (1988), Hinton *et al.* (1992), Weeks *et al.* (1992), Hinton and Lauren (1990), Cormier and Racine (1990), Yamashita *et al* (1990), and McMahon *et al.* (1990). A specific example of this approach was the study by Carr *et al.* (1991), which showed that the number and extent of hepatic lesions in winter flounder from the polluted waters of Boston Harbor were correlated with low tissue concentrations of ascorbic acid and hepatic glycogen (effects not observed in reference populations from clean sites). Grotesque deformities in smelt (*Osmerus eperlanus*) from the Elbe Estuary were observed by Pohl (1990).

Deformed fish had significantly higher lead and cadmium concentrations in their livers than did normal specimens, but the levels were still quite low in comparison to those at other polluted sites. Concentrations of each of the many pollutants in the Estuary were below critical levels, but when present together, the mixture of contaminants may have been related to the deformities.

7.16 SUMMARY AND CONCLUSIONS

The preceding account highlights the wide range of approaches available to assess the effects of pollutants in marine, estuarine, and brackish water environments. Identification of the type of injuries to be detected is necessary before selecting a measurement technique. No single approach is satisfactory. This review emphasizes the value of an integrated assessment approach involving several methods, each focusing at a different level of biological organisation. For an assessment useful in ecotoxicological decision-making, the fate of pollutants must be related to effects to which they give rise. Transplantation experiments and the use of biomarkers of toxicity that link responses to pollutants with changes in Darwinian fitness of organisms offer the greatest potential for future development as assessment tools.

7.17 REFERENCES

Aagaard, A., Andersen, B.B., and Depledge, M.H. (1990) Simultaneous monitoring of physiological and behavioural activity in aquatic organisms using non-invasive, computer-aided techniques. *Marine Ecol. Progr. Ser.* **73**, 277–282.

Al-Thaqafi, K., and White, K.N. (1991) Effect of shore position and environmental metal levels on body burdens in the barnacle *Elminius modestus*. *Environ. Pollut.* **69**, 89–104.

Anholdt, B.R. (1991) Measuring selection on a population of damselflies with a manipulated phenotype. *Evolution* **45**, 1091–1106.

Babich, H.R., Bewley, R.J.F., and Stotzky, G. (1983) Application of the ecological dose concept to the impact of heavy metals on some microbe-mediated ecological processes in soil. *Arch. Environ. Contam. Toxicol.* **12**, 421–426.

Ballan-Dufrancais, C., Jeantet, A.Y., and Coulon, J. (1990) Cytological features of mussels *(Mytilus edulis) in situ* exposed to an effluent of the titanium dioxide industry. *Ann. Inst. Oceanogr.* **66**, 1–18.

Bayne, B.L. (1989) Measuring the biological effects of pollution: the mussel watch approach. *Water Sci. Technol.* **21**, 1089–1100.

Bayne, B.L., Brown, D.A., Burns, K., Dixon, D.R., Ivanovici, A., Livingstone, D.R., Lowe, D.M., Moore, M.N., Stebbing, A.R.D., and Widdows, J. (1985) *The Effects of Stress and Pollution on Marine Animals.* Praeger, New York.

Begon, M., and Mortimer, M. (1986) *Population Ecology.* Blackwell Scientific, Oxford, 220 pp.

Berthou, F., Balouet, G., Bodennec, G., and Marchand, M. (1987) The occurrence of hydrocarbons and histopathological abnormalities in oysters for seven years following the wreck of the Amoco Cadiz in Brittany (France). *Marine Environ. Res.* **23**, 103–133.

Bickham, J.W., McBee, K., and Schlitter, D.A. (1986) Chromosomal variation among seven species of Myotis (Chiroptera: Vespertilionidae). *J. Mammals* **67**, 746–750.

Bjerregaard, P., and Depledge, M.H. (1992) Cadmium uptake in *Mytilus edulis*, *Littorina littorea* and *Carcinus maenas*: the influence of calcium ion concentration and salinity. *Marine Ecol. Progr. Ser.* **86**(1), 91–97.

Blackwood, L.G. (1992). The lognormal distribution, environmental data, and radiological monitoring. *Environ. Monitor. Assess.* **21**, 193–210.

Blanck, H., and Wangberg, S.-Å. (1988) Induced community tolerance in marine periphyton established under arsenate stress. *Can. J. Fish. Aquat. Sci.* **45**, 1816–1819.

Blanck, H., Wangberg, S.-Å, and Molander, S. (1988) Pollution-induced community tolerance—a new ecotoxicological tool. In: Cairns, J., and Pratt, J.R. (Eds.) *Functional Testing of Aquatic Biota for Estimating Hazards of Chemicals*, pp. 219–230. ASTM STP 988. American Society for Testing and Materials, Philadelphia, Pa.

Bright, D.A., and Ellis, D.V. (1990) A comparative survey of imposex in northeast Pacific neogastropods (Prosobranchia) related to tributyltin contamination, and choice of a suitable bioindicator. *Can. J. Zool.* **68**, 1915–1924.

Bryan, G.W. (1974) Adaptation of an estuarine polychaete to sediments containing high concentrations of heavy metals. In: Vernberg, F.J., and Vernberg, W.B. (Eds.) *Pollution and Physiology of Marine Organisms*, pp. 123–136. Academic Press, New York.

Bryan, G.W., and Hummerstone, L.G. (1971) Adaptation of the polychaete *Nereis diversicolor* to estuarine sediments containing high concentrations of heavy metals. I. General observations and adaptation to copper. *J. Marine Biol. Assoc. UK* **51**, 845–863.

Bryan, G.W., and Hummerstone, L.G. (1973) Adaptation of the estuarine polychaete *Nereis diversicolor* to estuarine sediments containing high concentrations of zinc and cadmium. *J. Marine Biol. Assoc. UK* **53**, 839–857.

Bryan, G.W., Gibbs, P.E., Hummerstone, L.G., and Burt, G.R. (1986) The decline of the gastropod *Nucella lapillus* around South-West England: evidence for the effect of tributyltin from antifouling paints. *J. Marine Biol. Assoc. UK* **68**, 733–744.

Bucke, D., and Waterman, B. (1988) Effects of pollutants on fish. In: Salomons, W., Bayne, B.L., Duursma, E.K., and Forstner, U. (Eds.) *Pollution of the North Sea—An Assessment*, pp. 612–623. Springer Verlag, Berlin and Heidelberg.

Calow, P. (1992) Can ecosystems be healthy? Critical consideration of concepts. *J. Aquat. Ecosys. Health* **1**, 1–6.

Carr, R.S., Hillman, R.E., and Neff, J.M. (1991) Field assessment of biomarkers for winter flounder. *Marine Pollut. Bull.* **22**, 61–67.

Chan, H.M., Rainbow, P.S., and Phillips, D.J.H. (1986) Barnacles and mussels as monitors of trace metal bio-availability in Hong Kong waters. In: Morton, B. (Ed.) *Proceedings of the Second International Marine Biological Workshop: The Marine Flora and Fauna of Hong Kong and Southern China*, pp. 1239–1268. Hong Kong University Press, Hong Kong.

Clark, R.B. (1986) *Marine Pollution.* Clarendon Press, Oxford, England. 215 pp.

Clark, J.R., Goodman, L.R., Borthwick, P.W., Patrick, J.M., Cripe, G.M., Moody, P.M., Moore, J.C., and Lores, E.M. (1989) Toxicity of pyrethroids to marine invertebrates and fish: a literature review and test results with sediment-sorbed chemicals. *Environ. Toxicol. Chem.* **8**, 393–401.

Connell, J.H., and Sousa, W.P. (1983) On the evidence needed to judge ecological stability and persistence. *Am. Natur.* **121**, 789–824.

Cossa, D. (1988) Cadmium in *Mytilus* spp.: world-wide survey and relationship between seawater and mussel content. *Marine Environ. Res.* **26**, 265–284.

Cossa, D. (1989) A review of the use of *Mytilus* spp. as quantitative indicators of cadmium and mercury contamination in coastal waters. *Oceanologica Acta* **12**, 417–432.

Coulon, J., Truchet, M., and Martoja, R. (1987) Chemical features of mussels (*Mytilus edulis*) *in situ* exposed to an effluent of the titanium dioxide industry. *Ann. Inst. Oceanogr.* **63**, 89–100.

Cormier, S.M., and Racine, R.N. (1990) Histopathology of Atlantic Tomcod: a possible monitor of xenobiotics in northeast tidal rivers and estuaries. In: McCarthy, J.F., and Shugart, L.R. (Eds.) *Biomarkers of Environmental Contamination*, pp. 59–72. Lewis, Boca Raton, Florida.

Depledge, M.H. (1984) Disruption of endogenous rhythms in *Carcinus maenas* (L.) following exposure to mercury pollution. *Compar. Biochem. Physiol.* **78A**, 375–379.

Depledge, M.H. (1987) Enhancement of copper toxicity resulting from environmental stress factor synergies. *Compar. Biochem. Physiol.* **87C**, 15–19.

Depledge, M.H. (1989) The rational basis for detection of the early effects of marine pollutants using physiological indicators. *Ambio* **18**, 301–302.

Depledge, M.H. (1990a) New approaches in ecotoxicology: can inter-individual physiological variability be used as a tool to investigate pollution effects? *Ambio* **19**, 251–252.

Depledge, M.H. (1990b) Interactions between heavy metals and physiological processes in estuarine invertebrates. In: Chambers, P.L., and Chambers, C.M. (Eds.) *Estuarine Ecotoxicology*, pp. 89–100. Japaga, Wicklow, Ireland.

Depledge, M.H. (1992) Ecotoxicological relevance and significance of ecophysiological pollution effects. In: Steinberg, C., and Kettrup, A. (Eds.) *Proceedings of the International Symposium on Ecotoxicology: Ecotoxicological Relevance of Test Methods, November 19–20, 1990*, pp. 1–16. GSF-Forschungszentrum, Neuherberg.

Depledge, M.H. (1993) The rational basis for the use of biomarkers as ecotoxicological tools. In:Fossi, M.C., and Leonzio, C. (Eds.) *Nondestructive Biomarkers in Higher Vertebrates*. Lewis, Boca Raton, Florida.

Depledge, M.H., and Andersen, B.B. (1990) A computer-aided physiological monitoring system for continuous, long-term recording of cardiac activity in selected invertebrates. *Compar. Biochem. Physiol.* **96A**, 474–477.

Depledge, M.H., and Bjerregaard, P. (1989) Explaining variation in trace metal concentrations in selected marine invertebrates: the importance of interactions between physiological state and environmental factors. In: Aldrich, J.C. (Ed.) *Phenotypic Response and Individuality in Aquatic Ectotherms*, pp. 121–126. Japaga, Wicklow, Ireland.

Depledge, M.H., and Rainbow, P.S. (1990) Models of regulation and accumulation of trace metals in marine invertebrates: a mini-review. *Compar. Biochem. Physiol.* **97C**, 1–7.

Depledge, M.H., Amaral-Mendes, J.J., Daniel, B, Halbrook, R.S., Kloepper-Sams, P., Moore, M.N., and Peakall, D.B. (1992) The conceptual basis of the biomarker approach. In: Peakall, D.B., and Shugart, L.R. (Eds.) *Strategy for Biomarker Research and Application in the Assessment of Environmental Health*. Lewis, Boca Raton, Florida.

Dorigan, J.V., and Harrison, F.L. (Eds.) (1987) *Physiological Responses of Marine Organisms to Environmental Stress*. US Department of Energy, Washington, D.C., 501 pp.

Fent, K., Fassbind, R., and Siegrist, H. (1988) Organotins in a municipal wastewater treatment plant. In: Løkke, H., Tyle, H., and Bro-Rasmussen, F. (Eds.) *Proceedings of the First European Conference on Ecotoxicology*, pp. 72–80. Copenhagen.

Ferraro, S.P., and Cole, F.A. (1992) Taxonomic level sufficient for assessing a moderate impact on macrobenthic communities in Puget Sound, Washington, USA. *Canadian J. Fisheries and Aquatic Sciences* **49**, 1184–1188.

Forbes, V.E., and Forbes, T.L. (1993) *Ecotoxicology in Theory and Practice: A Critique of Current Approaches.* Chapman and Hall, London.

Ford, J. (1989) The effects of chemical stress on aquatic species composition and community structure. In: Levin, S.A., Harwell, M.A., Kelly, J.R., and Kimball, K.D. (Eds.) *Ecotoxicology: Problems and Approaches*, pp. 99–144. Springer-Verlag, New York.

Fossi, M.C., and Leonzio, C. (Eds.) (1993). *Nondestructive Biomarkers in Higher Vertebrates.* Lewis, Boca Raton, Florida.

Futyuma, D.J. (1986) *Evolutionary Biology*, 2nd ed. Sinauer Associates, Sunderland, 600 pp.

GESAMP (1990) *The State of the Marine Environment.* Blackwell Scientific Publications, Oxford, pp. 146.

Gibbs, P.E., and Bryan, G.W. (1986) Reproductive failure in populations of the dogwhelk, *Nucella lapillus*, caused by imposex induced by tributyltin from antifouling paints. *J. Marine Biol. Assoc. UK* **66**, 767–777.

Gibbs, P.E., Pascoe, P.L., and Burt, G.R. (1988) Sex change in the female dog whelk, *Nucella lapillus*, induced by tributyltin from antifouling paints. *J. Marine Biol. Assoc. U.K.* **68**, 715–731.

Giddings, J.M. (1986) Protecting aquatic resources: an ecologist's perspective. In: Poston, T.M., and Purdy, R. (Eds.) *Aquatic Toxicol. Environ. Fate*, pp. 97–106. ASTM STP 921. American Society for Testing and Materials, Philadelphia, Pa.

Goksyor, A., Husoy, A.M., Larsen, H.E., Klungsoyr, J., Wilhelmsen, S., Brevik, E.M., Andersson, T., Celander, M., Pesonen, M., and Forlin, L. (1989). Evaluation of biochemical responses to environmental contaminants in flatfish from Hvaler Archipelago in Norway. *Marine Environ. Res.* **28**, 51–55.

Goksyor, A., Larsen, H.E., and Husoy, A. M. (1991) Application of cytochrome P-450 IA1-ELISA in environmental monitoring and toxicological testing of fish. *Compar. Biochem. Physiol.* **100C**, 157–160.

Grant, A., Hateley, J.G., and Jones, N.V. (1989) Mapping the ecological impact of heavy metals on the estuarine polychaete *Neries diversicolor* using inherited metal tolerance. *Marine Pollut. Bull.* **20**, 235–238.

Gray, J.S. (1974) Synergistic effects of three heavy metals on growth rates of a marine ciliate protozoan. In: Vernberg, F.J., and Vernberg, W.B. (Eds.) *Pollution and Physiology of Marine Organisms*, pp. 465–486. Academic Press, New York.

Gray, J.S. (1979) Pollution-induced changes in populations. *Philosoph. Transact. Roy. Soc.*, Series B, **286**, 545–561.

Gray, J.S. (1980) The measurement of effects of pollutants on benthic communities. *Rapp. P.-v Reun. Cons. Int. Explor. Mer.* **179**, 188–193.

Gray, J.S. (1981) *The Ecology of Marine Sediments.* Cambridge University Press, Cambridge, 185 pp.

Gray, J.S., Aschan, M., Carr, M.R., Clarke, K.R., Green, R.H., Pearson, T.H., Rosenberg, R., and Warwick, R.M. (1988) Analysis of community attributes of the benthic marcrofauna of Frierfjord/Langsundfjord and in a mesocosm experiment. *Marine Ecol. Progr. Ser.* **46**, 151–165.

Hinton, D.E., and Lauren, D.J. (1990) Liver structural alterations accompanying chronic toxicity in fishes: potential biomarkers of exposure. In: McCarthy, J.F., and Shugart, L.R. (Ed.) *Biomarkers of Environmental Contamination*, pp. 17–58. Lewis Publishers, Boca Raton, Florida.

Hinton, D.E., Baumann, P.C., Gardner, G.R., Hawkins, W.E., Hendricks, J.D., Murchelano, R.A., and Okihiro, M.S. (1992) Histopathological biomarkers. In: Huggert, R.J., Kimerle, R.A., Mehrle, P.M., and Bergman, H.L. (Ed.) *Biomarkers; Biochemical, Physiological and Histopathological Markers of Anthropogenic Stress*, pp. 155–210. Lewis Publishers, Boca Raton, Florida.

Holwerda, D.A., and Opperhuizen, A. (Eds.) (1991) Physiological and biochemical approaches to the toxicological assessment of environmental pollution. *Compar. Biochem. Physiol.* **100C**, 310.

Hopkin, S.P. (1993) *In situ* biological monitoring of pollution in terrestrial and aquatic ecosystems. In: Calow, P. (Ed.) *Handbook of Ecotoxicology*, Vol. 1. Blackwell, Oxford.

Howell, R. (1984) Acute toxicity of heavy metals to two species of marine nematodes. *Marine Environ. Res.* **11**, 153–161.

Huggert, R.J., Kimerle, R.A., Mehrle, P.M., and Bergman, H.L. (Eds.) (1992) *Biomarkers; Biochemical, Physiological and Histological Markers of Anthropogenic Stress*. Lewis Publishers, Boca Raton, Florida, 347 pp.

Ivanovici, A.M., and Weibe, W.J. (1981) Towards a working "definition" of "stress:" A review and critique. In: Barret, G.W., and Rosenberg, R. (Ed.) *Stress Effects in Natural Ecosystems*, pp. 13–28. John Wiley, Chichester.

Kelly, J.R., and Harwell, M.A. (1989) Indicators of ecosystem response and recovery. In: Levin, S.A., Harwell, M.A., Kelly, J.R., and Kimball, K.D. (Eds.) *Ecotoxicology: Problems and Approaches*, pp. 9–40. Springer Verlag, New York.

Kennelly, S.J. (1987) Physical disturbances in an Australian kelp community. 1. Temporal effects. *Marine Ecology Progress Series* **40**, 145–153.

Kuiper, J., and Gamble, J.C. (1988) Between test tubes and the North Sea: Mesocosms. In: Salomons, W., Bayne, B.L., Duursma, E.K., and Forstner, U. (Eds.) *Pollution of the North Sea—An Assessment*, pp. 638–654. Springer-Verlag, Berlin, Heidelberg.

Langston, W.J., Bryan, G.W., Burt, G.R., and Gibbs, P.E. (1990) Assessing the impact of tin and TBT in estuaries and coastal regions. *Functional Ecology* **4**, 433–443.

Lauenstein, G.G., Robertson, A., and O'Connor, T.P. (1990) Comparison of trace metal data in mussels and oysters from a mussel watch programme of the 1970s with those from a 1980s programme. *Marine Pollution Bulletin* **21**, 440–447.

Levine, S. (1989) Theoretical and methodological reasons for variability in the response of aquatic ecosystem processes to chemical stress. In: Levin, S.A., Harwell, M.A., Kelly, J.R., and Kimball, K.D. (Eds.) *Ecotoxicology: Problems and Approaches*, pp. 145–180. Springer-Verlag, New York.

Livingstone, D.R., Moore, M.N., and Widdows, J. (1988) Ecotoxicology: Biological effects measurements on molluscs and their use in impact assessment. In: Salomons, W., Bayne, B.L., Duursma, E.K., and Forstner, U. (Eds.) *Pollution of the North Sea—An Assessment*, pp. 624–637. Springer Verlag, London.

Lobel, P.B., Mogie, P., Wright, D.A., and Wu, B.L. (1982) Metal accumulation in four molluscs. *Marine Pollut. Bull.* **13**, 170–174.

Lobel, P.B., Belkhode, S.P., Jackson, S.E., and Longerich, H.P. (1989) A universal method for quantifying the residual variability of element concentrations in biological tissues using 25 elements in the mussel *Mytilus edulis* as a model. *Marine Biol.* **102**, 513–518.

Lobel, P.B., Belkhode, S.P., Jackson, S.E., and Longerich, H.P. (1990) Recent taxonomic discoveries concerning the mussel *Mytilus*: implications for biomonitoring. *Arch. Environ. Contam. Toxicol.* **19**, 508–512.

Luoma, S.N. (1977) Detection of trace contaminant effects inaquatic ecosystems. *J. Fish. Res. Board Can.* **34**, 436–439.

McCarthy, J.F., and Shugart, L.R. (Eds.) (1990) *Biomarkers of Environmental Contamination*. Lewis, Boca Raton, Florida, 457 pp.

McClusky, D.S., Bryant, V., and Campbell, R. (1986) The effects of temperature and salinity on the toxicity of heavy metals to marine and estuarine invertebrates. *Oceanogr. Marine Biol. Ann. Rev.* **24**, 481–520.

McMahon, G., Huber, L.J., Moore, M.N., Stegeman, J.J., and Wogan, G.N. (1990) c-K-*Ras* Oncogenes: prevalence in livers of winter flounder from Boston Harbour. In: McCarthy, J.F., and Shugart, L.R. (Eds.) *Biomarkers of Environmental Contamination*, pp. 229–238. Lewis, Boca Raton, Florida.

Mantoura, R.F.C., Dickson, A., and Riley, J.P. (1978) The complexation of metals with humic materials in natural waters. *Estuar. Coastal Marine Sci.* **6**, 387.

May, R.M. (1986) The search for patterns in the balance of nature: advances and retreats. *Ecology* **67**, 1115–1126.

Mayer, F.L., Versteeg, D.J., McKee, M.J., Folmar, L.C., Graney, R.L., McCume, D.C., and Rattner, B.A. (1992) Physiological and nonspecific biomarkers. In: Huggert, R.J., Kimerle, R.A., Mehrle, P.M., and Bergman, H.L. (Eds.) *Biomarkers; Biochemical, Physiological and Histological Markers of Anthropogenic Stress*, pp. 5–85. Lewis, Boca Raton, Florida.

Moore, P.G., Rainbow, P.S., and Hayes, E. (1991). The beach hopper *Orchestia gammarellus* (Crustacea, Amphipoda) as a biomonitor for copper and zinc: North Sea trials. *Sci. Total Environ.* **106**, 221–238.

Moran, P.J., and Grant, T.R. (1991) Transference of marine fouling communites between polluted and unpolluted sites: impact on structure. *Environ. Pollut.* **72**, 89–102.

Moriarty, F. (1983) *Ecotoxicology: The Study of Pollutants in Ecosystems*. Academic Press, London, 233 pp.

Morris, R.F. (1959) Single factor analysis in population dynamics. *Ecology* **40**, 580–588.

Nacci, D., and Jackim, E. (1989) Using the DNA alkaline unwinding assay to detect DNA damage in laboratory and environmentally exposed cells and tissues. *Marine Environ.* **28**, 333–337.

Newman, M.C., and McIntosh, A.W. (1989) Appropriateness of *Aufwuchs* as a monitor of bioaccumulation. *Environ. Pollut.* **60**, 83–100.

Pearson, T.H. (1970) The benthic ecology of Loch Linnhe and Loch Eil, a sea-loch system on the west coast of Scotland. I. Physical environment and the distribution of the macrobenthic fauna. *J. Exp. Marine Biol. Ecol.* **5**, 1–34.

Pearson, T.H. (1971) The benthic ecology of Loch Linnle and Loch Fil, a sea-lock system on the west coast of Scotland. III. The effect on the benthic fauna of the introduction of pulp mill effluent. *J. Exp. Marine Biol. Ecol.* **6**, 211–233.

Pearson, T.H. (1975) The benthic ecology of Loch Linnhe and Loch Eil, a sea-loch system on the west coast of Scotland. IV. Changes in the benthic fauna attributable to organic enrichment. *J. Exp. Marine Biol. Ecol.* **20**, 1–41.

Pearson, T.H., and Rosenberg, R. (1978) Macrobenthic succession in relation to organic enrichment and pollution of the marine environment. *Oceanogr. Marine Biol. Ann. Rev.* **16**, 229–311.

Phillips, D.J.H. (1980) *Quantitative Aquatic Biological Indicators.* Applied Science Publishers, London, 488 pp.

Pohl, C. (1990) Skeletal deformities and trace metal contents of European smelt, *Osmerus eperlanus*, in the Elbe estuary. *Meeresforschung* **33**, 76–89.

Powell, M.I., and White, K.N. (1990). Heavy metal accumulation by barnacles and its implications for their use as biological monitors. *Marine Environ. Res.* **30**, 91–118.

Rainbow, P.S., Moore, P.G., and Watson, D. (1989) Talitrid amphipods (Crustacea) as biomonitors for copper and zinc. *Estuarine Coast. Shelf Sci.* **28**, 567–582.

Rapport, D.J. (1992) Evaluating ecosystem health. *J. Aquat. Ecosys. Health* **1**, 15–24.

Remane, A., and Schlieper, C. (1971) *Biology of Brackish Water.* Wiley-Interscience, New York.

Rosenberg, R. (1972) Benthic faunal recovery in a Swedish fjord following the closure of a sulphite pulp mill. *Oikos* **23**, 92–108.

Rosenberg, R. (1973) Succession in the macrofauna in a Swedish fjord subsequent to the closure of a sulphite pulp mill. *Oikos* **24**, 1–16.

Sanders, B.M., Martin, L.S., Nelson, W.G., Phelps, D.K., and Welch, W. (1991) Relationships between accumulation of a 60 kDa stress protein and scope-for-growth in *Mytilus edulis* exposed to a range of copper concentrations. *Marine Environ. Res.* **31**, 81–97.

Scammell, M.S., Batley, G.E., and Brockbank, C.I. (1991) A field study of the impact on oysters of tributyltin introduction and removal in a pristine lake. *Arch. Environ. Contam. Toxicol.* **20**, 276–281.

Schimmel, S.G., Garnas, R.L., Patrick, J.M., and Moore, J.C. (1983). Acute toxicity, bioconcentration and persistence of AC222,705, benthiocarb, chlorpyrifos, fenvalerate, methyl parathion and permethrin in the estuarine environment. *J. Agric. Food Chem.* **31**, 104–113.

Schluter, D. (1988) Estimating the form of natural selection on a quantitative trait. *Evolution* **42**, 849–861.

Schulz-Baldes, M. (1989) The sea skater *Halobates micans*: an open ocean bioindicator for cadmium distribution in Atlantic surface waters. *Marine Biol.* **102**, 211–215.

Sericano, J.L., Atlas, E.L., Wade, T.L., and Brooks, J.M. (1990). NOAA's status and trends mussel watch program: chlorinated pesticides and PCB's in oysters (*Crassotrea virginica*) and sediments from the Gulf of Mexico, 1986–1987. *Marine Environ. Res.* **29**, 161–203.

Shugart, L.R., Bickham, J., Jackim, J., McMahon, G., Ridley, W., Stein, J., and Steinert, S. (1992) DNA alterations. In: Huggert, R.J., Kimerle, R.A., Mehrle, P.M., and Bergman, H.L. (Eds.) *Biomarkers; Biochemical, Physiological and Histopathological Markers of Anthropogenic Stress*, pp. 125–154. Lewis, Boca Raton, Florida.

Skwarzec, B., and Bojanowski, R. (1988) ^{210}Po content in seawater and its accumulation in southern Baltic plankton. *Marine Biol.* **97**, 301–307.

Smith, B.S. (1980). The estuarine mud snail, *Nassarius obseletus*: abnormalities in the reproductive system. *J. Molluscan Studies* **46**, 247–256.

Spence, S.K., Bryan, G.W., Gibbs, P.E., Masters, D., Morris, L., and Hawkins, S.J. (1990) Effects of TBT contamination on *Nucella* populations. *Functional Ecol.* **4**, 425–432.

Stegeman, J.J., Renton, K.W., Woodin, B.R., Zhang, Y., and Addison, R.F. (1990) Experimental and environmental induction of cytochrome P450E in fish from Bermuda waters. *J. Exp. Marine Biol. Ecol.* **138**, 49–67.

Stegeman, J.J., Brouwer, M., DiGuilio, R.T., Forlin, L., Fowler, B.A., Sanders, B.M., and Van Veld, P. (1992) Enzyme and protein synthesis as indicators of contaminant exposure and effects. In: Huggert, R.J., Kimerle, R.A., Mehrle, P.M., and Bergman, H.L. (Eds.) *Biomarkers; Biochemical, Physiological and Histopathological Markers of Anthropogenic Stress*, pp. 235–336. Lewis, Boca Raton, Florida.

Stickle, W.B., Sharp-Dahl, J.L., Rice, S.D., and Short, J.W. (1990) Imposex induction in *Nucella lima* (Gmelin) via mode of exposure to tributyltin. *J. Exp. Marine Biol. Ecol.* **143**, 165–180.

Tanabe, S., Tatsukawa, R., and Phillips, D.J.H. (1987) Mussels as bioindicators of PCB pollution: a case study on uptake and release of PCB isomers and congeners in green lipped mussels (*Perna viridis*) in Hong Kong waters. *Environ. Pollut.* **47**, 41–62.

Tedengren, M., Andre, C., Johannesson, K., and Kautsky, N. (1990) Genotypic and phenotypic differences between Baltic and North Sea populations of *Mytilus edulis* evaluated through reciprocal transplantations. III. Physiology. *Marine Ecol. Progr. Ser.* **59**, 221–227.

Thomas, P.G., and Ritz, D.A. (1986) Growth of zinc granules in the barnacle *Elminius modestus*. *Marine Biol.* **90**, 255–260.

Trim, A.H., and Marcus, J.M. (1990) Integration of long-term fish kill data with ambient water quality monitoring data and application to water quality management. *Environ. Management* **14**, 389–396.

Underwood, A.J. (1989) The analysis of stress in natural populations. *Biol. J. Linnean Soc.* **37**, 51–78.

Van Straalen, N.M. (1993) Soil and sediment quality criteria derived from invertebrate toxicity data. In: Dallinger, R., and Rainbow, P.S. (Eds.) *Ecotoxicology of Metals in Invertebrates*. Lewis, Boca Raton, Florida.

Varanasi, U., Stein, J.E., Reichert, W.L., Tilbury, K.L., and Chan, S.L. (in press) Chlorinated and aromatic hydrocarbons in bottom sediments, fish and marine mammals in US coastal waters: laboratory and field studies of metabolism and accumulation. In: Walker, C.H., and Livingstone, D.R. (Eds.) *Persistent Pollutants in the Marine Ecosystem*. SETAC Special Publication. Pergamon, New York.

Varley, G.C., and Gradwell, G.R. (1960) Key factors in population studies. *J. Animal Ecol.* **29**, 399–401.

Varley, G.C., Gradwell, G.R., and Hassell, M.P. (1975) *Insect Population Ecology*. Blackwell Scientific, Oxford.

Vernberg, F.J., and Vernberg, W.B. (Eds.) (1974) *Pollution and Physiology of Marine Organisms*. Academic Press, London, 492 pp.

Vernberg, F.J., Calabrese, A., Thurberg, F.P., and Vernberg, W.B. (Eds.) (1977) *Physiological Responses of Marine Biota to Pollutants*. Academic Press, London, 462 pp.

Vernberg, W.B., Calabrese, A., Thurberg, F.P., and Vernberg, F.J. (Eds.) (1982) *Physiological Mechanisms of Marine Pollution Toxicity*. Academic Press, London, 564 pp.

Via, S., and Lande, R. (1985) Genotype–environment interaction and the evolution of phenotypic plasticity. *Evolution* **39**, 505–522.

Waldock, M.J., Thain, J.E., and Miller, D. (1983) *The Accumulation and Depuration of bis(Tributyltin) Oxide in Oyster: A Comparison between the Pacific Oyster (Crassostrea gigas) and the European Flat Oyster (Ostrea edulis)*. Cooperative Research Report No. CM:1983/E:52. International Council for the Exploration of the Sea, Copenhagen.

Walker, C.H., and Johnston, G.O. (1989) Interactive effects of pollutants at the toxicokinetic level—implications for the marine environment. *Marine Environ. Res.* **28**, 521–525.

Warwick, R.M., Platt, H.M., Clarke, K.R., Agard, J., and Gobin, J. (1990). Analysis of macrobenthic and meiobenthic community structure in relation to pollution disturbance in Hamilton Harbour, Bermuda. *J. Exp. Marine Biol. Ecol.* **138**, 119–142.

Weeks, B.A., Anderson, D.P., DuFour, A.P., Fairbrother, A., Goven, A.J., Lahvis, G.P., and Peter, G. (1992) Immunological biomarkers to assess environmental stress. In: Huggert, R.J., Kimerle, R.A., Mehrle, P.M., and Bergman, H.L. (Eds.) *Biomarkers; Biochemical, Physiological and Pathological Markers of Anthropogenic Stress*, pp. 211–234. Lewis, Boca Raton, Florida.

Weeks, J.M., and Depledge, M.H. (1992) Inter-population variability in acute mercury toxicity in the amphipod, *Platorchestia platensis* and its significance for toxicity testing. In: *Collected Abstracts—International Symposium on Bio-Remediation, Toxicology, Environmental Fate and Ecology*, pp. D24-12. Joint Meeting of SETAC-Europe and the Aquatic Ecosystems and Health Management Society, 21–24 June, Potsdam, Germany.

Widdows, J., and Donkin, P. (1991) Role of physiological energetics in ecotoxicology. *Compar. Biochem. Physiol.* **100C**, 69–75.

Widdows, J., Burns, K.A., Menon, N.R., Page, D.S., and Soria, S. (1990). Measurement of physiological energetics (scope for growth) and chemical contaminants in mussels (*Arca zebra*) transplanted along a contaminant gradiet in Bermuda. *J. Exp. Marine Biol. Ecol.* **138**, 99–117.

Williamson, M.H. (1987) Are communities ever stable? In: Gray, A.J., Crawley, M.J., and Edwards, P.J. (Eds.) *Colonisation, Succession and Stability*, pp. 353–371. Blackwell Scientific, Oxford.

Wood, R.J., and Bishop, J.A. (1981) Insecticide resistance: populations and evolution. In: Bishop, J.A., and Cook, L.M. (Eds.) *Genetic Consequences of Man-Made Change*, pp. 97–128. Academic Press, London.

Woodwell, G.M. (1962) Effects of ionising radiation on terrestrial ecosystems. *Science* **138**, 572–577.

Yamashita, M., Kinae, N., Kimura, I., Ishida, H., Kumai, H., and Nakamura, G. (1990) The Croaker (*Nibea mitsukurii*) and the Sea Catfish (*Plotosus anguillaris*): useful biomarkers of coastal pollution. In: McCarthy, J.F., and Shugart, L.R. (Eds.) *Biomarkers of Environmental Contamination*, pp. 73–86. Lewis, Boca Raton, Florida.

8 Assessment of Effects of Chemicals on Wetlands

William H. Queen and Donald W. Stanley
East Carolina University, USA

8.1 INTRODUCTION

From the late 1950s, scientific studies have conclusively shown that wetlands contribute substantially to human health and welfare. Specifically, wetlands have been shown to improve water quality, provide food for commercially and recreationally important fish and shellfish living in nearby water bodies, serve as critical habitat for numerous species of wildlife, reduce shoreline erosion, and act as flood control devices (Hemond and Benoit, 1988). As important functions of wetlands were being identified, data were also being assembled that suggested that wetlands were being lost or degraded at alarming rates by either development or chemical pollution (Mitsch and Gosselink, 1986). These data raised questions in both the scientific and environmental management communities about which wetlands are most valuable, current rates of wetland loss or degradation, and the effects of chemical pollutants on wetland characteristics and functions. These and similar questions have been difficult to answer, because wetlands differ enormously in terms of their basic characteristics (i.e., vegetation type, soil type, areal extent, spatial configuration, hydrology), their geographical distribution, and their abundance in different state or provincial or regional watersheds.

Unlike changes resulting from development projects that are easily detected, adverse chemical impacts on wetlands are especially difficult to detect and evaluate. First, wetlands can be damaged by a variety of substances (e.g., high nutrient concentrations, heavy metals, and a vast array of organic compounds). Second, these chemicals may reach wetlands as a result of either point or non-point source discharge, and they may be transported by either water or air from nearby or distant discharge points. Third, negative impacts of particular chemical pollutants may be limited to the wetland vegetation or to specific faunal species. Fourth, chemical pollutants in wetlands often cause chronic rather than acute problems for the affected organisms, making detection far more difficult. Other factors may further complicate the chemical impact picture. For example, chemical pollutants are often released at very irregular intervals, and, therefore, may move to a wetland site, adversely affect wetland organisms, and not be detected even with a sensitive

Methods to Assess the Effects of Chemicals on Ecosystems
Edited by R. A. Linthurst, P. Bourdeau, and R. G. Tardiff

monitoring program in place.

For these and other reasons, considerable attention over the past few decades has been given to the development and improvement of methodologies to assess the effects of chemicals on both plants and animals and on entire wetland systems. As a result, an array of methodologies is now available. These methodologies are summarized in Table 8.1.

Table 8.1. Methods to assess the effects of chemicals on wetlands

Method	Application
Plant and animal methods	
Bioaccumulation methods	Assess toxicity of specific chemicals to individual species
Fertilization and sewage addition	Assess capacity of wetlands to remove (field experiments) nutrients from sewage, and impact of fertilization on wetland system
Microcosm/mesocosm	Provide controlled experiment (laboratory experiments) conditions for testing nutrient and toxics additions
System methods	
Inventories	Identify threatened wetlands and the threatening pollutants
Chemical loadings and mass balance determinations	Quantify degree of threat, assess relative importance of sources, and compare different wetlands
Modelling	
energy/nutrient models	Track energy flow through or nutrient cycling within wetland ecosystem
Hydrodynamic models	Simulate pollutant transport through wetlands
spatial ecosystem models	Combination of ecosystem and hydrodynamic transport models
Process models	Analyse individual wetlands functions
Monitoring	
Abiotic monitoring	Track changes in chemical concentrations and provide data for interpreting biomonitoring results
Biomonitoring	Detect impacts from many sources and integrate intermittent stressor effects

The discussion of these methodologies begins with a review of those that are routinely employed in bioaccumulation and experimental studies that are undertaken to evaluate the effects of chemicals on wetland plants and animals. Then attention is devoted to the chemical loading and modelling methodologies being developed to determine the effects of chemicals on wetland systems. Finally, wetland monitoring is discussed.

8.2 METHODS FOR STUDYING PLANTS AND ANIMALS

For the past three decades, water quality functions of wetlands have been a major concern of wetland scientists. Among these functions are their capacity to remove or transform excess nutrients, organic compounds, trace metals, sediment, and refractory chemicals from water as it moves downstream (Hemond and Benoit, 1988). In addressing questions related to these water quality functions, wetland scientists have: (1) developed sophisticated protocols for collecting samples, (2) refined methodologies for both field and laboratory experiments as well as methodologies to analyse biological materials, and (3) developed mathematical models for water quality studies. Those specifically applicable to studying wetland plants and animals are discussed in terms of either bioaccumulation determinations or experimental studies.

8.2.1 BIOACCUMULATION DETERMINATIONS

Measurement of chemical accumulation in tissues of organisms (bioaccumulation) is a methodology widely used to assess contamination in aquatic ecosystems. Bioaccumulation is assessed in the laboratory by first exposing organisms to a chemical, mixed in water or sediments, and then measuring the concentration of the chemical in the tissues (body burden) as a function of the duration of the exposure. The results may be expressed as a bioaccumulation rate—the rate at which the body burden increases—or a bioaccumulation factor—the unitless quotient of the equilibrium body burden divided by the average exposure concentration. A fairly large number of bioaccumulation studies have been reported in the literature, including works by Anthony and Kozlowski (1982), Aulio (1980), Behan *et al.* (1979), Larsen and Schierup (1981), McIntosh *et al.* (1978), Mouvet (1985), Niethamner *et al.* (1985), Schierup and Larsen (1981), and Taylor and Crowder (1983).

Laboratory measurements of bioaccumulation permit chemicals to be tested one at a time. Another attractive attribute of this methodology is that it is not as affected by short-term fluctuations in contaminant levels as are measurements of sediment, and particularly water-column concentrations of the chemicals. Longer-lived organisms integrate exposure levels over longer periods of time, so selection of a species, or age class, can influence results. Aquatic macrophytes, macroinvertebrates, amphibians, bird eggs, and small mammals may be preferable for detecting short-term exposures, whereas woody plants, fish, turtles, large adult birds, and fur-bearing mammals may be better for monitoring long-term exposure (Leibowitz *et al.*, 1991).

The two major drawbacks to bioaccumulation studies are cost and the difficulty of quantitative interpretation of the results. Interpretation is difficult because variability in tissue residues can be high, even within a species, because they are affected by many factors, including age, tissue type, diet, season, and the length of exposure. Therefore, a sufficient number of individuals from each sampling location has to be collected to overcome the limitation imposed by this variability. This requirement is regrettable, because tissue contaminant levels are costly to measure, especially if analyses are made for more than a few of the wide array of contaminants present in some environments (Leibowitz *et al.*, 1991). Even more serious, however, is the difficulty of evaluating the consequence of a measured body burden to the individual organism, the population of which it is a part, or the higher trophic level consumers that feed on it. The problem is that few studies have attempted to correlate body burden with adverse effects on the organism, or to determine the elimination rates of accumulated chemicals from tissues upon cessation of exposure. Therefore, the longevity of the body burden and any associated harmful effects are unknown.

8.2.2 EXPERIMENTAL STUDIES

Experimental studies of the impacts of chemicals on wetlands have been conducted in the field and in laboratories, greenhouses, microcosms, and mesocosms. Most of these studies have emphasized nutrients, but some have been directed at heavy metals. Salt marsh plants, especially *Spartina* and *Juncus*, have been used frequently in experimental research projects.

Numerous field experiments to identify the effects of nitrogen and phosphorus fertilization on North American salt marshes (Chalmers, 1982; Morris, 1991) have been conducted. Typically, plots ranging in size from <1.0 m^2 to 100 m^2 were established in *Spartina* or *Juncus* dominated communities. A few plots were used as controls, while the others were enriched with various levels of inorganic fertilizer materials in a single application at the beginning of the growing season, or at weekly or monthly intervals. Response to the fertilization was often measured by calculating net primary productivity from changes in standing biomass. Examples of studies of this type include those of Tyler (1967), Valiela and Teal (1974), Mendelssohn (1979), de la Cruz *et al.* (1981), and Morris (1988). This procedure was modified by Patrick and DeLaune (1976) and Buresh *et al.* (1980) with the use of ^{15}N-depleted nitrogen to permit differentiation between soil-derived and fertilizer-derived nitrogen in plant shoots. Van Raalte *et al.* (1976) and Sullivan (1981) investigated the effects of nitrogen enrichment on salt marsh diatom communities. Wetland nutrient addition experiments have also been made in North American arctic tundra (Shaver and Chapin, 1980), in European fens and wet grasslands (Vermer, 1986), and in wet heathlands in the Netherlands (Aerts and Berendse,

1988). Several laboratory studies have been made to test the influence of nitrogen levels on the growth response of *Spartina alterniflora* (Haines and Dunn, 1976; Linthurst and Seneca, 1981) and on species of *Jungus, Littorella*, and *Sphagnum* (Roelofs *et al.*, 1984). Response was measured by following changes: in plant height, total dry weight, rhizome length and weight, and root weight.

Research on the capacity of wetlands to remove nutrients and other chemicals from wastewater has used nutrient enrichment experiments carried out in the field (Godfrey *et al.*, 1985; Nichols, 1983). The most complete study of the effects of sewage on a salt marsh is the multiyear study by Valiela and Teal in Massachusetts, US, in which they made biweekly additions of a commercial fertilizer made from sewage sludge (Valiela and Teal, 1974; Valiela *et al.*, 1975, 1976, 1985). Similar studies were conducted by Chalmers (1979) and Haines (1979). Sewage treatment capacity has also been studied in several freshwater wetland types (Sloey *et al.*, 1978; Nichols, 1983; Fetter *et al.*. 1978; and Bayley *et al.*, 1985). The retention of nutrients and heavy metals by a tidal freshwater wetland receiving urban stormwater runoff was estimated by comparing inputs with outputs (Simpson *et al.*, 1983). Lan *et al.* (1992) described the use of a pond with cattails (*Typha latifolia*) for treating wastewater from a lead and zinc mine in China.

The uses of laboratory microcosms and mesocosms for marine and freshwater ecosystem research have been described in volumes edited by Giesy (1980) and Grice and Reeve (1982), and a few wetland systems studies have been reported that made use of these enclosures (Kitchens, 1979; Portier and Meyers, 1982). The flume approach was used by Windom (1977) in Georgia to measure the uptake by the marsh of nutrients and heavy metals contained in effluent from dredged material disposal. For the most part, however, wetlands work has emphasized greenhouse-type studies in which various kinds of plants have been grown in different soils under varying conditions. For example, in a greenhouse experiment designed to differentiate between acid and nitrogen effects, mixtures of different wetland plant species were exposed to simulated rain containing various combinations of inorganic sulphur and nitrogen (Schuurkes *et al.*, 1986). Manipulations and additions involving complete portions of wetlands with sediments, many plants and animal species, and water flows have been rare (Nixon and Lee, 1986).

8.3 METHODS TO STUDY THE EFFECTS OF CHEMICALS ON THE SYSTEM ITSELF

8.3.1 INVENTORIES

Development of methodologies to address the broader issue of the toxicity of chemicals on wetland systems should begin with the realization that all wetland - organisms, processes, and functions cannot be studied, or even monitored, on a regular basis for even a single chemical, much less for the vast array of chemical pollutants found in wetlands. Hence, a major problem for both the wetland scientific and management communities is that of selecting those wetland sites that should be studied (i.e., monitored), and the specific chemical pollutants that should receive special attention in regulatory, monitoring, and research programmes. The resolution of this important problem requires comprehensive inventories of both - wetlands and the sources of potential chemical pollutants. Moreover, these inventories need to be prepared on both national and local scales (i.e., city or county jurisdiction, watershed, state or province).

The national perspective is essential for several reasons. First, wetlands are an important natural resource for many nations, and issues concerning them are issues of national interest. Second, funding for either wetlands protection or control of chemical discharges that adversely impact wetlands is generally provided by national governments. Hence, these funds should be used to: (1) protect a nation's most valuable wetlands and (2) control those chemical discharges that are most threatening to their health. The local inventories are essential, because most public decisions involving wetlands focus on specific wetland sites and on particular developmental activities or chemical discharges that threaten the specific sites. The following summarizes the essential features and considerations concerning preparation of the four inventories:

1. National wetlands inventories. A national inventory should show the distribution of all wetlands by location, type, and areal extent; and data on wetland loss over time should be included if available (Office of Technology Assessment; OTA, 1984; Turner *et al.*, 1981; Gosselink and Baumann, 1980; Tiner 1984). Assembly of the national wetland inventories should be based on the local wetland inventories described below.
2. Local wetlands inventories. Local wetland inventories should be developed in conformance with national guideline, and should be used to assemble the national inventory of wetlands. Local inventories should contain information in much greater detail than the national inventories, because the local inventories can be used for site-specific decisions on wetland development projects.
3. National inventories of sources of potential chemical pollutants. This inventory should identify both point and non-point sources of chemical pollution (Pait *et al.*, 1992; Quinn *et al.*, 1989). Data on nutrients, heavy metals, and organic compounds should be included. The inventory should be assembled from data contained in the local inventories.
4. Local inventories of sources of potential chemical pollutants. The local inventories should be developed according to national guidelines, and the information collected used to assemble the national inventory on sources of chemical pollution. Like the national inventory, the local inventories should also contain data on nutrients, heavy metals, and organic compounds.

These inventories by themselves will not establish the adverse impacts of chemical pollutants on wetland systems. They can, however, be used to: (1) identify the wetlands within a system that may be threatened by chemical pollutants, (2) identify the threatening pollutant(s), (3) permit generalization to be made about the severity of the threat, and (4) suggest specific wetland sites and particular chemicals for subsequent studies designed to evaluate specific threats.

8.3.2 CHEMICAL LOADINGS AND MASS BALANCE DETERMINATIONS

Once specific wetland sites and chemical pollutants are selected for study, chemical loading estimates can be developed as indicators of the degree to which the wetland is threatened by toxicity or eutrophication. Loading data can also be used to assess the relative importance of various chemical sources (point and non-point) (Stanley, 1992) and to compare loadings among wetlands. By combining loading data with export values, a chemical budget or mass balance for the wetland can be generated. Although difficult and, therefore, costly to obtain in most situations, mass balances provide a measure of the capacity of wetlands to perform their normal water quality functions (e.g., immobilization of toxic pollutants, plant nutrient removal, removal of biological oxygen demand) (Hemond and Benoit, 1988). Anyone interested in the mass balance approach should consult the excellent review by Nixon and Lee (1986). Although this review is limited to the US literature, it provides a critical assessment of all research between 1970 and 1985 on the role of wetlands as sources, sinks, and transformers of nutrients and heavy metals. The authors cite hundreds of examples of loading and mass balance studies, and argue that developing the ability to predict responses to changes in inputs of these materials depends on a knowledge of the various cycles and flows within the system, and that a mass balance can provide the framework for developing this knowledge. Another useful review by Johnston (1991) references numerous studies of internal nutrient fluxes among wetland compartments.

While conceptually straightforward, a complete budget is difficult to develop because, as Brinson *et al.* (1980) noted, most wetlands are very open ecosystems by virtue of lateral inflows and outflows during flooding. Thus, obtaining accurate water flow data for the computation of fluxes and mass balances is difficult. Consequently, researchers have frequently resorted to estimating fluxes across wetland boundaries indirectly by measuring various internal storages and flows. The result, concluded Nixon and Lee (1986), is a considerable amount of confusion about whether specific wetland systems act as sources, sinks, or transformers of nutrients, metals, etc., at different times of the year.

8.3.3 MODELLING

Wetland modelling, although relatively new compared to modelling of other types of ecosystems, can also be employed to provide insights into the effects of chemicals on the system itself. Wetland modellers have borrowed from the more developed lake and estuary modelling techniques (Henderson-Sellers, 1984;

Straskraba and Gnauck, 1985; Jorgensen, 1988) and from terrestrial models. Reviews of models of freshwater wetlands are summarized by Mitsch *et al.* (1982, 1988), Mitsch (1983), and Costanza and Sklar (1985).

Several types of models have been developed that incorporate the effects of chemicals on wetland biological structure and processes. Mitsch *et al.* (1983) summarized this modelling in the following way:

1. Energy/nutrient ecosystem models. Energy, nutrients, or other materials flow through or cycle among biotic and abiotic components and exchange with the surroundings. These models are often an outgrowth of chemical loadings and mass balance studies, or studies of the ability of wetlands to process wastewaters (Jorgensen, 1988). Examples include a model of energy flow and nutrients in an Everglades marsh system in Florida (US) (Bayley and Odum, 1976), a salt marsh carbon flow model (Wiegert *et al.*, 1981), an energy-nutrient model of water hyacinth (Mitsch, 1976), models for cypress swamps in Florida (Ewel, 1976), a model of hydrology and nutrient cycling in forested wetlands (Mitsch, 1988a, 1988b), models of nutrient retention in a freshwater wetland (Mitsch and Reeder, 1992; Kadlec and Hammer, 1988), and a model dealing with acid mine drainage in a freshwater wetland (Mitsch *et al.*, 1983).
2. Hydrodynamic transport models. These models, frequently used by engineers for stream flow and runoff calculations, have been applied to simulate wetland hydrology and pollutant transport. Hopkinson and Day (1980) described the application of one of these models (the Storm Water Management Model, SWMM, developed by the USEPA) to predict nutrient dynamics and eutrophication effects in a Louisiana (US) swamp forest resulting from anticipated increases in nutrient loading over a twenty-year period.
3. Spatial ecosystem models. These models combine features of ecosystem models with hydrodynamic transport models. The entire ecosystem is separated into discrete blocks or nodes, with each node having the characteristics of an ecosystem model. The nodes are connected by hydrodynamic processes. A suite of spatial models has been used to investigate various consequences associated with using freshwater wetlands for disposal of treated sewage (Kadlec and Tilton, 1979).
4. Process models. This type of model emphasizes individual processes, such as photosynthesis and nutrient cycling processes, rather than emphasizing interactions among compartments as in ecosystem models. Among the relatively few wetland examples of this model type are process models of primary production and vascular plant growth for wet arctic tundra (Miller *et al.*, 1976, 1978a; 1978b; Stanley 1976).

8.4 MONITORING

8.4.1 ABIOTIC MONITORING

Programs designed to monitor chemicals in aquatic ecosystems have as their primary goal the identification of long-term trends in the quality of water and sediments of these systems. These programs also have secondary goals that include the provision of data for interpreting the results of other approaches, such as biomonitoring studies. In addition, monitoring is required to estimate chemical loadings and mass balances, and to validate wetland process models. However, field monitoring is inherently retrospective, allowing effects of an activity to be observed, but not predicted. Many works provide information on strategies and practices for environmental monitoring. One of the best is by the National Research Council (NRC, 1990). In addition, a USEPA Environmental Monitoring and Assessment Program report is entitled *Research Plan for Monitoring Wetland Ecosystems* (Leibowitz *et al.*, 1991).

Wetlands pose unusual challenges for monitoring programs. As transitional areas, they are located between uplands and deep water areas. Thus, they exhibit extreme spatial and temporal variability, which requires the collection of many samples for adequate characterization of the wetland. Only a minority of wetlands have permanent surface water, so access may be difficult, and sampling techniques developed for other surface waters are not always applicable. In some cases, downstream sampling may provide appropriate spatial averaging. Unless a relatively frequent sampling protocol can be followed, sampling for chemical contaminants in sediments should provide more consistent results than samples from the water column.

Several standard chemical measurement techniques developed for water and upland soil samples (American Public Health Association, 1989; USEPA, 1979; Parsons *et al.*, 1984; Black, 1965) have been adapted for use in wetlands. Wetland pore waters are often anoxic, so that substantial chemical changes may occur between sampling and analysis if water samples are not rigorously protected. For example, oxidation can cause iron to precipitate and adsorb transition metals, while sulphide may oxidize and release metals. Also, high concentrations of humic substances in wetland surface and pore waters interfere with the chemistry of many standard analyses; nitrate and sulphate are prime examples in this regard. Apparently, no compilation or critical evaluation of chemical analytical methods used in wetlands has been produced (Hemond and Benoit, 1988).

8.4.2 BIOMONITORING

A growing body of literature exists which suggests that monitoring of biological - community structures provides cost-effective diagnostic information about the health of ecosystems. Biological monitoring (or biomonitoring) (1) directly addresses the result of pollution, not its cause, (2) integrates intermittent stressor conditions, and (3) can detect impacts from many sources for which chemical criteria are poorly developed. If biomonitoring suggests that a chemical stress is occurring, then traditional methods (e.g., field monitoring of chemicals, toxicity testing) can be used to help determine the cause(s).

Adamas and Brandt (1990) have prepared a valuable synopsis of the uses of biomonitoring to measure responses to chemicals and other anthropogenic stresses in wetlands. This USEPA-sponsored report first describes general considerations in the design of wetlands biomonitoring studies. Then, for each major taxonomic group (e.g., algae, herbaceous vegetation, birds, fishes), the authors discuss the group's responses to various stressors, and the sampling protocols and equipment appropriate for monitoring those responses. The following general considerations are discussed in the report:

1. Monitoring of multiple indicators—having both short and long lifespans, and both localized and broad ranges—is necessary, because some chemical effects may be brief (e.g., biodegradable pesticides), while others occur over long time periods (e.g., bioaccumulation of metals).
2. The choice of species to be used as indicators must take into account both policy and scientific considerations. Many of the scientific criteria point to the use of so-called keystone species, which physically alter the landscape so profoundly that they create or destroy habitat for a much larger group of species over a wide area. Examples of wetland keystone animals include beavers, muskrats, alligators, and some herbivorous birds. However, the choice of indicator species is an unsettled issue; some researchers point out that changes in community-level measures give a clearer indication of chemical stress than does the presence or absence of a single indicator species, regardless of its reputation as a keystone (Browder, 1988; Karr, 1987; Kelley and Harwell, 1989).
3. A wide range of qualitative and quantitative biological sampling methods is available. Qualitative methods include the use of chronological vegetation maps to identify changes in plant composition, species lists and estimates of the presence or absence of indicator species, and the Habitat Evaluation Procedures (HEP) and the Wetland Evaluation Technique (WET), that do not directly measure biological communities but assume biological community structure or wetland function using information on habitat structure (Schroeder, 1987). Quantitative methods are detailed in the Adamus and Brandt (1990) report and in Erwin (1989), Hellawell (1986), Welcomme (1979), and Woods (1975).
4. Data analysis and interpretation involve the selection of appropriate measures (e.g., abundance, biomass, species richness, and similarity indices). To optimize detection of impacts, several procedures should be used in combination (Schindler, 1987). The utility and sensitivity of some procedures may vary by wetland type. For example, those that depend on species-level data (richness, ordination, and similarity indices) may be ineffective at describing the condition of wetlands that characteristically have low species richness (e.g., breeding bird richness in salt marshes and fish richness in montane wetlands).

8.5 CONCLUSIONS AND RECOMMENDATIONS

For the past three decades, wetland scientists have developed a wide array of methodologies to study the effects of chemicals on wetland organisms, processes, and functions. These methodologies include sampling protocols, field and laboratory experimental methods, and mathematical models. Use of these methodologies has greatly enhanced understanding of impacts of particular chemicals on wetland plants and animals. These methodologies to study plants and animals are now being coupled with information collected in inventories of wetlands and of chemical pollutant sources to characterize the effects of chemicals on wetland systems. Inventories are generally of national and local wetlands, and of national and local sources of chemical pollutants. They are undertaken to identify critical wetlands being threatened by either development projects or chemical pollutants, and to pinpoint pollution sources. The national inventories can also be used to either develop or justify: (1) regulatory restrictions on chemical pollutant discharges that may adversely affect wetlands, (2) the structure and funding of monitoring programs, (3) the allocation of funds for mitigation or restoration activities, and (4) the establishment of chemical pollutant and wetland ecology research priorities. Because most public decisions involving wetlands focus on specific wetland sites, the local inventories can also be used in public policy decisions concerning development projects or chemical discharges that threaten specific wetland sites.

8.6 REFERENCES

Adamas, P.R., and Brandt, K. (1990) *Impacts on Quality of Inland Wetlands of the United States: A Survey of Indicators, Techniques, and Application of Community-Level Biomonitoring Data.* Report No. EPA/600/3-90/093. US Environmental Protection Agency, Environmental Research Laboratory, Corvallis, Oregon, 396 pp.

Aerts, R., and Berendse, F. (1988) The effect of increased nutrient availability on vegetation dynamics in wet heathlands. *Vegetation* **76**, 63–69.

American Public Health Association. (1989) *Standard Methods for the Examination of Water and Wastewater*, 17th edn. American Public Health Association, Washington, D.C., 1550 pp.

Anthony, R.G., and Kozlowski, R. (1982) Heavy metals in tissues of small mammals inhabiting waste-water-irrigated habitats. *J. Environ. Qual.* **11**, 20–22.

Aulio, K. (1980) Accumulation of copper in alluvial sediments and yellow water lilies (*Nuphar lutea*) at varying distances from a metal processing plant. *Bull. Environ. Contam. Toxicol.* **25**, 713–717.

Bayley, S., and Odum, H.T. (1976) Simulation of interrelations of the Everglades' marsh, peat, fire, water, and phosphorus. *Ecol. Modelling* **2**, 169–188.

Bayley, S.E., Zoltek, Jr, J., Hermann, A.J., Dolan, T.J., and Tortora, L. (1985) Experimental manipulation of nutrients and water in a freshwater marsh: Effects on biomass, decomposition, and nutrient accumulation. *Limnol. Oceanogr.* **30**, 500–512.

Behan, M.J., Kinraide, T.B., and Seiser, W.I. (1979) Lead accumulation in aquatic plants from metallic sources, including shot. *J. Wildl. Manage.* **43**(1), 240–244.

Black, C.A. (Ed.) (1965) *Methods of Soil Analysis*. American Society of Agronomy, Madison, Wisconsin.

Brinson, M., Bradshaw, H.D., Holmes, R.N., and Elkins, J.B., Jr (1980) Litterfall, stemflow, and throughfall nutrient fluxes in an alluvial swamp forest. *Ecology* **61**, 827–835.

Browder, J.A. (1988) Introduction: aquatic organisms as indicators of environmental pollution. *Water Resources Bull.* **24**(5), 927–929.

Buresh, R.J., DeLaune, R.D., and Patrick, W.H., Jr (1980) Nitrogen and phosphorus distribution and utilization by *Spartina alterniflora* in a Louisiana Gulf Coast marsh. *Estuaries* **3**, 111–121.

Chalmers, A.G. (1979) The effects of fertilization on nitrogen distribution in a *Spartina alterniflora* salt marsh. *Estuar. Coastal Shelf Sci.* **8**, 327–337.

Chalmers, A.G. (1982) Soil dynamics and the productivity of *Spartina alterniflora*. In: Kennedy, V.S. (Ed.) *Estuarine Comparisons*, pp. 231–241. Academic Press, New York.

Costanza, R., and Sklar, F.H. (1985) Articulation, accuracy, and effectiveness of mathematical models: a review of freshwater wetland applications. *Ecol. Modelling* **27**, 45–68.

de la Cruz, A.A., Hackney, C.T., and Stout, J.P. (1981) Above ground net primary productivity of three Gulf Coast marsh macrophytes in artificially fertilized plots. In: Nielson, B.J., and Cronin, L.E. (Eds.) *Estuaries and Nutrients*, pp. 437–445. Humana Press, Clifton, New Jersey.

Erwin, K. (1989) Wetland evaluation for restoration and creation. In: Kusler, J., and Kentula, M. (Eds.) *Wetland Creation and Restoration: The Status of the Science*, pp. 15–46. Report No. EPA/600/3-89/038. US Environmental Protection Agency, Cincinnati, Ohio.

Ewel, K.C. (1976) Effects of sewage effluent on ecosystem dynamics in cypress domes. In: Tilton, D.L., Kadlec, R.H., and Richardson, C.J. (Eds.) *Freshwater Wetlands and Sewage Effluent Disposal*, pp. 169–195. University of Michigan, Ann Arbor.

Fetter, C.W., Jr. C.W., Sloey, W.E., and Spangler, F.L. (1978) Use of a natural marsh for wastewater polishing. *J. Water Pollut. Control Fed.* **50**, 290–307.

Giesy, J.P., Jr (Ed.) (1980) *Microcosms in Ecological Research.* DOE Symposium No. 52. US Department of Energy, Savannah River Ecology Laboratory. Available from National Technical Information Service (NTIS), Springfield, Virginia 22161, as publication no. CONF-781101.

Godfrey, P.J., Kaynor, E.R., Pelczarski, S., and Benforado, J. (1985) *Ecological Considerations in Wetlands Treatment of Municipal Wastewaters*. Van Nostrand Reinhold, New York, 474 pp.

Gosselink, J.G., and Baumann, R.H. (1980) Wetland inventories: wetland loss along the United States coast. *Z. Geomorphol., N.F. Suppl.-Bd.* **34**, 173–187.

Grice, G.D., and Reeve, M.R. (1982) *Marine Mesocosms*. Springer-Verlag, New York, 430 pp.

Haines, B.L., and Dunn, E.L. (1976) Growth and resource allocation responses of *Spartina alterniflora* Loisel. to three levels of NH_4–N, Fe, and NaCl in solution culture. *Botan. Gaz.* **137**, 224–230.

Haines, E.B. (1979) Growth dynamics of cordgrass, *Spartina alterniflora* Loisel. on control and sewage sludge fertilized plots in a Georgia salt marsh. *Estuaries* **2**(1), 50–53.

Hellawell, J.M. (1986) *Biological Indicators of Freshwater Pollution and Environmental Management*. Elsevier, New York. 515 pp.

Hemond, H.F., and Benoit, J. (1988) Cumulative impacts on water quality functions of wetlands. *Environ. Manage.* **12**(5), 639–653.
Henderson-Sellers, B. (1984) *Engineering Limnology*. Pitman, London, 356 pp.
Hopkinson, C.S., Jr and Day, J.W., Jr (1980) Modeling hydrology and eutrophication in a Louisiana swamp forest ecosystem. *Environ. Manage.* **4**, 325–335.
Johnston, C.A. (1991) Sediment and nutrient retention by freshwater wetlands: Effects on surface water quality. *CRC Crit. Rev. Environ. Control* **21**, 491–565.
Jorgensen, S.E. (1988) Modelling of wetlands. In: Marani, A. (Ed.) *Advances in Environmental Modelling*, pp. 535–543. Elsevier, New York.
Kadlec, R.H., and Hammer, D.E. (1988) Modelling nutrient behavior in wetlands. *Ecol. Modelling* **40**, 37–66.
Kadlec, R.H., and Tilton, D.L. (1979) The use of freshwater wetlands as a tertiary wastewater treatment alternative. *CRC Crit. Rev. Environ. Control* **9**, 185–212.
Karr, J.R. (1987) Biological monitoring and environmental assessment: a conceptual framework. *Environ. Manage.* **11**, 249–255.
Kelly, J.R., and Harwell, M.A. (1989) Indicators of ecosystem response and recovery. In: Levin, S.A., Harwell, M.A., Kelly, J.R., and Kimball, K.D. (Eds.) *Ecotoxicology: Problems and Approaches*, pp. 9–35. Springer Verlag, New York.
Kitchens, W.M. (1979) Development of a saltmarsh microcosm. *Int. J. Environ. Studies* **13**, 109–118.
Lan, C., Chen, G., Li, L., and Wong, M.H. (1992) Use of cattails in treating wastewater from a Pb/Zn mine. *Environ. Manage.* **16**, 75–80.
Larsen, V.J., and Schierup, H.H. (1981) Macrophyte cycling of zinc, copper, lead, and cadmium in the littoral zone of a polluted and a non-polluted lake: seasonal changes in heavy metal content of above-ground biomass and decomposed leaves. *Aquat. Bot.* **9**, 26–34.
Leibowitz, N.C., Squires, L., and Baker, J.P. (1991) *Research Plan for Monitoring Wetland Ecosystems.* Report No. EPA/600/3-91/010. US Environmental Protection Agency, Environmental Research Laboratory, Corvallis, Oregon, 191 pp.
Linthurst, R.A., and Seneca, E.D. (1981) Aeration, nitrogen and salinity as determinants of *Spartina alterniflora* Loisel. growth response. *Estuaries* **4**, 53–63.
McIntosh, A.W., Shephard, B.K., Mayes, R.A., Atchison, G.J., and Nelson, D.W. (1978) Some aspects of sediment distribution and macrophytes cycling of heavy metals in a contaminated lake. *J. Environ. Qual.* **7**, 301–305.
Mendelssohn, I.A. (1979) The influence of nitrogen level, form and application method on the growth response of *Spartina alterniflora* in North Carolina. *Estuaries* **2**, 106–112.
Miller, P.C., Stoner, W.A., and Tieszen, L.L. (1976) A model of stand photosynthesis for the wet meadow tundra at Barrow, Alaska. *Ecology* **57**, 411–430.
Miller, P.C., Oechel, W.C., Stoner, W.A., and Sveinbjornsson, B. (1978a) Simulation of CO_2 uptake and water relations of four arctic bryophytes at Point Barrow, Alaska. *Photosynthetica* **12**, 7–20.
Miller, P.C., Stoner, W.A., Tieszen, L.L., Allessio, M., McCown, B., Chapin, F.S., and Shaver, G. (1978b) A model of carbohydrate, nitrogen, and phosphorus allocation and growth in tundra production. In: Tieszen, L.L. (Ed.) *Vegetation and Production Ecology of an Alaskan Arctic Tundra*, pp. 577–598. Springer Verlag, New York.
Mitsch, W.J. (1976) Ecosystem modeling of waterhyacinth management in Lake Alice, Florida. *Ecol. Modelling* **2**, 69–89.

Mitsch, W.J. (1983) Ecological models for management of freshwater wetlands. In: Jorgensen, S.E., and Mitsch, W.J. (Eds.) *Application of Ecological Modeling in Environmental Management*, Part B, pp. 283–310. Elsevier, Amsterdam.

Mitsch, W.J. (1988a) Ecological engineering and ecotechnology with wetlands applications of systems approaches. In: Marani, A. (Ed.) *Advances in Environmental Modelling*, pp. 565–580. Elsevier, Amsterdam.

Mitsch, W.J. (1988b) Productivity-hydrology-nutrient models for forested wetlands. In: Mitsch, W.J., Straskraba, M., and Jorgensen, S.E. (Eds.) *Wetland Modelling*, pp. 115–132. Elsevier, Amsterdam.

Mitsch, W.J., and Gosselink, J.G. (1986) *Wetlands*. Van Nostrand Reinhold, New York, 539 pp.

Mitsch, W.J., and Reeder, B.C. (1992) Nutrient and hydrologic budgets of a Great Lakes coastal freshwater wetland during a drought year. *Wetlands Ecol. Manage.* **1**, 211–222.

Mitsch, W.J., Day, J.W., Jr, Taylor, J.R., and Madden, C. (1982) Models of North American freshwater wetlands. *Int. J. Ecol. Environ. Sci.* **8**, 109–140.

Mitsch, W.J., Taylor, J.R., Benson, K.B., and Hill, P.L., Jr (1983) Wetlands and coal surface mining in western Kentucky—a regional impact assessment. *Wetlands* **3**, 161–179.

Mitsch, W.J., Straskraba, M., and Jorgensen, S.E. (Eds.) (1988) *Wetland Modelling*. Elsevier, Amsterdam.

Morris, J.T. (1988) Pathways and controls of the carbon cycle in salt marshes. In: Hook, D.D. *et al.* (Eds.) *The Ecology and Management of Wetlands*, pp. 497–510. Croom Helm, London.

Morris, J.T. (1991) Effects of nitrogen loading on wetland ecosystems with particular reference to atmospheric deposition. *Ann. Rev. Ecol. Syst.* **22**, 257–279.

Mouvet, C. (1985) The use of aquatic bryophytes to monitor heavy metals pollution of freshwaters as illustrated by case studies. *Verh. Int. Verein. Limnol.* **22**, 2420–2425.

NRC (National Research Council) (1990) *Managing Troubled Waters: The Role of Marine Environmental Monitoring*. National Academy Press, Washington, D.C., 140 pp.

Nichols, D.S. (1983) Capacity of natural wetlands to remove nutrients from wastewater. *J. Water Pollut. Control Fed.* **55**, 495–505.

Niethamner, K.R., Atkinson, R.D., Baskett, T.S., and Samson, F.B. (1985) Metals in riparian wildlife in the lead mining district of southeastern Missouri. *Arch. Environ. Contam. Toxicol.* **14**, 213–223.

Nixon, S.W., and Lee, V. (1986) *Wetlands and Water Quality: A Regional Review of Recent Research in the United States on the Role of Freshwater and Saltwater Wetlands as Sources, Sinks, and Transformers of Nitrogen, Phosphorus, and Various Heavy Metals.* Technical Report Y-86-2 to US Army Engineer Waterways Experiment Station, Vicksburg, Mississippi. University of Rhode Island, Kingston, 229 pp.

Office of Technology Assessment (1984) *Wetlands: Their Use and Regulation.* Report No. OTA-O-206. US Congress, Office of Technology Assessment, Washington, D.C., 208 pp.

Pait, A.S., DeSouza, A.E., and Farrow, D.R.G. (1992) *Agricultural Pesticides in Coastal Areas: A National Summary.* National Oceanic and Atmospheric Administration, Strategic Environmental Assessments Division, Rockville, Maryland, 112 pp.

Parsons, T.R., Maita, Y., and Lalli, C.M. (1984) *A Manual of Chemical and Biological Methods for Seawater Analysis*. Pergamon Press, New York, 173 pp.

Patrick, W.H., Jr, and DeLaune, R.D. (1976) Nitrogen and phosphorus utilization by *Spartina alterniflora* in a salt marsh in Barataria Bay, Louisiana. *Estuar. Coastal Marine Sci.* **4**, 59–64.

Portier, R.J., and Meyers, S.P. (1982) Use of microcosms for analyses of stress-related factors in estuarine-wetland ecosystems.In: Gopal, B., Turner, R.E., Wetzel, R.G., and Whigham, D.F. (Eds.) *Wetlands Ecology and Management*, pp. 375–387. National Institute of Ecology and International Scientific Publications. Lucknow Publishing, Lucknow, India.

Quinn, H.A., Tolson, J.P., Klein, C.J., Orlando, S.P., and Alexander, C. (1989) *Strategic Assessment of Near Coastal Waters: Susceptibility and Status of Gulf of Mexico Estuaries to Nutrient Discharges. Summary Report.* National Oceanic and Atmospheric Administration, Strategic Assessment Branch, Rockville, Maryland, 35 pp.

Roelofs, J.G.M., Schuurkes, J.A.A.R., and Smits, A.J.M. (1984) Impact of acidification and eutrophication on macrophyte communities in soft water. II. Experimental studies. *Aquat. Bot.* **18**, 389–411.

Schierup, H.H., and Larsen, V.J. (1981) Macrophyte cycling of zinc, copper, lead and cadmium in the littoral zone of a polluted and a non-polluted lake. 1. Availability, uptake, and translocation of heavy metals in *Phragmites australis*. *Aquat. Bot.* **11**, 197–210.

Schindler, D.W. (1987) Detecting ecosystem response to anthropogenic stress. *Can. J. Fish. Aquat. Sci.* **44**, 6–25.

Schroeder, R.L. (1987) *Community Models for Wildlife Impact Assessment: A Review of Concepts and Approaches.* US Fish and Wildlife Service, National Ecology Center, Resource Evaluation and Modelling Section, Fort Collins, Colorado.

Schuurkes, J.A., Kok, C.J., and den Hartog, C. (1986) Ammonium and nitrate uptake by aquatic plants from poorly buffered and acidified waters. *Aquat. Bot.* **24**, 131–146.

Shaver, G.R., and Chapin III, F.S. (1980) Response to fertilization by various plant growth forms in an Alaskan tundra: Nutrient accumulation and growth. *Ecology* **61**(3), 662–675.

Simpson, R.L., Good, R.E., Walker, R., and Frasco, B.R. (1983) The role of Delaware River freshwater tidal wetlands in the retention of nutrients and heavy metals. *J. Environ. Qual.* **12**(1), 41–48.

Sloey, W.E., Spangler, F.L., and Fetter, C.W., Jr (1978) Management of freshwater wetlands for nutrient assimilation. In: Good, R.E., Whigham, D.F., and Simpson, R.L. (Eds.) *Freshwater Wetlands: Ecological Processes and Management Potential*, pp. 321–340. Academic Press, New York.

Stanley, D.W. (1976) A carbon flow model of epipelic algal productivity in Alaskan tundra ponds. *Ecology* **57**, 1034–1042.

Stanley, D.W. (1992) *Historical Trends: Water Quality and Fisheries, Albemarle-Pamlico Sounds, with Emphasis on the Pamlico River Estuary*. Report No. UNC-SG-92-04. University of North Carolina, Sea Grant College, Raleigh, 215 pp.

Straskraba, M., and Gnauck, A. (1985) *Freshwater Ecosystems: Modeling and Simulation. Developments in Environmental Modeling*, Vol. 8. Elsevier, New York, 309 pp.

Sullivan, M.J. (1981) Effects of canopy removal and nitrogen enrichment on a *Distichlis spicata*-edaphic diatom complex. *Estuar. Coastal Shelf Sci.* **13**, 119–130.

Taylor, G.J., and Crowder, A.A. (1983) Accumulation of atmospherically deposited metals in wetland soils of Sudbury, Ontario. *Water Air Soil Pollut.* **19**, 29–42.

Tiner, R.W. (1984) *Wetlands of the United States: Current Status and Recent Trends.* US Department of Interior, Fish and Wildlife Service, National Wetlands Inventory, Washington, D.C., 58 pp.

Turner, R.E., Forsythe, S.W., and Craig, N.J. (1981) Bottomland hardwood forest land resources of the southeastern United States. In: Clark, J.R., and Benforado, J. (Eds.) *Wetlands of Bottomland Hardwood Forests*, pp. 13–28. Elsevier, Amsterdam.

Tyler, G. (1967) On the effect of phosphorus and nitrogen, supplied to Baltic shore-meadow vegetation. *Bot. Notiser* **120**, 433–447.

USEPA (US Environmental Protection Agency) (1979) *Methods for Chemical Analyses of Water and Wastes.* US Environmental Protection Agency, Washington, D.C.

Valiela, I., and Teal, J.M. (1974) Nutrient limitation in salt marsh vegetation. In: Reimold, R.J., and Queen, W.H. (Eds.) *Ecology of Halophytes*, pp. 543–565. Academic Press, New York.

Valiela, I., Teal, J.M., and Sass, W.J. (1975) Production and dynamics of salt marsh vegetation and the effects of experimental treatment with sewage sludge. *J. Appl. Ecol.* **12**, 973–981.

Valiela, I., Teal, J.M., and Persson, N.Y. (1976) Production and dynamics of experimentally enriched salt marsh vegetation: below-ground biomass. *Limnol. Oceanogr.* **21**, 245–252.

Valiela, I., Teal, J.M., Cogswell, C., Hartman, J., Allen, S., Van Etten, R., and Goehringer, D. (1985) Some long-term consequences of sewage contamination in salt marsh ecosystems. In: Godfrey, P.J., Kaynor, E.R., Pelczarski, S., and Benforada, J. (Eds.) *Ecological Considerations in Wetland Treatment of Municipal Waste Waters*, pp. 301–316. Van Nostrand Reinhold, New York.

Van Raalte, C.D., Valiela, I., and Teal, J.M. (1976) The effect of fertilization on the species composition of salt marsh diatoms. *Water Res.* **10**, 1–4.

Vermer, J.G. (1986) The effect of nutrients on shoot biomass and species composition of wetland and hayfield communities. *Acta Ecol. Acta Plant.* **7**, 31–41.

Welcomme, R.L. (1979) *Fisheries Ecology of Floodplain Rivers.* Longman, New York, 318 pp.

Wiegert, R.G., Christian, R.R., and Wetzel, R.L. (1981) A model view of the marsh. In: Pomeroy, L.R., and Wiegert, R.G. (Eds.) *The Ecology of a Salt Marsh*, pp. 183–218. Springer Verlag, New York.

Windom, H.L. (1977) *Ability of Salt Marshes to Remove Nutrients and Heavy Metals from Dredged Material Disposal Area Effluents.* Technical Report D-77-37. US Army Engineers, Waterways Experiment Station, Vicksburg, Mississippi.

Woods, R.D. (1975) *Hydrobotanical Methods.* University Park Press, Baltimore. 173 pp.

9 Methods to Assess the Effects of Chemicals on Forests

M. Bonneau
Institut National de la Recherche Agronomique, France

S. Fink
Institüt für Forstbotanik und Baumphysiologie, Germany

H. Rennenberg
Institüt für Forstbotanik und Baumphysiologie, Germany

9.1 INTRODUCTION

Despite growing in generally fairly clean air, forests may be subjected to a variety of pollutant: salts (Na, Ca, Mg, sulphates, chlorides, or bicarbonates), gases (SO_2, NO_2, and ozone), nitrogen compounds (HNO_3, NH_3, and NH_4^+), heavy metals (Cd, Cu, Ni, Pb, and Zn), acids (H_2SO_4 and HNO_3). Weed control agents and other pesticides appear to have a negligible impact on forests, because, at most, they are used for only a few years in a rotation of a hundred years or more, and some countries have banned their use completely.

Based on the nature of these substances, their effects may be identified either at either of two levels: that of the trees and that of the forest ecosystem.

9.2 METHODS USED TO STUDY TREES

9.2.1 CHEMICAL STUDIES

9.2.1.1 Stress reactions at the metabolic level

Plants are exposed in their environment to numerous stress factors that may act either simultaneously or consecutively with different intensities and frequencies. Plants experience stress from several environmental influences such as (1) contact with chemicals of anthropogenic origin, (2) climate (e.g., high light intensity, drought, high temperature, and low temperature), (3) nutrient deficiency, (4) excess availability of trace elements, and (5) pathogens. At the metabolic level, many

Methods to Assess the Effects of Chemicals on Ecosystems
Edited by R. A. Linthurst, P. Bourdeau, and R. G. Tardiff

factors generate oxidative stress to plants (Polle and Rennenberg, 1993a). Therefore, metabolic effects caused by exposure of plants to oxidative chemicals cannot easily be distinguished from effects caused by other factors.

The site(s) of initial generation of oxidative stress in plants may be different for airborne chemicals and other stress factors. Atmospheric substances enter the shoot of plants predominantly via the stomata of the leaves, wherein they first contact the aqueous phase of the cell walls (apoplastic space). Reactions in this compartment are considered initiating events. Metabolic alterations in the cytoplasm may be caused by products of initial reactions in the apoplastic space or by a fraction of the parent substance. By contrast, oxidative stress from climate or nutrient deficiency may predominantly take place inside the cell with the chloroplast as the most important initial target. Compartmental analysis, therefore, may be used as a diagnostic tool to distinguish the influence of alternative stresses.

The metabolic responses of trees to air pollutants (Bytnerowicz and Grulke, 1992) and oxidative stresses (Polle and Rennenberg, 1993a, 1993b) have been reviewed recently in detail. This chapter discusses metabolic reactions in trees as diagnostic tools of damage caused by atmospheric sulphur and nitrogen compounds, photochemical oxidants, and atmospheric xenobiotics.

9.2.1.2 Metabolic effects of photochemical oxidants

During atmospheric transport of pollutants, radical reaction chains are initiated by light, producing highly reactive hydroxyl radicals and peroxides, hence the name photochemical oxidants. Ozone is believed to be the most important final product of these reaction chains. By this mechanism, emissions of volatile organic compounds contribute to the ozone troposphere. Among the many photochemical oxidants, ozone is the most studied for its toxic properties on vegetation; hence, the discussion of the metabolic effects of photochemical oxidants is focused on ozone.

Like other atmospheric pollutants ozone enters the leaves predominantly via the stomata, and first contacts the aqueous phase of the cell walls (apoplastic space). Ozone's high oxidation potential ozone will lead either to its rapid degradation to other reactive oxygen species or to its reaction with natural components in the apoplastic space. Ascorbate is known to protect against injury from ozone, is found in mM concentrations in the apoplastic space, and, therefore, plays a major part in the neutralization of ozone absorbed by leaves. When the dehydroascorbate produced in this reaction is regenerated, this reaction protects against damage from ozone (Polle and Rennenberg, 1993a). Ozone appears unlikely to pass the aqueous phase of the cell wall and enter the symplasm. Hence, ozone will damage simplistic structures and alter metabolic processes only in the apoplastic space inside the cell.

Ozone-mediated changes in apoplastic ascorbate content and peroxidase activity have been observed in fumigation experiments and field studies. However, even the direction of these changes varied between species and even for the same species in different studies (Castillo *et al.*, 1987; Castillo and Greppin, 1988; Ogier *et al.*, 1991; Polle *et al.*, 1991; Polle and Rennenberg, 1992). Therefore, apoplastic markers cannot be recommended for the identification of perturbations by ozone stress.

Likewise, biochemical markers in the symplasm are unavailable. Though ozone is known to affect photosynthesis, stomatal aperture, respiration, allocation of photosynthetase, substrate levels, and enzyme activities of the antioxidant system, and ethylene biosynthesis, none of these effects are specific for ozone (Bytnerowicz and Grulke, 1992). Ozone-mediated induction of polyamine and stilbene biosynthesis and pathogen-related proteins has been observed in plants (Ernst *et al.*, 1992; Rosemann *et al.*, 1991; Schraudner *et al.*, 1992); however, the specificity of these biochemical responses to ozone remains to be elucidated.

9.2.1.3 Metabolic effects of sulphur compounds

Trees growing in regions of volcanic or geothermal activity or in heavily industrialized areas are subjected to high levels of atmospheric sulphur compounds, most predominantly sulphur dioxide (SO_2) and hydrogen sulphide (H_2S), which are both rapidly absorbed by plant shoots where it enters the plant's metabolic pathways of sulphur reduction and association (DeKok, 1990). Foliar deposition of sulphur at high concentrations can result in damage to plants, but, at low concentrations, may also be beneficial for plant growth and development.

Sulphur dioxide

Herbaceous and woody plants exposed to SO_2 have increased sulphur content in the shoots (Bytnerowicz and Grulke, 1992). The extent of sulphur accumulation varies between species, and depends on suffix nutrition to the roots, atmospheric SO_2 concentration, duration of exposure, and climatic conditions during exposure (DeKok, 1990). Shoot sulphur content cannot be used as a specific diagnostic marker of SO_2 exposure, when plants grow in high sulphur soil or are in contact with other sulphur-containing gases in high concentrations in the air (Rennenberg, 1984).

Elevated sulphur levels in shoots can predominantly be ascribed to the accumulation of sulphate in the vacuole (Rennenberg, 1984; DeKok 1990; Cram, 1990). Though a decrease in the sulphate content of shoots has been observed upon cessation of SO_2 fumigation (Maas *et al.*, 1987a), sulphate accumulation may still be a diagnostic tool for atmospheric sulphur exposure especially in trees, because mobilization of sulphate stored in the vacuole proceeds at a low rate (Cram, 1990) and dilution of accumulated sulphate by growth may only cause injury during specific developmental stages.

Some of the SO_2 absorbed by plant shoots is reduced to the level of sulphide (DeKok, 1990), and up to 10–15 percent of absorbed SO_2 can be re-emitted into the atmosphere in the form of H_2S (Rennenberg, 1991). The sulphide is incorporated into cysteine, and used for the synthesis of other low molecular weight sulphur compounds, such as glutathione, and for protein synthesis. Since elevated levels of low molecular weight thiols are also present from various stresses besides air pollutants (Smith *et al.,* 1990), thiol levels provide no specific diagnostic measure for SO_2 stress in plants. However, the ratio of sulphate-S/organic-S may provide a relatively specific tool to diagnose SO_2 stress in plants. In pine, this ratio varied from 0.29 at a reference location to 0.88 in an area with high SO_2 deposition (Legge *et al.*, 1988). No such changes have been observed from other stress factors including exposure to other atmospheric sulphur compounds or to excess sulphate in the soil.

Exposure of foliage to atmospheric SO_2 results in a transport of SO_2-derived sulphur to the roots in the form of sulphate and organic sulphur compounds like cysteine and glutathione (Garsed, 1985; Rennenberg and Polle, 1992). Sulphate and organic sulphur compounds are normal constituents of both xylem and phloem, and their long-distance transport in these pathways is important for the inter-organ regulation of sulphur nutrition (Herschbach and Rennenberg, 1991; Schupp *et al.*, 1992). Consequently, atmospheric SO_2 interacts in plants with processes involved in the regulation of sulphur nutrition at the whole plant level, but no diagnostic tools has been developed to benefit from this observation.

SO_2 and its metabolic products affect numerous physiological processes in plant cells, including stomatal aperture and its control (Black, 1985), carbon balance and allocation of photosynthetate (Winner *et al.,* 1985; Bytnerowicz and Grulke, 1992), stromal and thylakoid functions (Wellburn, 1985), and antioxidant defence mechanisms (Rennenberg and Polle, 1992). None are specific for SO_2 exposure, or to that of other air pollutants. Therefore, they are not suitable as diagnostic measures.

Likewise, the analysis of apoplastic processes is not helpful diagnostically. SO_2 that has entered the leaves rapidly dissolves in the aqueous phase of the cell walls and reacts with water to form bisulphite. Finally, determination of apoplastic sulphate can presently not be recommended for the diagnosis of atmospheric SO_2 exposure.

Hydrogen sulphide

Fumigation of plants with H_2S results in elevated sulphur contents of the shoot. Only a minor part of this increase can be ascribed to enhanced levels of sulphate. The bulk of the H_2S absorbed by leaves is directly incorporated into reduced sulphur compounds. Consequently, elevated levels of sulphydryl compounds, mainly glutathione and cysteine, are found in shoots fumigated with H_2S (DeKok, 1990). Several other stress factors cause elevated levels of sulfhydryl compounds, mainly glutathione (Smith *et al.*, 1990) Therefore, this finding is unsuitable as a diagnostic measure of exposure to H_2S. In spinach, prolonged H_2S exposure results in an accumulation of sulphate in the roots (Maas *et al.,* 1987b), and such accumulation in the roots may be used as an indication of H_2S exposure, provided the phenomenon can be generalized.

9.2.1.4 Metabolic effects of nitrogen compounds

Anthropogenic activities have led to high emissions of gaseous nitrogen compounds into the atmosphere—specifically nitrogen oxides released during the combustion of fossil fuels by power plants and by automobiles. Agricultural activities result in the emission of high amounts of ammonia and ammonium into the atmosphere. Consequently, critical loads for nitrogen to forests are exceeded in many regions (Hadwiger-Fangmeier *et al.*, 1992). Although excess nitrogen input may initially be beneficial to forests growing under nitrogen limitation, prolonged exposure will result in nutritional disorders and enhanced susceptibility to other stresses. Furthermore, the species composition of forest ecosystems will change dramatically during long-term exposure to excess nitrogen (Hadwiger-Fangmeier *et al.*, 1992).

Nitrogen oxides

The combustion of fossil fuel produces nitrogen monoxide (NO) as the primary oxide of nitrogen. In the presence of ozone, NO is rapidly oxidized to nitrogen dioxide (NO_2). Both gases can contact the stomata of the leaves, a major path of entry in plants. Deposition velocities are one order of magnitude higher for NO_2 than for NO, due to differences in solubility. Both NO_2 and NO react with water to form nitrate and nitrite; for NO, this reaction is much slower than for NO_2 (Wellburn, 1990). Having entered a leaf and been solubilized in the aqueous phase of the cell wall, each species reacts with other constituents of the apoplastic space (e.g., with ascorbate), perhaps a very significant group of interactions (Thoene, *et al.*, 1991).

As reaction products of NO and NO_2 in the aqueous phase of the cell wall, nitrite and nitrate may provide a diagnostic tool to identify exposure of trees to nitrogen oxides. In herbaceous plants, nitrite levels rarely rise, and changes in the level of total nitrate in response to NO_2 depend on the *N* supply to the roots. When red spruce seedlings were fumigated with NO_2 in the winter, nitrite accumulated in the apoplastic space, whereas nitrate levels remained unchanged (Wolfenden *et al.*, 1991). In the same study, the catabolism of nitrite was slower than its production from NO_2. The finding of nitrate in the xylem sap of control plants in this study was unexpected, and may be an indication of insufficient mycorrhization of the seedlings. Nitrate reduction in mycorrhized trees is supposed to proceed in the fungi associated with the roots. From these fungi amino acids are transferred to the roots and transported to the leaves with the transpiration stream (Martin and Botton, 1993). Therefore, the occurrence of nitrate in leaf extracts from mycorrhized trees may be an indication of exposure to nitrogen oxides. Whether the fungi withhold the nitrate because of the high concentrations in soil is unknown. Furthermore, total nitrogen content of the leaves provides no diagnostic measure, since it often declines in response to exposure of plants to nitrogen oxides, presumably due to enhanced export of nitrogen from the leaves (Wellburn, 1990).

For many trees, the assumption is incorrect that the conversion of NO and NO_2 to nitrite and nitrate in the aqueous phase of the cell wall (passed the plasmalemma) indicates a normal pathway of nitrate reduction followed by synthesis of amino acids and protein. Nitrate reduction and amino acid synthesis may take place in the mycorrhizal fungi of the roots and not in the leaves. Thus, nitrate reduction and amino acid synthesis are not considered normal processes in the leaves of mycorrhized trees, although their leaves remain capable of reducing nitrate and synthesizing amino acids. Nitrate reductase (NaR), which catalyses the reduction of nitrate to nitrite, is induced by its substrate; therefore, the level of its activity is determined by the nitrate supply. Studies of several tree species lacking significant nitrate reduction in the leaves have shown that exposure to atmospheric NO_2 concentrations results in increased NaR activity of the shoots (Thoene *et al.*, 1991). While nitrite reductase activity has been shown to increase, the change is less at normal atmospheric NO_2 concentrations (Wellburn, 1990; Thoene *et al.*, 1991). Therefore, the elevated NaR activity is postulated to be a reliable diagnostic tool to identify nitrogen oxide exposure to mycorrhized trees.

By eliciting oxidative stress, NO and NO_2 produce numerous physiological changes similar to those found with other oxidants. The responses include membrane damage due to inhibition of lipid biosynthesis and lipid peroxidation and reduction in net photosynthesis (Wellburn, 1990; Bytnerowicz and Grulke, 1992). Since none of these effects is specific for NO and NO_2, they are considered to be diagnostic measures.

Ammonia and ammonium

Stomatal aperture provides the only resistance for the influx of gaseous ammonia or ammonium into plant shoots, a phenomenon that may be due to the high water solubility of ammonia or ammonium that facilitates penetration through the aqueous phase of the cell wall. Dry deposition of these compounds accounts for 60 percent of the total, and takes place in close proximity to the source, because of the high deposition velocity (i.e., 65 percent within a distance of 30 km and 90 percent within 90 km from the source (Hadwiger-Fangmeier *et al.*, 1992). Therefore, the forests adjacent to agricultural activities are a large sink for atmospheric ammonia and ammonium.

Exposure (e.g., soil fertilization) of herbaceous and woody plants to atmospheric ammonia and ammonium increases the concentrations of these compounds in foliage and shoots. Such exposures also increased the amino acid content, especially arginine, in the leaves, and glutamine synthetase activity was found to be enhanced (Hadwiger-Fangmeier *et al.*, 1992). Physiological reactions of the exposure to atmospheric ammonia and ammonium included increased stomatal conductance connected with increased photosynthesis, enhanced transpiration, reduction of the water potential in water stressed plants, and mineral disorders such as reduced K and P content in the leaves (Hadwiger-Fangmeier *et al.*, 1992). These physiological changes are related to differences in susceptibility to other biotic and abiotic stresses such as drought, frost resistance, attack by herbivores, and fungi. Mycorrhization was found to be reduced (Hadwiger-Fangmeier *et al.*, 1992). Apparently, increased ammonia and ammonium contents of the shoots provides a reliable measure of atmospheric ammonia and ammonium input into plants. Elevated total nitrogen of the shoots may be the result of atmospheric as well as pedospheric input. Reduced K and P contents of the leaves may be used to diagnose atmospheric ammonia input at sufficient K and P supply in the soil.

9.2.1.5 Metabolic effects of airborne xenobiotics

Halogenated hydrocarbons and pesticides are found in the atmosphere of highly industrialized areas, in the vicinity of agricultural regions, and in remote sites, such as alpine forests (Koval *et al.*, 1986; Elling *et al.*, 1988; Herterich and Herrmann, 1990). Xenobiotics can be transported in the atmosphere over long distances, and may be deposited preferentially in forests of mountain slopes. Trees growing in this environment are already exposed to various natural stresses like extreme temperatures, elevated ozone concentrations, high irradiation, and drought. The deposition of xenobiotics, as evident from the accumulation of xenobiotics in conifer needles (Gaggi and Bacci, 1985; Reichl *et al.*, 1987; Hinckel *et al.*, 1990), provide additional stress to these trees. Numerous crops, weeds, and even the needles of conifer species are known to contain glutathione S-transferases (GSTs), a group of constitutive and inducible isoenzymes capable of detoxifying halogenated and nitrated xenobiotics (Lamoureux and Rusness, 1989; Schroder *et al.*, 1990a, 1990b; Schroder and Rennenberg, 1992). GSTs may be induced by exposure to xenobiotics that they metabolize (Lamoureux and Rusness, 1989); thus, analysis of the GST isoenzyme pattern may be a useful means of identifying xenobiotic exposure of trees. Such an approach could be misleading, if GST activity was also induced by oxidative stresses such as ozone (Price *et al.*, 1990), since protection from oxidative stress may be considered a natural function of this group of enzymes.

9.2.1.6 Conclusions

Numerous biochemical parameters and physiological processes have been studied in herbaceous and woody plant species to ascertain the impact of exposures to photochemical oxidants, atmospheric sulphur and nitrogen compounds, and airborne xenobiotics. Though many effects of these pollutants have been identified, only few have proven to be sufficiently specific to serve as diagnostic measures for the identification of exposure to an individual pollutant. The ratio of sulphate-

S/organic–S may be used to identify sulphur dioxide exposure; apoplastic or cellular nitrate and nitrite as well as nitrate reductase activity of the leaves may be suitable to identify exposure of mycorrhized tree species to nitrogen oxides. Several biochemical parameters may be useful to identify atmospheric or pedospheric ammonia and ammonium input into trees. GST isoenzyme analysis for xenobiotics remains a research interest. A reliable diagnostic measure of exposure to photochemical oxidants is unavailable.

9.2.2 ANATOMICAL STUDIES

9.2.2.1 Microscopic diagnosis of damage to needles and leaves

The sudden occurrence of widespread "new type forest decline" in Central Europe and North America during the early 1980s caused an intensive search for diagnostic criteria to evaluate possible causal pathogenic factors. Although abiotic agents seemed to be the main toxic agents, discussions focused on whether symptoms seen in the field were caused by the direct impact of acid precipitation on the foliage or of gaseous pollutants on the leaves, or by indirect disturbances in mineral nutrition from the increased deposition of acidity and nitrogen (Roberts *et al.*, 1989).

Several approaches have been developed to construct criteria that could identify the actual causative factors. These criteria were first deduced from comparisons of healthy control trees with those subjected to artificial acid precipitation, controlled fumigation in chambers, or controlled experiments with varying mineral nutrition. These criteria were then applied in the field, using material from healthy and declining trees. Most diagnostic approaches use quantitative biochemical analysis of homogenates of needles or leaves, looking for carbohydrates and associated metabolites, lipids, antioxidants, or enzymes. Limitations of these indicators include unspecific reactions to numerous stress factors and only occasionally the determination of the impact of specific stresses.

To improve the possibility of diagnosing damage in needles, the reactions to stress of individual cells in the complex tissue of conifer needles and angiospermous leaves is an appropriate starting point for exploring more detailed diagnostic approaches. Qualitative changes within specialized cells and tissues were chosen as a criterion instead of quantitative changes in a homogenate of all cell types. The approach, originally developed in plant hormone research (Osborne, 1978, 1984), focused on finding target cells in needles and leaves that react differentially to external influences. The needles of Norway spruce *(Picea abies* (L.) Karst.) have been intensively studied for this purpose. The regular structures in these needles and leaves and in those from other species can be compared to alterations induced in controlled experiments to establish causal relationships, and the results then can be applied to damage patterns observed in the field.

9.2.2.2 Results from controlled experiments

Acid precipitation

The influence of acid precipitation is first investigated at the leaf/needle surface with epidermis, cuticle, and epicuticular waxes. In controlled experiments with simulated acid rain or mist, the amorphous and fibrillar crystalline waxes (especially in the epistomatal chambers) become partially fragmented and eroded, and partially baked together to amorphous masses, which is especially pronounced on long-lived leaves such as conifer needles. On one side, formation of these waxes on young leaves is retarded by the acidic conditions, whereas on the other side acid erosion is accelerated by 2–5 times the rate without acid; the regeneration capability is reduced by acid (Haines *et al.*, 1985; Riding and Percy, 1985; Crossley and Fowler, 1986; Grill *et al.*, 1987; Mengel *et al.*, 1987, 1989; Schmitt *et al.*, 1987; Tuomisto, 1988; Rinallo *et al.*, 1986; Turunen and Huttunen, 1989, 1991; Percy and Baker, 1990).

Acid precipitation may also lead to punctual necrosis of epidermal cells on the adaxial surface, frequently close to leaf veins or stomata, and at the bases of trichomes or glandular hairs. If cells adjacent to stomata collapse, the stomata may stay open continuously. At pH values below about 3.5, larger epidermal necroses may occur, which then may spread to adjoining cells of the palisade parenchyma and eventually spongy mesophyll. The neighbouring undamaged mesophyll cells then react with hypertrophy; the cells divide again; and, by hyperplasia, they may finally lead to circular or ring-like (with necrotic tissue in the centre) proliferations of the leaves, somehow resembling insect galls. These tissues form a delineation, resembling early stages of wound periderm formation, and may thus limit further progress of damage. In extreme cases, the damage may also proceed to the vascular bundles after most of the mesophyll has already collapsed. In general, developing young leaves are more affected by acid rain than older leaves (Evans *et al.*, 1978; Evans and Curry, 1979; Swiecki *et al.*, 1982; Paparozzi and Tukey, 1983; Adams *et al.*, 1984; Evans, 1984; Crang and McQuattie, 1986; Rinallo *et al.*, 1986). This pattern is mainly confined to angiospermous leaves. At pH-values below 2.5, similar epidermal necroses may occur in young needles (Whitney and Ip, 1991; Zobel and Nighswander, 1991).

After acid rain-treatment, non-lethal changes may be seen in chloroplast structures. Chloroplasts in outer mesophyll cells are more irregular in form with fewer thylakoids and larger starch grains. Sometimes thus treated needles show more severe changes in chloroplast structure after they have been subjected to cold treatments (Holopainen and Nygren, 1989).

Fumigation with gaseous air pollutants

Within a leaf or needle, gaseous pollutants primarily attack the mesophyll, especially that directly adjacent to the substomatal cavities; they sometimes directly attack the guard cells where the pollutants contact the stomata. The vascular bundle is mainly unaffected during initial exposure. This observation has been confirmed for many conifer needles and deciduous leaves subjected to SO_2, O_3, NO_x, HF, Cl_2, singly or in combination; in long-living conifer needles or evergreen leaves cumulative damage proceeds through additional necrosis in subsequent years (Hartig, 1896; Evans and Miller, 1972, 1975; Stewart *et al.*, 1973; Percy and Riding, 1981; Soikkeli 1981a, 1981b, 1981c; Carlson and Gilligan, 1983; Fink, 1988, 1989; Kozlowski and Constandinidou, 1986; Ruetze *et al.*, 1988; Hafner *et al.*, 1989; Ebel *et al.*, 1990). The damaged mesophyll cells show plasmolysis and shrinkage, breakdown of internal compartmentalization, and finally necrosis and discolorations; the mesophyll cells directly bordering necrotic cells may react with slight discolorations, thickenings of their cell walls, or hypertrophy; a specific reaction in conifer needles is the hypertrophy of epithelial cells of the resin sacs leading to their occlusion. Collapse of mesophyll may lead to depressions in the overlying epidermis. The vascular bundles generally remain unaffected during an initial phase, and only after extensive damage to the mesophyll the sieve elements may also show some necrosis, accompanied by a hypertrophy of adjacent cells of the phloem and transfusion parenchyma. Damage by O_3 preferentially affect the palisade parenchyma, even if this is located on the upper side of a leaf, i.e., opposite the stomata that are on the lower side in hypostomatic leaves; O_3 hardly affects the vascular tissue. HF seems to induce first injury to the spongy mesophyll, which spreads to the vascular tissue. SO_2 induces a more general damage to all mesophyll cells, spreading only later to the vascular tissues.

Epicuticular waxes can be affected when SO_2 dissolves in moisture upon the leaf surface, and then causes the same reactions as acid precipitation. Ozone seems to have no effect on the structure of the already developed waxes, but obviously reduces the biosynthesis of waxes in the young needle (Gwrthardt-Goerg and Keller, 1987; Schmitt *et al.*, 1987; Percy *et al.*, 1992).

Before lethality in the mesophyll, the most pronounced changes produced by gaseous pollutants are alterations in the structure of chloroplasts: granulation of stroma, curling and swelling of thylakoids, reduction of grana, accumulation of plastoglobuli, and small, round chloroplasts with occasional irregular outlines. Most changes are rather non-specific signs of premature senescence, and only few specific changes for the different pollutants may be distinguished speculatively. SO_2 seems to lead more to reduced grana, swollen thylkoids, and light plastoglobuli; O_3 and other oxidants produce generally smaller and flattened chloroplasts that may have irregular outlines with granulated stroma partially containing crystaloids or fibrils, and accumulation of electron-dense material between the two membranes of the chloroplast envelope; NO_2 produces tubular protrusions from the chloroplast envelope; and BF produces stretched chloroplast envelopes and swollen or curled thylakoids (Thomson *et al.*, 1966, 1974; Thomson, 1975; Athanassious *et al.*, 1978; Athanassious, 1980; Soikkeli, 1981a, 1981b, 1981c; Soikkeli and Karenlampi, 1984; Karenlampi and Houpis, 1986; Sutinen, 1987c; Ruetze *et al.*, 1988; Ebel *et al.*, 1990; Sutinen *et al.*, 1990).

Induced mineral deficiencies

The widespread mineral deficiencies typical of many declining forests have structural impacts different from those caused by gaseous air pollutants. In conifer needles, deficiencies of magnesium, potassium, and calcium cause initial changes in the vascular bundle with early necrosis of the sieve cells and hypertrophy and hyperplasia of cambial cells and adjacent parenchyma cells, as well as alterations of the transfusion parenchyma. The mesophyll cells are only affected at a later stage, and may continue to live, even though their organelles may be severely modified (Fink, 1988, 1989, 1990a; Holopainen and Nygren, 1989; Holopainen *et al.*, 1992). These changes may be explained by the importance of these cations for the stabilization of membranes and cell walls, by their role in osmotic regulations and hormone metabolism, and by their part in modulating enzyme activities (e.g., Mg-dependent ATPase responsible for loading of sugars into the phloem). For deficiencies of Mg, K, Ca, the sequence of degeneration (with regard to necrosis) is as follows: vascular bundle, then mesophyll, and finally epidermis which remains mainly unaffected through development of damage (Fink, 1988).

Deficiencies of Mg and K lead to a reduction and swelling of the thylakoids in chloroplasts, which are frequently also distorted because of the occurrence of large starch grains; reduced and irregularly organized grana compartments are typical, as are increased amounts of plastoglobuli (Whatley, 1971; Hamzah and Gomez, 1979; Fink, 1989; Holopainen and Nygren, 1989; Holopainen *et al.*, 1992).

9.2.2.3 Patterns from damaged foliage in the field

The first visible and most pronounced damage in declining trees is seen macroscopically as alterations of colour (turns yellow), partial necrosis, or premature abscises. By applying results from controlled experiments to the field, more conclusive microscopic criteria were sought to discriminate among possible causes of such alterations. Several microscopic criteria are discussed below.

Integrity of epicuticular waxes

In current research on forest decline, several studies have reported on possible changes in wax structure in field material as an indicator of the direct action of acid rain. Heavy degradations, accompanied by contamination by particulates, were indeed found in pine needles from urban areas of Berlin (Hafner, 1986; Hafner *et al.*, 1989). Under much less polluted conditions, fused stomatal waxes and degraded surface waxes were reported from declining trees in various areas in Central Europe (Sauter and Voss, 1986; Badot *et al.*, 1988). A close correlation of wax structure exists with the vitality of the trees, but not necessarily with type or amount of emissions. Some conflicting data exist, e.g., for the Black Forest, since other researchers could find no significant difference in wax structure between healthy and declining trees, because of the large variation of wax structure even within a single tree. A perplexing result is that the healthiest trees in Scotland have the least intact waxes (highest wettability), while the most diseased trees in the Black Forest have the most intact ones and the slowest weathering rates (Cape *et al.*, 1988). These findings strongly suggest that, outside "classical" heavily polluted industrial areas, the structure of surface waxes may be mainly determined by natural (climatic) influences, water stress, mineral nutrition, and indirectly by the general vitality of the trees (Guth and Frenzel, 1989; Ylimartino *et al.*, 1992; van Gardeningen *et al.*, 1991).

Mesophyll necrosis

The types of primary mesophyll damage caused by gaseous pollutants have been found widespread in areas with well-established ozone damage to susceptible species, mainly in the northeastern and southwestern parts of the USA (Costonis and Sinclair, 1969; Costonis, 1970; Karenlampi, 1986; Evans and Leonard, 1991). Concerning forest decline in Germany, some instances of this type of damage were found close to the borders to CSSR and GDR, where with easterly winds occasionally very high SO_2 concentrations occur (Cape *et al.*, 1988; Forschner *et al.*, 1989; Hasemann and Wild, 1990). Furthermore, they were encountered in heavily polluted urban regions of Berlin, where dead mesophyll cells already appear in green needles (Hafner, 1986; Hafner *et al.*, 1989). In leaves from sensitive hardwoods (*Fraxinus, Prunus*), necrotic palisade parenchyma cells (pointing to ozone damage) were claimed to have been found during the summer of 1989 in the Black Forest and the Alps with that region's unusually high ozone concentrations. No mesophyll necrosis, however, is associated with the most widespread yellowing of conifer (especially spruce) needles, which may stay living and turgescent for 2–4 years without necrosis of mesophyll cells, being totally yellow. Besides limited reports on mesophyll necrosis in some areas (Sutinen, 1987b; Forschner *et al.*, 1989) only little evidence exists that widespread direct damage by gaseous pollutants is involved with the typical "montane yellowing."

Some problems arise to avoid confounding pollution-caused necroses with necroses of other causes. This is especially a problem for several necrotic flecks that frequently occur on many leaves and needles in the field. In conifer needles (especially at higher altitudes), an increasing density of necrotic flecks is found on the upper surfaces as the needles age. No clear relationship has been observed between the occurrence of such flecks and either air quality or the vitality of the trees. A possible link between these necrotic spots on conifer needles and damage by acid rain has been hypothesized (Whitney and Ip, 1991). These necrotic spots are characterized by initial damage to mesophyll cells, frequently beneath the hypodermis, whereas the epidermis continues to live, thereby excluding the possibility of damage by acid precipitation (Fink, 1990b). Hypotheses that these spots had been caused by ozone (Lang and Holdenrieder, 1985; Sutinen, 1986) are not in accordance with this damage pattern, so that the explanation remains speculative, "winter fleck injury" remaining the most likely explanation (Daniker, 1923; Schmidt, 1936; Miller and Evans, 1974; Brennan, 1988; Fink 1990b); however, acidity and ozone should be rejected as possible causative agents.

Phloem alterations

Yellow needles from declining conifers that show symptoms of "montane yellowing" (associated mainly with Mg-deficiency) frequently exhibit symptoms characterized by drastic changes in their vascular bundles (sieve cell necrosis, hypertrophy and hyperplasia of cambium and phloem parenchyma), and less alterations in their still living mesophyll cells. These symptoms have been found in decaying yellow spruces and firs in the Black Forest (Fink, 1983, 1988, 1989; Parameswaran *et al.*, 1985) and the Hunsruck mountains (Forschner *et al.*, 1989; Hasemann and Wild, 1990). The observations of Schmitt *et al.* (1986) on early changes in the phloem, transfusion parenchyma, and endodermis (while mesophyll cells were still intact) already in green needles from decaying trees in northern Germany are in accordance with these observations of damage starting in the vascular bundle. These observations can be exclusively attributed to mineral deficiencies, which are probably caused by a combination of soil acidification, increased input of nitrogen, and climatic stresses. Further evidence for this phenomenon arises from the fact that at least moderately yellow needles may repair their chloroplasts and start to produce functional sieve cells as soon as the deficient minerals are added as fertilizers (Huttl and Fink, 1988).

Chloroplast ultrastructure

Alterations of chloroplast ultrastructure have been studied mainly in yellowing needles from declining trees in the field, and consist of reduction of thylakoids and accumulation of numerous plastoglobuli (Parameswaran *et al.*, 1985; Sutinen, 1987b; Jung and Wild, 1988; Zellnik and Gailhofer, 1989). These changes are comparable with those induced by controlled nutrient deficiencies. The type of swelling and distortion produced experimentally by ozone or SO_2 has been described as that from heavily polluted industrial and urban areas (Soikkeli, 1981b, 1981c; Hafner, 1986). However, these changes are only partly pollutant-specific and partly signs of senescence, either the natural rate of ageing or accelerated senescence caused by air pollutants or mineral deficiencies (Dodge, 1970; Whatley, 1971; Hecht-Buchholz, 1972, 1983; Chabot and Chabot, 1975; D'Agostino and Pennazio, 1981; Soikkeli, 1981 a, 1981b, 1981c; Cunninghame *et al.*, 1982; Sutinen, 1987a).

Swelling, thylakoid disorganization, formation of vesicles, and partial accumulation of large starch grains and lipid bodies have been found in needles from declining conifers to be caused by ozone, SO_2, water deficit, or mineral deficiencies (Thomson *et al.*, 1966, 1974; Whatley, 1971; Thomson, 1975; Pell and Weissberger, 1976; Vapaavuori *et al.*, 1984; Fink, 1988, 1989, 1992).

Observations on separation and dilatation of the double membranes of the thylakoids, forming electron-translucent spaces and inducing a swelling of the chloroplasts to roundish bodies, have stimulated hypotheses about the possible effects of organic atmospheric pollutants; however, no evidence yet supports this contention (Meyberg *et al.*, 1988). Similar symptoms have been reported with SO_2, O_3, NO_x, and mineral deficiencies (Wellburn *et al.*, 1972; D'Agostino and Pennazio, 1981; Soikkeli, 1981a, 1981b; Hecht-Buchholz, 1983; Platt-Aloia *et al.*, 1983; Soikkeli and Karenlampi, 1984). Sutinen (1986, 1987b, 1987c) and Sutinen *et al.* (1990) suggested that smaller and flatter chloroplasts with reduced thylakoids and more plastoglobuli may be typical for impacts of ozone (alone or together with SO_2), but the same changes are reported for different mineral deficiencies (Fink, 1988, 1989). Other types of chloroplast changes originally were thought to be typical for ozone or acid rain, but apparently may also be caused by water stress, such as crooking of the chloroplasts and the formation of thylakoid-free parts of the stroma, in which crystalline structures may occur (Thomson *et al.*, 1966, 1974; Athanassious *et al.*, 1978; Athanassious, 1980; Vapaavuori *et al.*, 1984; Crang and McQuattie, 1986; Fink, 1989).

Generally, chloroplast changes seem to be non-specific, and may offer little help in determining the cause of damage in needles from declining trees (Fink, 1988). However, such studies may help to identify early damage in needles (without determining the causes), while these still may look green and healthy from the outside (Grill *et al.*, 1989; Barsig *et al.*, 1990). Application of qualitative and quantitative morphometric measurements at the cell and chloroplast levels may help in more detailed analysis, as currently applied to a large-scale project in Finland (Jokela *et al.*, 1992).

Histochemical tests

The structural changes described above represent relatively late stages of damage, since metabolic changes usually start long before they finally become manifest in the form of structural consequences. The search for early indicators of damage should be investigations combining biochemical and structural (cell, tissue) measures such as histochemical tests for the localization and quantification of substances (e.g., starch) and enzyme activities.

Gaseous pollutants and mineral deficiencies can induce similar pathological starch accumulations (Hanson and Stewart, 1970; Fischer and Bussler, 1988; Ebel *et al.*, 1990; Luethy-Krause and Landolt, 1990). However, with gaseous pollutants, the inhibition of carbohydrate translocation from the chloroplasts seems to be inhibited at the level of the single cells, whereas with mineral deficiencies translocation apparently is inhibited at the more central level of the phloem. These changes lead to patchy, heterogeneous distribution of starch grains in the first case and a homogeneous distribution in the second case (Fink, 1992). Thus, the pattern of starch distribution may give more information than a measure of the concentration, although other factors (e.g., insect damage or viruses) may lead to similar patterns in starch accumulation. Starch occurrence in hardened mesophyll cells in late autumn and winter may reflect disturbances in the hardening process due to impact of acid precipitation (Back and Huttunen, 1992).

The distribution of calcium oxalate in conifer needles may be a sensitive indicator for pathogenic influences (Fink, 1991 a, 1991b, 1991c). In the epidermis, minute crystals are normally distributed within the epidermal cell walls and especially in the cuticular layer between the cuticle proper and the cell walls. Under the impact of acid precipitation, this pattern of distribution now becomes altered, and only very small amounts of crystals occur in the cuticular layer and the outer epidermal cell walls, whereas most of them remain restricted to the inner epidermal cell wars. This effect probably reflects "leaching" of Ca^{2+} from the outer cell walls by the impact of acid precipitation, whereas the inner parts of the walls are unaffected by the acidity. Already existing crystals of calcium oxalate are unlikely to become dissolved by the acid; rather the cations are leached out of the apoplast before they are precipitated as oxalate. Physiologically, whether Ca^{2+} is leached away from the apoplast and is lost to the needle or would otherwise become precipitated as insoluble calcium oxalate and thus also disappear physiologically matters little. This observation is consistent with the microscopical detection of crystals of calcium sulphate on the surface of needles subjected to acid rain (Bosch *et al.*, 1983; Nebe *et al.*, 1988; Fiedler *et al.*, 1990; Huttunen *et al.*, 1990).

Ozone alters the localization of the calcium oxalate as well. In the epidermis, numerous small crystals suddenly occur within the lumen of the cell, whereas the cell walls remain free of them and in the mesophyll cells where the crystals usually occur outside the cell walls towards the intercellular spaces, these crystals also grow from cell wall projections into the cells. In some mesophyll cells, free crystals can be found within the vacuoles, although these seem to become dissolved again quite rapidly. Generally, the mesophyll and epidermal cells in conifer needles seem to have a strong tendency to keep excess Ca^{2+} outside in the apoplast and to prevent entrance into the symplast and the vacuole. Under the impact of ozone, however, permeability of the plasma membranes may increase, and calcium may leak into the cells. This effect could provoke defense reactions such as callous deposition, which further can develop into wall ingrowths, or can lead to a breakthrough into the vacuole. Historically, the effects of ozone on mineral nutrition have been attributed to enhanced leaching of minerals and afflux out of the needles; however, recent findings suggest that the main effect may consist in a disturbance of the internal mineral compartmentalization and, at least for calcium, an increased influx from the apoplast into the symplast.

Early accumulation of phenolic compounds in cells close to the stomata have been regarded as possible early indicators for the direct impact of acid rain on conifer needles (Zobel and Nighswander, 1991).

Enzyme activities have been measured by histochemical means in only a few cases so far. However, such approaches might offer quite sophisticated tools for detailed analysis of pathological changes in affected needles or leaves.

9.2.2.4 Conclusions

Compound-specific reactions of target cells hypothetically offer an improved differentiated diagnostic approach over the quantitative analysis of parameters in tissue homogenates. For conifer needles, the main target cells of acid precipitation are those in the epidermis; for gaseous pollutants, the mesophyll cells; and for Mg^- or K^- deficiency, the sieve cells. Histochemical techniques offer promise for the localization of early metabolic changes in affected cells. Nevertheless, additional investigations are needed on the influence of natural factors (e.g., cold shocks, insects, water deficiency) on the structure of tissues under controlled conditions. Neither of the mentioned structural approaches offer a totally secure diagnosis, but always has to be combined with other techniques (e.g., biochemical analysis or nutrient analysis).

9.3 METHODS TO IDENTIFY EFFECTS ON FOREST ECOSYSTEMS

9.3.1 INTRODUCTION

Measurable effects of chemicals in trees are signs of on-going change, or perhaps even damage, as chemicals transported by air or water are deposited on leaves or needles or enter the soil and are absorbed by roots. They may accumulate in soils or in soil animals or micro-organisms with no immediate effects in trees, but perhaps changes in soil composition or constituents. These changes may be signs of an excessive chemical load and of a hazard to be manifest in the future. To the extent that they confirm assessments made directly on trees or animals, they may be useful in detecting or confirming the chemical load on an ecosystem.

Methods to study the effects of chemicals on forest ecosystems include those of soil solid phase, soil solutions, species composition, and chemical composition of ground vegetation, microfauna, and microflora.

9.3.2 SOIL SOLID PHASE STUDIES

9.3.2.1 Total soil chemical composition

Total element concentrations in soil are easy to determine by acid digestion. In several areas and forest types, concentrations of cadmium, copper, nickel, lead, and zinc have been measured, shown to vary widely according to region and soil layer, and are higher in industrial regions (Southern Norway: Steinnes *et al.*, 1989; Austria, Weiner Wald, beech: Kazda and Glatzel, 1984; Germany, Black Forest: Zottl, 1985; Germany, Solling, spruce: Heinrichs and Mayer, 1980; Germany, Nordrhein Westfalien, beech: Neite, 1989; Germany Hoglwald (Bavaria), spruce: Schierl and Kreutzer, 1991; USA, Vermont, Camel Humps, Boreal forest: Friedland *et. al*, 1984; Canada, boreal spruce fir forest: Geballe *et al.*, 1990; Canada, Turkey Lake, Ontario, maple and beech: Morrison and Hogan, 1986; USA, New England, Great Smoky Mountains: Turner *et al.*, 1985; USA, North-East, Coniferous forest 1000 m above sea level: Herrick and Friedland, 1990). Such data are of limited utility, because potential effects of chemicals depend on their form and availability. Besides, even the highest concentrations (e.g., in solling area) are too low to cause toxicity. Nonetheless, the measurement of total heavy metal concentrations, particularly in the forest floor where most of these metals accumulate, may be of value. If the concentrations are not of the same order as the highest values, a low risk of toxicity exists; but periodic measurements assess the evolution of concentrations, because heavy metals (especially Pb) accumulate because the rate at which they are drained from the soil is slower than that at which they are deposited.

The total concentrations considered to be without injurious consequences in garden soils (Kloke, 1981) can be used to estimate acceptable levels in forest soils. In mg/kg, these tolerable concentrations are As, 20; B, 25; Be, 10; Br, 10; Cd, 3; Co, 50; Cr, 100; Cu, 100; F, 200; Hg, 2; Mo, 5; Ni, 50; Pb, 100; Sb, 5; Se, 10; Sn, 50; U, 5; V, 50; and Zn, 300. These values take into account risks to human health. If phytotoxicity alone were to be addressed, the following values in mg/kg would be obtained: water soluble B, 3; total Cu, 130; Cu (EDTA), 50; total Ni, 70; exchangeable Ni, 20; total Zn, 300; and exchangeable Zn, 130.

9.3.2.2 Available forms of elements

Several extraction processes allow one to obtain fractions of cadmium, copper, lead, nickel, or zinc in soil that are considered available to trees (ammonium acetate, acetic acid, NH_4Cl, EDTA) (Austria, Wiener Wald-beech: Kazda and Glatzel, 1984; several European beech forests: Wittig and Neite, 1989; Germany, Nordhein, Westfalien, beech forests: Neite, 1989). Such values are more useful than total concentrations, as they can be compared to quantities experimentally added to soil samples where seedlings are grown. Toxic concentrations (mg/kg) of several heavy metals have been reported for trees (Kahle *et al.*, 1989; Geballe *et al.*, 1990; Seiler and Paganelli, 1987; Chappelka *et al.*, 1991; Kristodorov *et al.*, 1989; Smith and Brennan, 1984; Denny and Wilkins, 1987; Brown and Wilkins, 1985; Burton *et al.*, 1984; Patterson and Olson, 1983; Nakos, 1979).

Some of these toxicity values are probably incorrect. For instance, when seedlings are grown in sand, the determined toxic value is probably lower than when grown in soil, because sand has a lower exchange capacity; thus the solution in equilibrium with sand is more concentrated than soil solution in the presence of clay and organic matter. Furthermore, only results with mycorrhizal seedlings are valid, because fungi are a barrier against toxic element uptake. Moreover, most of the published results of experiments in soil samples suppose that the whole quantity of a toxic metal added in the form of salt (mainly chloride) remains in available form, which is incorrect most of the time. Such experiments should be completed by the analysis of the actual available contents of toxic metals at the end of the experiment, as done by Burton *et al.* (1986); otherwise estimated toxic levels may be higher than the actual ones.

Regrettably, research efforts have focused predominantly on cadmium and lead. The toxicity levels of each vary according to species and authors. The toxic level of lead is in the area of 50–300 mg/kg; toxicity of cadmium begins at 0.3-5 mg/kg. In some areas, present levels of available lead may be toxic to beech seedlings since they are higher than 55 mg/kg, the value which was found toxic by Kahle *et al.* (1989). No data have been reported for nickel, copper, and zinc.

For aluminium, Devevre (1990) found that the yellowing of young Norway spruce seedlings, the death of roots, and poor mycorrhization occurred in French and German soils containing more than 80 meq/kg (720 mg/kg) Al extracted by NH_4Cl. However, different salts used in extracting exchangeable *Al* give different results; KCl and NH_4Cl generally extract more Al than does $BaCl_2$. Also, the level of exchangeable Al must be assessed according to the exchangeable Ca and Mg contents. Ca/Al and Mg/Al ratios of 0.05 and 0.02, respectively, seem to be threshold values below which adult coniferous trees such as *Picea abies* or *Abies pectinate* may suffer from aluminium toxicity (Mohamed, 1992; Bonneau *et al.*, 1992). The corresponding values for young spruce plantations, where roots are less developed, are 0.15 and 0.04 (Bonneau *et al.*, 1992).

The heterogeneous distribution of heavy metals in a forest stand must be ascertained according to the distance of the sampling point from the tree stems, since heavy metals are leached from tree crowns and concentrated in the vicinity of the trunks (Kazda and Glatzel, 1984).

9.3.3 SOIL SOLUTION STUDIES

Concentrations of cadmium, copper, nickel, lead, and zinc in soil solution have been measured (mg/L) and reported (Germany, Solling: Godbold and Huttermann, 1985; USA, Turkey Lake, drainage water: Foster and Nicolson, 1986; USA, Northern Vermont, drainage water: Friedland and Johnson, 1985). However, some studies were conducted on freely drained water. Heavy metal concentrations should be determined in micropore water extracted by centrifugation or by displacement, i.e., applying a water flux at low pressure on a soil column, because tree uptake occurs mainly from this water that is representative of risks of injury. Other works have attempted to determine levels of heavy metals corresponding to toxicity thresholds (Godbold and Huttermann, 1985; Burton *et. al.*, 1986; Chappelka *et al.*, 1991; Lozano and Morrison, 1982; Clarke and Brennan, 1980; Russo and Brennan, 1979; Mitchell and Fretz, 1977; Smith and Brennan, 1984). Results are often incomparable, because many were obtained by adding nutrient solutions with varied heavy metal concentrations to plants grown in soils or with non-mycorrhizal seedlings. Realistic toxicity thresholds can only be obtained from experiments where nutrition solutions are applied to sand cultures. Toxic element absorption by soil colloids and the reduction of their concentration in the solution can hence been avoided. It is also necessary to experiment with mycorrhizal seedlings as indicated for available elements. Threshold levels greatly vary according to tree species and to elemental interactions; for instance Pb toxicity is significantly increased by the addition of 2 ppm Cd (Kahle *et al.*, 1989). Toxicity thresholds are of the order of 0.5–5 mg/L Cd, 10–100 mg/L Cu, 10–50 mg/L Ni, more than 200 mg/L Pb and 0.4–2 mg/L Zn. Such levels are far from being reached in soil solutions for Cd, Cu, and Pb, but might be reached for Zn.

For aluminium, toxicity levels seem to be controversial, and vary according to tree species. In species well adapted to acid soils, such as *Picea abies,* root growth is hampered only by high concentrations: 800–1200 micromoles/L (27 mg/L), while Ca uptake is depressed by lower concentrations (100 micromoles/L, i.e. 2.7 mg/L). Sugar maple is much more sensitive: its growth is depressed by about 3 mg/L; such concentrations are rare in forest soils. In coniferous forests in the Vosges, a concentration of 4 mg/L was rarely found in drainage water (Gras *et al.*, 1989). Under usual conditions, the concentrations are in the order of 1–3 mg/L in drainage and in micropore water (extracted by the displacement technique). As for the exchangeable elements, the Ca:Al and Mg:Al ratios should be taken into account. The value of *Ca*:*Al* equivalent ratios corresponding to toxicity varies from 0.06–0.12 in very tolerant species (*Fagus sylvatica, Pseudotsuga menziesii*) to 0.6 in rather tolerant species (*Picea abies*) (1 in molar ratios) (Ulrich, 1984). An *Mg*:*Al* equivalent ratio of 0.7 in micropore water in the Vosges corresponds to a very healthy spruce stand, while values of 0.3–0.5 correspond to yellow, Mg-deficient stands (Mohamed, 1992; Ranger *et al.*, 1993).

When comparing metal soil solution concentrations to toxicity thresholds, only ionic forms of metals are toxic, while metals complexed in organic molecules are not. The extracted soil solution must be percolated through exchange resins to assess toxicity (Schierl, 1989). The composition of soil water may be indicative of tree sensitivity to salts: very sensitive species do not tolerate more than 2 g/L NaCl, the resistant ones up to 9 g/L.

The following conclusions can be drawn from this information:

1. Aluminium's toxicity, which differs by tree species, can be assessed by determining exchangeable Al concentrations and Ca:Al and Mg:Al ratios of exchangeable elements, or Ca:Al and Mg:Al ratios in the micropore water; and only the ionic forms of Al must be considered.
2. To assess heavy metal toxicity, the concentrations of exchangeable metals in the soil can be measured and compared to toxicity levels determined in appropriate experiments, in which different quantities of heavy metal salts are added to soil samples containing seedlings supplied with mycorrhizae, and which are incubated several days. Exchangeable concentrations are then measured at the end of the experiment.

3. To assess heavy metal toxicity, the concentrations of ionic forms of heavy metals must be measured in soil micropore water in an adequate season (end of the rainy season) and compared to toxicity levels. These toxicity levels have to be determined by percolating a sand culture (of mycorrhizal seedlings) with nutrient solutions containing different concentrations of heavy metals.

9.3.4 ELEMENT CYCLING STUDIES

Soil composition data reflect a static measurement of potential toxicity. Estimating dynamic aspects of the processes and evolution of risks in the future is not possible. The stress effects of nitrogen mineral forms (NO_3^- and NH_4^+) cannot be investigated by static soil studies, since they are very variable with time, and do not accumulate in the system, (particularly NO_3^-). Thus, knowledge of the cycling of several elements (mainly protons and nitrogen, but also heavy metals) is essential to obtain valid views on the ecosystem function.

9.3.4.1 Protons

Proton cycling can be established only indirectly. The first difficulty is input determination, because of the interference of the forest canopy. Acidic dust or aerosols are deposited on forest leaves or needles and leached by the rain. Simultaneously, acidity is exchanged on the canopy with base cations (Ca, Mg, K) from the inside of the foliar tissues ("recretion"). Thus the total acidity input must be measured in the throughfall and is the sum of the H^+ and of the recreted base cations (Ca, Mg, K). The excess of the flux of the base cations in the throughfall is the sum of wet, dry, and occult deposition and of the recreted base cations (Ca, Mg, K). Thus for an exact determination of acidity, a correct estimation would be required either of dry and occult deposition or of recreted basic cations. Presently, a method to estimate canopy exchanges as a function of rainfall quantity and composition does not exist. Several methods of estimating dry and occult deposition of cations exist (Lovett and Lindberg, 1984; Mayer and Ulrich, 1974, 1978), but none is totally satisfactory. Thus, the acidity added to the ecosystem cannot be estimated exactly.

Global effects of the acidity flux in the soil may be estimated from composition of the drainage water and the fluxes of elements at the base of the soil. If only basic cations (Na, Ca, Mg, K) were in the drainage water, the soil is able to buffer acidity input by weathering or exchange processes. In these conditions, the ecosystem may be considered to be stable, but its "acid neutralizing capacity" (ANC) is consumed little by little, and risks of acidification and release of aluminium or heavy metals may arise later (Van Breemen *et al.*, 1984). If the drainage water contains high quantities of H^+ and Al^{3+} (only in the case of acid soils), the soil is subjected to excessive stress; the acid input is too high compared to the quantity of exchangeable cations and to quantities and weathering rates of the soil minerals; the uptake of nutrients is difficult, and the toxicity of aluminium and heavy metals may ensue.

A dynamic estimate of the acid stress may be achieved by models that calculate the critical load for the ecosystem. The "critical load" of a soil is defined as the flux of acidity acceptable without ionic aluminum being in too high a concentration in the drainage water. Several models exist that take into account quantity, nature, size, and weathering rates of minerals in the soil, the quantity of exchangeable cations and the rate of their exchangeability against protons, and the potential of hydroxides for fixing SO_4^{2-} anions, a process that consumes protons (Sverdrup *et al.*, 1988; Hettelingh and de Vries, 1991; Henriksen *et al.*, 1990; Hettelingh *et al.*, 1991). Comparison of the total acidity to which a soil is submitted with a calculated critical load may give a fairly accurate idea of the acidity stress, but to calculate the total acidity to which a soil is submitted, the external proton sources (estimated in the throughfall as indicated above) and the internal sources (uptake of ions by roots and acidity coming from litterfall, humus, root, and micro-organism decomposition and from nitrogen cycling) must be added (Van Breemen *et al.*, 1984).

The total acidity to which a soil is submitted may also be estimated as the sum of the total output of basic cations in excess of the input, plus lack of SO_4^{2-} in comparison with SO_4^{2-} input, (if drainage of SO_4^{2-} is higher than input this difference must be put in the proton sources), plus the protons and aluminum in the drainage water plus the difference ($C^+ - A^-$) of the cations and anions immobilized annually in wood. This sum is called "the proton sinks." Comparing all the proton sinks and all the proton sources estimated as above is called the "proton balance." This balance, theoretically equalling zero, allows verification of the estimate of the proton sources and more accurate comparisons between the estimated critical load and the total proton sources.

9.3.4.2 Nitrogen

When the nitrogen supply to forest trees is in excess, mineral nitrogen is a stress factor. Independent of the fact that NO_3^- anions are often distributed to the forest ecosystems together with a proton, an excessive N uptake creates an imbalance in nutrient composition, and induces deficiencies in P, Ca, and Mg (Roelofs and Van Dijk, 1986; Nys, 1989) that may be detected easily by leaf or needle analysis. When N is brought in the form of NH_4^+, its nitrification (that may be very active even in acid soils) supplies protons to soil and increases the acidification rate. The proton sources linked to nitrogen cycling may be calculated from input and output fluxes of NH_4^+ and NO_3^- using the formula (Van Breemen *et al.*, 1984):

$$(NH_3^- \textit{ output} - NH_3^- \textit{ input}) - (NH_4^+ \textit{ output} - NH_4^+ \textit{ input})$$

Nitrogen cycling specialists hypothesize that, in a sound forest ecosystem, nearly no nitrogen loss occurs in the drainage water (Schulze, 1987). When this latter contains fairly large quantities of nitrogen, namely in the form of nitrate, stress from an excessive nitrogen supply or from other factors that depress stand health and reduce N uptake below the normal value takes place. In addition, such a high N drainage may cause problems in ground or surface water.

Comparing N output to its permits judgement as to the origin of the excess supply of N. N input of 10 to 15 kg/ha–yr and N drainage of about 10 kg/ha–yr may be considered normal. An N drainage much lower than that input means that N brought into the ecosystem from outside is consumed by trees, plants, or micro-organisms. When N drainage is as high as, or higher than, N input, the ecosystem is called "nitrogen saturated." When N drainage is higher than its input, the quantity of N mineralised in the soil is higher than uptake by trees or other living organisms. The causes of such a situation may be an unhealthy state for the forest, leading to restricted uptake or, if the stand is healthy, mineralization in the soil of excess organic nitrogen coming from an accumulation in a previous ecosystem. Feger *et al.* (1992) give examples in the Black Forest, Germany, of excessive organic nitrogen mineralization in a spruce stand (Schluchsee) following a beech forest of which the root system was deeper, and of an unsaturated spruce ecosystem (Viflingen) where the total N input is consumed—a phenomenon that tends to compensate for a previous excess of N exportation by litter grubbing. Nys (1987, 1989) gives an example of a spruce stand in the French Ardennes where an excessive N drainage is linked to high N input and where the spruce stand suffers severe Mg and Ca deficiencies and severe defoliation.

9.3.4.3 Heavy metals

The drainage and input may also be compared for heavy metals. In most current forest conditions, inputs exceed outputs, mainly for Pb which is absorbed strongly in soils. Zottl (1985) has shown the input and drainage of several heavy metals in the Black Forest in Germany. Such comparisons demonstrate the rate of accumulation of these metals in forest soils and the risk of a short-term excess. For instance, Friedland and Johnson (1985) think that Pb quantities may be doubled in 40 years in the forest soils of Vermont.

The input–output balances of acidity, nitrogen, or heavy metals can be made at soil or at watershed level. If the functioning and the health of the forest ecosystem is considered, the soil level is the only valid one. If the impact of forest condition on aquatic ecosystems is paramount, the watershed level has to be taken into account together with the soil level, because many phenomena may occur in subsoil layers: nitrate denitrification, weathering of minerals in the regolith, and the resaturation of water by basic cations together with the precipitation of aluminum and heavy metals.

9.3.5 FLORA, FAUNA, MICROFLORA SPECIES AND THEIR COMPOSITION

9.3.5.1 Species composition

Species composition may indicate the stresses that a forest ecosystem is undergoing. For instance, an abundance of N-liking species in the ground flora of an acid forest ecosystem indicates an excess of nitrogen and often of nitrates. Presently, such a situation is often described in a European forest (Becker *et al.*, 1992; Thimonier *et al.*, 1992). This abundance may be a result of tree defoliation that permits more light and heat to reach the soil surface and that increases the litter mineralization rate. Abundance of acidophilous species, absence of lumbricidae, abundance of collembolas and acatridae are normal in acid forests, even in a good state of health; it is difficult to interpret this as an acidification stress. Generally, the identification of stress by species composition is very uncertain, and such methods are more relevant for comparing the evolution of ecosystems over years and decades.

9.3.5.2 Chemical composition

Chemical composition of plant or animal species might be used to detect a high availability of heavy metals in the upper layers of soils. Mapping of heavy metal concentrations in the mosses *Hylocomium spendens* and *Pleurozium Schreberi* is being carried out in Europe under the supervision of Swedish researchers. High levels of metals in these mosses can localize areas with high deposition rates, but do not indicate whether the concentration of heavy metals really is a current stress agent for those ecosystems. A similar method has been used by other researchers to analyze meso- and microfauna (Zottl and Lamparski, 1981; Roth-Holzapfel *et al.*, 1992), but great attention must be paid to the choice of animal groups to be analyzed, since some are far more able than others to accumulate heavy metals, and the variability within species is high (Roth-Holzapfel *et al.*, 1992). Much advance in understanding is needed before the determination of the chemical composition of microfauna species can be used for detecting real heavy metal stress.

9.3.6 CONCLUSION

Many different ways exist to detect the effects of chemicals on forest ecosystems. At the present state of knowledge, the most efficient and simple ones seem to be analyses of exchangeable heavy metals and aluminium, soil solution analyses and cycling studies, eventually in connection with the determination of the critical loads of acid input.

9.4 MONITORING ECOSYSTEM CONDITIONS

9.4.1 INTRODUCTION

All of the investigations indicated above provide information on the state of an ecosystem at a defined time; they are difficult to interpret in terms of ecosystem evolution. Monitoring characteristics of a forest stand and properties of soil or flora and fauna composition together at defined time intervals provide opportunities to improve knowledge of toxicity thresholds and of the effects of chemicals on the long-term evolution of forest ecosystems. Many countries in the world have set up permanent assessment plots since forest decline appeared in the 1980s. Roughly three kinds of plots exist according to the intensity of monitoring.

9.4.2 THE LOWEST INTENSITY

At the lowest intensity, systematic assessment networks like the 16 × 16 km ECC network exist. In the corresponding plots, there are only annual assessments of stand health, simple analyses of soil upper horizons about every tenth year: texture, pH (water and KCl or $CaCl_2$), cation exchange capacity, exchangeable base cations, exchangeable Al and Mn at soil pH, total C and N. Determination of the concentration of the major elements in needles or leaves is foreseen every fifth year by sampling at least five trees and analysing a composite sample.

The aim of such networks is to monitor the evolution of stand health and to detect possible relationships with the state and changes in the main soil properties and in mineral nutrition. This type of network has many major drawbacks. The position of the plots is systematic, and thus they are not very representative of the regional forest types; in addition, the stands are of any age and health state. Furthermore, the analysis of only a composite soil sample does not allow any statistical comparison of properties to be made between dates.

9.4.3 INTERMEDIATE-INTENSITY OBSERVATION PLOTS

Intermediate-intensity observation plots, generally less numerous than low-intensity plots, are chosen in order to be highly representative of the regional forest types. In addition, they have to be of adult age to be observed over several decades without being young or senile, i.e., without risk of confusing age change and stress effects. Each observation plot (about 0.5 ha) must be very homogeneous concerning soil and stand characteristics.

Soil must be sampled at a minimum of twenty or twenty-five locations, and at each point with more detail than in the low-intensity network. For instance, the organic horizons have to be taken up from a defined area (about 0.1 m^2) and weighed. Weight increase over the observation period may be a sign of a decrease in the mineralization rate of the litterfall. Below the organic horizons, A_1 horizons and two or three mineral horizons must be sampled. Sampling must be carried out at defined depths to compensate for operator differences. The soil horizons must be collected at a minimum of 25 points, and sampling points must be defined according to a precise geometric configuration. A defined extrapolation must be anticipated from the initial points to other locations to avoid taking a second sample in a previously perturbed location. Each sample may be analysed individually, or a composite of five samples from five locations may be assembled. These precautions permit valid statistical comparisons of soil properties among sampling dates. To avoid site waste and because soil properties change very slowly, the time interval between two successive sampling times should be at least ten years. The flora beneath the stand has to be inventoried at the beginning of the work and every tenth year; this causes disturbances, and must be made only in a separate plot.

In these intermediate intensity plots, climatic data (temperature and precipitation) must be collected daily to distinguish climatic stresses from chemical ones.

In addition to observations of defoliation and colour changes, several other stand characteristics must be noted every few years (e.g., circumference) or yearly (e.g., date of bud burst and date of leaf fall). The determination of litterfall quantity is of great interest in deciduous stands as an index of primary production. In evergreen stands, interpreting litterfall is difficult, because it is a function of yearly foliage production together with duration of persistence.

Ideally, needle or leaf analysis should be performed yearly to permit detection of long-term trends. Foliage composition is submitted to annual fluctuations with the result that analyses at five-year intervals might mask long-term trends. Since yearly foliage sampling and analysis are somewhat expensive, the frequency of these operations requires professional judgement. If foliage is sampled yearly, damage to the tree crown must be prevented by previewing two sets of ten trees to be sampled alternatively every second year, or by sampling in an additional plot around the observation plot. The same precaution must be taken to set up water collection devices in high-intensity plots.

9.4.4 HIGH-INTENSITY OBSERVATION PLOTS

In these high intensity plots, the following are determinations to be made:

1. Collection of precipitation in an open field, as near as possible to the observation plot, and measurement of pH, NH_4^-, NO_3^-, SO_4^{2-}, Cl^-, Na^+, and CO_3H^-.
2. Collection of throughfall inside (or around, if it is a large homogeneous site) the plot and analysis of the same elements as in precipitation with the possible addition of Al and heavy metals. Throughfall must be collected in a manner representative of the conditions throughout the whole plot (i.e., below-crown and between-crown areas);

3. Collection of drainage water at the base of the soil at three points inside or around the plot and analysis of the same elements as in the throughfall with the possible addition of SO_2, O_3, and NO_x in air in or around these plots.

In France a national systematic observation network (16 km × 1 km) and the ECC systematic network (16 km × 16 km) are already functioning. A network of a hundred intermediate and high intensity observation plots, of which 25 are high intensity plots (only 15 of which have drainage water collection), is being set up. Such networks of permanent monitoring plots are very expensive; therefore, their installation must proceed with particular care.

9.5 CHEMICAL STRESSES IN FORESTS TODAY AND IN THE FUTURE

Some chemical stresses are tightly localized. An illustration is salt stress: NaCl near the highways in temperate or northern countries. After many studies of forest decline in temperate areas, direct effects of SO_2 in its gaseous form have been rather rare and localised in regions where many factories are burning coal with a high S-content without a smoke scrubbing system. Some damage by NH_3 or HF are known in areas around large chemical industries. In developed countries such chemical effects are less common after the installation of filtering systems.

The most widespread chemical stresses are acidification in temperate and tropical forests, in the former as a consequence of pollution, and in the latter as a result of the strong weathering of soil materials, with the consequent elimination of most basic cations.

Acidification of forest soils is widespread because of HNO_3 and H_2SO_4 input. While H_2SO_4 pollution is decreasing in developed countries as a result of such countervailing measures as burning low S-content fuels, adding scrubbing systems, or changing to nuclear power, HNO_3 increases with expanding automobile traffic. The total acid pollution remains reasonably constant. In several forest areas with poor parent rocks and soils, damage to trees is clear and severe (defoliation, Mg and Ca deficiencies). Other effects are still more severe in water downstream or in lakes where aquatic life is disappearing due to high aluminium concentrations. Damage due to acidification is fortunately easy to stop by spreading limestone or dolomitic limestone. Many experiments have demonstrated that such fertilization improves forest health and drainage water quality, but soil pH is increased only in the upper soil horizons, and risks are present of increasing the nitrate concentration in drainage water in temperate forests where the N organic content is generally high. However, fertilization is very expensive, because, in many areas, the helicopter is the only way of spreading fertilizers and, therefore, is very costly to apply in developing countries.

Effects of ozone are still rather low except in particular regions (California), where a combination of high quantities of emitted NO_x from auto traffic and a sunny climate favours ozone synthesis. In temperate areas, this effect is equivocal.

Claustres (1992) in an open top chamber experiment in the South-West of France found a 25 percent depression in *Picea abies* by a summer ozone content of air of 90–100 μg/m^3. Many Mediterranean regions, with climate conditions like in California, have not been well investigated for this effect; in these regions, increased auto traffic eventually may increase ozone stress.

Except for narrow areas near industrial regions, heavy-metals are not yet a problem for tree health. However, as many heavy metals are retained in the soil and acidification favours the liberation of certain of these metals (Cd, Zn) in soil solution, this risk is increasing gradually (Kahle and Breckle, 1992). Liming forest soils may diminish the free ion concentration for several metals, but increased organic complexation may increase the concentration of others (Cu, Pb) in drainage water, and contribute to the load in the waters downstream. In addition to the health of forest trees, heavy metals, particularly Cd, Pb, and Hg in wild fruits and mushrooms, raise concerns for human health. These metals may become increasingly important in some forests with increasing acidification and spreading of sewage sludge.

With the progressive reduction of SO_2 emissions in developed countries and the possibility of liming the most acidified forests, the major problems for these countries eventually will become those effects produced by heavy metals and ozone (the latter, mainly in Mediterranean regions), while in tropical areas of developing countries, soil acidification with aluminum toxicity and with difficulty in Ca, Mg and K nutrition may eventually alter forest production along with the many other threats that exist in the forests of these regions.

9.6 REFERENCES

Adams, C.M., Dengler, N.G., and Hutchinson, T.C. (1984) Acid rain effects on foliar histology of *Artendsia tilesii*. *Can. J. Bot.* **62**, 463–474.

Athanassious, R., Klyne, M.A., and Phan, C.T. (1978) Ozone effects on radish (*Raphanus sativus* L. cv. Cherry Belle): morphological and cellular damage. *Z. Pflanzenphysiol.* **90**, 183–187.

Athanassious, R. (1980) Ozone effects on radish (*Raphanus sativus* L. cv. Cherry Belle): gradient of ultrastructural changes. *Z. Pflanzenphysiol.* **97**, 227–232.

Back, J., and Huttunen, S. (1992) Structural responses of needles of conifer seedlings to acid rain treatment. *New Phytol.* **120**, 77–88.

Badot, P.M., Garrec, J.P., Millet, B., Badot, M.J., and Mercier, J. (1988) Deperissement et etat hydrique des aiguifles chez le *Picea abies*. *Can. J. Bot.* **66**, 1693–1701.

Barsig, M., Endler, W., Weese, G., and Hafner, L. (1990) Cytomorphologische Untersuchungen an Nadeln von Kieferm (*Pinus silvestris* L.) eines Ballungsgebietes. II. Das Spektrum feinstruktureller Schadphanomene. *Angew. Bot.* **64**, 303–315.

Becker, M., Bonneau, M., and Le Tacon, F. (1992) Long–term vegetation changes in an *Abies alba* forest: natural development compared with the response to fertilization. *J. Veg. Sci.* **3**, 467–474.

Black, V.J. (1985) SO_2 effects on stomatal behavious. In: Winner, W.E., Mooney, H.A., and Goldstein, R.A. (Eds.) *Sulfur Dioxide and Vegetation: Physiology, Ecology, and Policy*

Issues, pp. 96–117. Stanford University Press, Palo Alto, California.

Bonneau, M., Landmann, G., and Adrian, M. (1992) La fertilisation comme remede au deperissement des forets en sol acide: essais dans les Vosges. *Rev. For. Fr.* **44**(3), 207–223.

Bosch, C., Pfannkuch, E., Baum, U., and Rehfuess, K.E. (1983) Uber die Erkrankung der Fichte *(Picea abies* Karst.) in den Hochlagen des Bayerischen Waldes. *Forstw. Centralbl.* **102**, 167–181.

Brennan, E. (1988) Winter spot: a common symptom on coniferous evergreens in New Jersey. *Shade Tree* **61**, 44–45.

Brown, M.T., and Wilkins, D.A. (1985) Zinc tolerance of mycorrhizal Betula. *New Phytol.* **99**, 101–106.

Bucher, J.B., and Landolt, W. (1985) Zur Diagnose von Ozonsymptomen auf Waldbaumen. *Schweiz. Z. Forstwes.* **136**, 863–865.

Burton, K.W., Morgan, E., and Roig, A. (1984) The influence of heavy metals upon the growth of Sitka spruce in South Wales forests. *Plant and Soil* **78**, 271–282.

Burton, K.W., Morgan, E., and Roig, A. (1986) Interactive effects of cadmium, copper and nickel on the growth of Sitka spruce and studies of metal uptake from nutrient solutions. *New Phytol.* **103**, 537–549.

Bytnerowicz, A., and Grulke, N.E. (1992) Physiological effects of air pollutants on western trees. In: Olson, R.K., Binkley, D., and Bohm, M. (Eds.) *The Response of Western Forests to Air Pollution*, pp. 183–232. Springer-Verlag, New York.

Cape, J.N., Paterson, I.S., Wellburn, A.R., Wolfenden, J., Mehlhorn, H., Freer-Smith, P.H., and Fink, S. (1988) *Early Diagnosis of Forest Decline.* Institute of Terr. Ecology, Edinburgh, 68 pp.

Carlson, C.E., and Gilligan, C.J. (1983) *Histological Differentiation among Abiotic Causes of Conifer Needle Necrosis.* USFS Research Paper No. INT–298. US Department of Agriculture, Forest Service, Washington, D.C., 17 pp.

Castillo, F., and Greppin, H. (1988) Extracellular ascorbic acid and enzyme activities related to ascorbic acid metabolism in *Sedum album* L. leaves after ozone exposure. *Environ. Exp. Bot.* **28**, 231–238.

Castillo, F., Miller, P., and Greppin, H. (1987) Waldsterben: Extracellular biochemical markers of photochemical oxidant air pollution damage to Norway spruce. *Experiential* **43**, 111–115.

Chabot, J.F., and Chabot, B.F. (1975) Developmental and seasonal patterns of mesophyll ultrastructure in *Abies* balsamea. *Can J. Bot.* **53**, 295–304.

Chakrabarti, K., Rennenberg, H., and Polle, A. (1991) Interaction of plants and atmospheric hydrogen peroxide. In: Borren, P., Borrell, P.M., and Seiler, W. (Eds.) *Proceedings of EUROTRAC Symposium, 1990*, pp. 119–121. SPB Acad. Publ., The Hague.

Chappelka, A.H., Kush, J.S., Runion, G.B., Meier, S., and Kelley, W.D. (1991) Effects of soil–applied lead on seedling growth and ectomycorrhizal colonisation of Loblolly Pine. *Environ. Pollut.* **72**, 307–316.

Clarke, B.B., and Brennan, E. (1980) Evidence for a cadmium and ozone interaction on *Populus tremuloides. J. Arboric.* **6**, 130–134.

Claustres, J.P. (1992) Influence de la Pollution Atmospherique sur les Echanges Gazeux des Epiceas. Resultats d'une Nouvelle Methodologie. These de Doctorat, Universite de Nancy I, Specialite "Biologie Vegetale et Forestiere," 96 pp.

Costonis, A.C. (1970) Acute foliar injury of Eastern White Pine induced by sulfur dioxide and ozone. *Phytopathology* **60**, 994–999.

Costonis, A.C., and Sinclair, W.A. (1969) Relationship of atmospheric ozone to needle blight of Eastern White Pine. *Phytopathology* **59**, 1566–1574.

Cram, W.J. (1990) Uptake and transport of sulfate. In: Rennenberg, H., Brunold, C., DeKok, L.J., and Stulen, I. (Eds.) *Sulfur Nutrition and Sulfur Assimilation in Higher Plants. Fundamental, Environmental and Agricultural Aspects*, pp. 3–11. SPB Acad. Publ., The Hague.

Crang, R.E., and McQuattie, C.J. (1986) Qualitative and quantitative effects of acid misting and two air pollutants on fohar structures of *Liriodendron tulipifera*. *Can. J. Bot.* **64**, 1237–1243.

Crossley, A., and Fowler, D. (1986) The weathering of Scots pine epicuticular wax in polluted and clean air. *New Phytol.* **103**, 207–218.

Cunninghame, M.E., Hillman, J.R., and Bowes, B.G. (1982) Ultrastructural changes in mesophyl cells of *Larix decidua* × *kaempferi* during leaf maturing and senescence. *Flora* **172**, 161–172.

D'Agostino, P.S. (1981) An ultrastructural study of the senescence induced in *Gonmphrena globosa* leaves by mineral deficiency. *J. Submicrosc. Cytol.* **13**, 373–384.

Daniker, A. (1923) Biologische Studien uber Baum- und Waldgrenze. Vierteljahrschr. *Naturforsch. Ges. Zurich* **68**, 1–102.

DeKok, L.J. (1990) Sulfur metabolism in plants exposed to atmospheric sulfur. In: Rennenberg, H., Brunold, Ch., DeKok, L.J., and Stulen, I. (Eds.) *Sulfur Nutrition and Sulfur Assimilation in Higher Plants. Fundamental, Environmental and Agricultural Aspects*, pp. 111–130. SPB Acad. Publ., The Hague.

Denny, H.J., and Wilkins, D.A. (1987) Zinc tolerance in *Betula* spp. I: effect of external concentrations of zinc on growth and uptake. *New Phytol.* **106**, 517–524.

Devevre, O. (1990) Mise en Evidence Experimentale d'une Microflore Rhizosphe–Rique Deletere Associee au Depdrissement de I'Epicda en France et en Allemagne. DEA de Biologic Forestiere. Universite de Nancy, I/INRA, Dijon, France, 65 pp.

Dodge, J.D. (1970) Changes in chloroplast fine structure during the autumnal senescence of *Betula* leaves. *Ann. Bot. N.S.* **34**, 817–824.

Ebel, B., Rosenkranz, J., Schiffgens, A., and Liitz, C. (1990) Cytological observations on spruce needles after prolonged treatment with ozone and acid mist. *Environ. Poll.* **64**, 323–335.

Elling, W., Huber, S.J., Bankstahl, B., and Hock, B. (1988) Atmospheric transport of atrazine: a simple device for its detection. *Environ. Pollut.* **48**, 77–82.

Ernst, D., Schraudner, M., Langebartels, C., Sanderrnann, H., Jr (1992) Ozone-induced changes of mRNA levels of b-1,3-glucanase, chitinase and "pathogen-related" protein lb in tobacco plants. *Plant Molec. Biol.* **20**, 673–682.

Evans, L.S. (1984) Botanical aspects of acidic precipitation. *Bot. Rev.* **50**, 449–490.

Evans, L.S., and Curry, T.M. (1979) Differential responses of plant foliage to simulated acid rain. *Am. J. Bot.* **66**, 953–962.

Evans, L.S., and Leonard, M.R. (1991) Histological determination of ozone injury symptoms of primary needles of giant sequoia (*Sequoiadendron giganteum* Bucch.). *New Phytol.* **117**, 557–564.

Evans, L.S., and Miller, P.R. (1972) Ozone damage to ponderosa pine: a histological and histochemical appraisal. *Am. J. Bot.* **59**, 297–304.

Evans, L.S., and Miller, P.R. (1975) Histological comparison of single and additive O_3 and SO_3 injuries to elongating pine needles. *Am. J. Bot.* **62**, 416–421.

Evans, L.S., Gmur, N.F., and DaCosta, F. (1978) Foliar response of six clones of hybrid poplar. *Phytopathology* **68**, 847–856.

Feger, K.H., Brahmer, G., and Zottl, H.W. (1992) *Projekt ARINUS. VI. Stickstoffumsatz und Auswirkungen der experimentellen Ammonsulfatgabe*, pp. 199–211. 8 Status Cofloquium des PEF. Kernforschungszentrum Karlsruhe.

Fiedler, H.J., Baronius, G., and Ehrig F. (1990) Rasterelektronenmikroskopische Untersuchungen gruner und chlorotischer Nadeln eines immissionsgeschaldigten Kiefernbestandes. *Flora* **184**, 91–101.

Fink, S. (1983) Histologische und histochemische Untersuchungen an Nadeln erkrankter Tannen und Fichten im Sudschwarzwald. *Allg. Forstz.* **38**, 660–663.

Fink, S. (1988) Histological and cytological changes caused by air pollutants and other abiotic factors. In: SchulteHostede, S., Darrall, N.M., Blank, L.W., and Wellburn, A.R. (Eds.) *Air Pollution and Plant Metabolism*, pp. 36–54. Elsevier, London, New York.

Fink, S. (1989) Pathological anatomy of conifer needles subjected to gaseous pollutants or mineral deficiencies. *Aquilo Ser. Bot.* **27**, 1–6.

Fink, S. (1990a) Structural changes in conifer needles due to Mg and K deficiency. *Fertil. Res.* **27**, 23–27.

Fink, S. (1990b) Todliche Saureflecken auf Fichtennadeln? *Allg. Forstz.* **45**, 989–992.

Fink, S. (1991a) The micromorphological distribution of bound calcium in the needles of Norway spruce, *Picea abies* (L.) Karst. *New Phytol.* **119**, 33–40.

Fink, S. (1991b) Unusual patterns in the distribution of calcium oxalate in spruce needles and their possible relationships to the impact of pollutants. *New Phytol.* **119**, 41–51.

Fink, S. (1991c) Comparative microscopical studies on the patterns of calcium oxalate distribution in the needles of various conifer species. *Bot. Acta* **104**, 306–315.

Fischer, E.S., and Bussler, W. (1988) Effects of magnesium deficiency on carbohydrates in *Phaseolus vulgaris. Z.* Pflanzenernahr. *Bodenk.* **151**, 295–298.

Forschner, W., Schmitt, V., and Wild, A. (1989) Investigations on the starch content and ultrastructure of spruce needles relative to the occurrence of novel forest decline. *Bot. Acta* **102**, 208–221.

Foster, N.W., and Nicolson, J.A. (1986) Trace elements in the hydrologic cycle of a tolerant hardwood forest. *Water Air Soil Pollut.* **31**, 501–508.

Friedland, A.J., and Johnson, A.H. (1985) Lead distribution and fluxes in a high elevation forest in northern Vermont. *J. Environ. Qual.* **14**, 332–336.

Friedland, A.J., Johnson, A.H., Siccama, T.G., and Mader, D.L. (1984) Trace metals profiles in the forest floor of New England. *Soil Sci. Soc. Am. J.* **48**, 422–425.

Gaggi, C., and Bacci, E. (1985) Accumulation of chlorinated hydrocarbon vapour in pine needles. *Chemosphere* **14**, 451–456.

Garsed, S.G. (1985) SO_2 uptake and transport. In: Winner, W.E., Mooney, H.A., and Goldstein, R.A. (Eds.) *Sulfur Dioxide and Vegetation. Physiology, Ecology and Policy Issues*, pp. 75–95. Stanford University Press, Palo Alto, California.

Geballe, G.T., Smith, W.H., and Wargo, P.H. (1990) Red spruce seedling health: an assessment of acid fog deposition and heavy metal soil contamination as interactive stress factors. *Can. J. For. Res.* **20**, 1680–1683.

Godbold, D.L., and Huttermann, A. (1985) Effect of zinc, cadmium and mercury on root elongation of *Picea abies* (Karst.) seedlings, and the significance of these metals to forest dieback. *Environ. Pollut.* **38**, 375–381.

Gras, F., Boudot, J.P., Becquer, T., Merlet, D., and Rouiller, T. (1989) Acidification et liberation de L'aluminium dans un sol forestier vosgien, role de la nitrification et des apports atmospheriques de soufre. *Journées de Travail DEFORPA (INRA, Nancy)* **4**, 841–848.

Grill, D., Pfeifhofer, H., Halbwachs, G., and Waltinger, H. (1987) Investigations on epicuticular waxes of differently damaged spruce needles. *Eur. J. For. Pathol.* **17**, 246–255.

Grill, D., Guttenberger, H., Zellnik, G., and Bermadinger, E. (1989) Reactions of plant cells on air pollution. *Phyton* **29**, 277–290.

Guth, S., and Frenzel, B. (1989) Epicuticularwachs der Tanne *(Abies alba* Mill.) und Walderkrankung. I. Die Wachsstrukur. *Angew. Botanik* **63**, 241–258.

Gwrthardt-Goerg, M.S., and Keller, T. (1987) Some effects of long–term ozone fumigation on Norway spruce. II. Epicuticular wax and stomata. *Trees* **1**, 145–150.

Hadwiger-Fangmeier, A., Fangmeier, A., and Jager, H.-J. (1992) *Ammoniak in der bodennahen Atmosphar—Emission, Immission und Auswirkungen auf terrestrische Okosysteme.* Forschungsberichte zum Forschungsprogramm des Landes Nordrhein-Westfalen "Luftverunreinigungen und Waldschalden" Nr. 28. Ministerium fur Umwelt, Raumordnung und Landwirtschaft des Landes Nordrhein–Westfalen, Dusseldorf.

Hafner, L. (1986) Zur Feinstruktur der geschadigten Kiefernnadel. *Allg. Forstz.* **41**, 1119–1121.

Hafner, L., Endler, W., Wendering, R., and Weese, G. (1989) Ultrastructural studies on the needles of damaged Scotch pine trees from forest dieback areas in West Berlin. *Aquilo Ser. Bot.* **27**, 7–14.

Haines, B.L., Jernstedt, J.A., and Neufeld, H.S. (1985) Direct foliar effects of simulated acid rain. II. Leaf surface characteristics. *New Phytol.* **99**, 407–416.

Hamzah, S.B., and Gomez, J.C. (1979) Ultrastructure of mineral deficient leaves of Hevea. I. Effects of macronutrient deficiencies. *J. Rubber Res. Inst. Malaysia* **27**, 132–142.

Hanson, G.P., and Stewart, W.S. (1970) Photochemical oxidants: effects on starch hydrolysis in leaves. *Science* **168**, 1223–1224.

Hartig, R. (1896) Uber die Einwirkung des Hutten- und Steinkohlenrauches auf die Gesundheit der Nadelwaldblume. *Forstl.-naturwiss. Z.* **5**, 245–290.

Hasemann, G., and Wild, A. (1990) The loss of structural integrity in damaged spruce needles from locations exposed to air pollution. I. Mesophyll and central cylinder. *J. Phytopathol.* **128**, 15–32.

Hecht-Buchholz, C. (1972) Wirkung der Mineralstoffernshrung auf die Feinstruktur der Pflanzenzelle. *Z. Pflanzenern. Bodenkd.* **132**, 45–69.

Hecht-Buchholz, C. (1983) Light and electron microscopic investigations of the reactions of various genotypes to nutritional disorders. *Plant Soil* **72**, 151–165.

Heinrichs, H., and Mayer, R. (1980) The role of forest vegetation in the biogeochemical cycle of heavy metals. *J. Environ. Qual.* **9**, 111–118.

Henriksen, A.J., Lien, L., and Traaen, T.S. (1990) *Critical Loads for Surface Waters. Chemical Criteria for Inputs of Strong Acids.* Report No. D 89210 NIVA. Norwegian Institute for Water Research, Oslo.

Herrick, G.T., and Friedland, A.J. (1990) Patterns of trace metal concentrations and acidity in montane forest soils of the northeastern United States. *Water Air Soil Pollut.* **53**, 151–157.

Herschbach, C., and Rennenberg, H. (1991) Influence of glutathione on sulphate influx, xylem loading and exudation in excised tobacco roots. *J. Exp. Bot.* **42**, 1021–1029.

Herterich, R., and Herrmann, R. (1990) Comparing the distribution of nitrated phenols in the atmosphere of two German hill sites. *Environ. Technol. Lett.* **11**, 961.

Hettelingh, J.P., and de Vries, W. (1991) *Mapping Vademecum.* National Institute of Public Health and Environmental Protection, Coordination Centre for Effects, Bilthoven, The Netherlands.

Hettelingh, J.P., Downing, R.J., and Smedt, P.A.M. (1991) *Mapping Critical Loads for Europe.* Technical Report No. 1. National Institute of Public Health and Environmental Protection, Coordination Centre for Effects, Bilthoven, The Netherlands, 183 pp.

Hibben, C.R. (1969) The distinction between injury to tree leaves by ozone and mesophyll-feeding leafhoppers. *For. Sci.* **15**, 154–157.

Hinckel, M., Reichl, A., Schramm, K.-W., Trautner, F., Reissinger, M., and Hutzinger, O. (1990) Concentration levels of nitrated phenols in conifer needles. *Chemosphere* **18**, 2433–2439.

Holopainen, T., and Nygren, P. (1989) Effects of potassium deficiency and simulated acid rain, alone and in combination, on the ultrastructure of Scots pine needles. *Can. J. For. Res.* **19**, 1402–1411.

Holopainen, T., Anttonen, S., Wulff, A., Palomaki, V., and Karenlampi, L. (1992) Comparative evaluation of the effects of gaseous pollutants, acidic deposition and mineral deficiencies: structural changes in the cells of forest plants. *Agric. Ecosyst. Environ.* **42**, 365–398.

Huttl, R.F., and Fink, S. (1988) Diagnostische Dungungsversuche zur Revitalisierung geschadigter Fichtenbestande *(Picea abies* Karst.) in Sudwestdeutschland. *Forstw. Centralbl.* **107**, 173–183.

Huttunen, S., Turunen, M., and Reinikainen, J. (1990) Scattered $CaSO_4$–crystallites on needle surfaces after simulated acid rain as an indicator of nutrient leaching. *Water Air Soil Pollut.* **54**, 169–173.

Jokela, A., Back, J., and Huttunen, S. (1992) Applicability of microscopic methods to forest damage research (Abstr.). In: *Proceedings of the 15th International Meeting on Air Pollution and Interactions between Organisms in Forest Ecosystems, 9–11 September, 1992*, p. 25. Tharandt, Dresden.

Jung, G., and Wild, A. (1988) Electron microscopic studies of spruce needles in connection with the occurrence of novel forest decline. I. Investigations of the mesophyll. *J. Phytopathol.* **122**, 1–12.

Kahle, H., and Breckle, S.W. (1992) Blei und Cadmium. Zeithombe in unseren Waldboden. *Biol. unserer Zeit.* **22**, 21–27.

Kahle, H., Bertels, H., Noack, G., Roder, U., Ruther, P., and Breckle, S.W. (1989) Wirkungen von Blei und Cadmium auf Wachstum und Mineralstoffhaushalt von Baehenjungwuchs. *Allg. Forstz.* **29/30**, 783–788.

Karenlampi, L. (1986) Relationships between macroscopic symptoms of injury and cell structural changes in needles of ponderosa pine exposed to air pollution in California. *Ann. BOL Fenn.* **23**, 255–264.

Karenlampi, L., and Houpis, J.L.J. (1986). Structural conditions of mesophyll cells of *Pinus ponderosa* var. *scopulorum* after SO_2 fumigation. *Can. J. For. Res.* **16**, 1381–1385.

Kazda, M., and Glatzel, G. (1984) Schwermetalleinreicherung und Schwermetallverfugbarkeit im Einsicherungsbereich von Stammablaufwasser in Buchenwalders *(Fagus sylvatica)* der Wienerwald. *Z. Pflanzenernahr. Bodenk.* **147**, 743–752.

Kelly, J.M., Parker, G.R., and McFee, W.W. (1979) Heavy metal accumulation and growth of seedlings of five forest species as influenced by soil cadmium level. *J. Environ. Qual.* **8**, 361–364.

Khristodorov, V., Kamenova-Yukhimenko, S., and Merakchiiska-Nikolova, M. (1989) Effects of pollution with cadmium (Cd^{2+}) on the survival and growth of one–year–old seedlings of Scots pine *(Pinus silvestris). Nauka-za-Gorata* **2**, 18–26.

Kloke, A. (1981) Sollen Richtwerte fur tolerierbare Schwermetallgehalte in landwirtschafdich/gartnerish genutzten Boden auch fur Forstboden gelten? *Mitteil. Forstl. Bundesversuchsanstalt Wien* **137**, 241–248.

Koval, M., Giehl, H., Matuska, P., Scharffe, D., Scheel, H.E., Schlitt, K., Seiler, W., and Werhahn, J. (1986) *Bestimmung der horizontalen und vertikalen Verteilung verschiedener Spurengase im Rahmen des TULLA-Projekts.* CEC Report. Ispra.

Kozlowski, T., and Constantinidou, H. (1986). Responses of woody plants to environmental pollution. *For. Abst.* **47**, 5–51.

Lamoureux, G.L., and Rusness, D.G. (1989) The role of glutathione and glutathione S-transferases in pesticide metabolism, selectivity, and mode of action in plants and insects. In: Dolphin, D., Poulson, R., and Avranovic, O. (Eds.) *Glutathione: Chemical, Biochemical and Medical Aspects*, Vol. III. *Series: Enzymes and Cofactors*, pp. 153–196. John Wiley & Sons, New York.

Lang, K.J., and Holdenrieder, O. (1985) Nekrotische Flecken an Nadeln von *Picea abies*—ein Symptom des Fichtensterbens? *Eur. J. For. Pathol.* **15**, 52–58.

Langebartels, Ch., Kemer, K., Leonardi, S., Schraudner, M., Trost, M., Heller, W., and Sandermann, H. (1990) Biochemical plant responses to ozone. I. Differential induction of polyamine and ethylene biosynthesis in tobacco. *Plant Physiol.* **95**, 882–889.

Legge, A.H., Bogner, J.C., and Krupa, S.V. (1988) Foliar sulphur species in pine: a new indicator of a forest ecosystem under air pollution stress. *Environ. Pollut.* **55**, 15–27.

Lovett, G.M., and Lindberg, S.E. (1984) Dry deposition and canopy exchange in a mixed oak forest as determined by analysis of throughfall. *J. Appl. Ecol.* **21**, 1013–1027.

Lozano, F.C., and Morrison, I.K. (1982) Growth and nutrition of white pine and white spruce seedlings in solutions of various nickel and copper concentrations. *J. Environ. Qual.* **11**, 437–441.

Luethy-Krause, B., and Landolt, W. (1990) Effects of ozone on starch accumulation in Norway spruce *(Picea abies). Trees* **4**, 107–110.

Maas, F.M., DeKok, L.J., Strik-Trimmer, W., and Kuiper, P.J.C. (1987a) Plant responses to H_2S and SO_2 fumigation. II. Differences in metabolism of H_2S and SO_2 in spinach. *Physiol. Plant.* **70**, 722–728.

Maas, F.M., DeKok, L.J., Hoffmann, I., and Kuiper, P.J.C. (1987b) Plant responses to H_2S and SO_2 fumigation. I. Effects on growth, transpiration and sulfur content of spinach. *Physiol. Plant.* **70**, 713–721.

Martin, F., and Botton, B. (1993) Nitrogen metabolism of ectomycorrhizal fungi and ectomycorrhizas. *Adv. Plant Pathol.* **9**.

Mayer, R., and Ulrich, B. (1974). Conclusions on the filtering action of forests from ecosystem analysis. *Oecol. Plant.* **9**, 157–168.

Mayer, R., and Ulrich, B. (1978) Input of atmospheric sulphur by dry and wet deposition into two central European forest ecosystems. *Atmos. Environ.* **12**, 375–377.

Mengel, K., Lutz, H.-J., and Breininger, M.-T. (1987) Auswaschung von Nahrstoffen aus jungen intakten Fichten *(Picea abies). Z. Pflanzenern. Bodenk.* **150**, 61–68.

Mengel, K., Hogrebe, A.M.R., and Esch, A. (1989) Effect of acidic fog on needle surface and water relations of *Picea abies. Physiol. Plant.* **75**, 201–207.

Meyberg, M., Lockhausen, J., and Kristen, U. (1988) Ultrastructural changes in mesophyll cells of spruce needles from a declining forest in northern Germany. *Eur. J. For. Pathol.* **18**, 169–175.

Miller, H.G., Cooper, J.M., and Miller, J.D. (1976) Effect of nitrogen supply in litter fall and crown leaching in a stand of Corsican pine. *J. Appl. Ecol.* **13**, 233–248.

Miller, P.R., and Evans, L.S. (1974) Histopathology of oxidant injury and winter fleck injury on needles of western pines. *Phytopathology* **64**, 801–806.

Mitchell, C.D., and Fretz, T.A. (1977) Cadmium and zinc toxicity in white pine, red maple, and Norway spruce. *J. Am. Soc. Hortic. Sci.* **102**, 81–84.

Mohamed, A.D. (1992) Role du Facteur Edaphique dans le Fonctionnement Biogeochimique et l'Etat de Santé de Deux Pessieres Vosgiennes. Effet d'un Amendement Calci-Magnesien. These de Doctorat, Université de Nancy, I/INRA Nancy, France, 136 pp.

Morrison, T.K., and Hogan, H.D. (1986) Trace element distribution within the tree phytomass and forest floor of a tolerant hardwood stand, Algoma, Ontario. Acid precipitation, Part 2. *Water Air Soil Pollut.* **51**, 493–500.

Nakos, G. (1979) Lead pollution. Fate of lead in the soil and its effect on *Pinus halepensis. Plant Soil* **53**, 427–443.

Nebe, W., Schierhorm, E., and Ilgen, G. (1988) Rasterelektronenmikroskopische und chemische Untersuchungen von immissionsgeschadigten Fichtennadeln (*Picea abies* [L.] Karst.). *Flora* **181**, 409–414.

Neite, H. (1989) Gehalte und Losligkeit von Blei, Zink, und Kupfer in den Boden ausgewahlter Buchenwalder Nordrhein-Westfalien. *Allg. Forstz.* **29/30**, 778–780.

Nys, C. (1987) Fonctionnement du Sol d'un Ecosysteme Forestier: Etude des Modifications Dues a la Substitution d'une Plantation d'Epicea Commun (*Picea abies*) a une Foret Feuillue Melangee des Ardennes. These de Doctorat, Université de Nancy INRA, Nancy, France, 207 pp.

Nys, C. (1989) Fertilisation, deperissement et production de l'epicea commun *(Picea abies)* dans les Ardennes. *Rev. For. Fr.* **41**, 336–347.

Ogier, G., Greppin, H., and Castillo, F. (1991) Ascorbate and guaiacol peroxidase capacities from apoplastic and cell material extracts of Norway spruce needles after long-term ozone exposure. In: Lobarzewski, J., Greppin, H., Penel, C., and Gaspar, T. (Eds.), *Molecular, Biochemical and Physiological Aspects of Plant Peroxidases*, pp. 391–400. University Press, Geneva.

Osborne, D.J. (1978) Target cells—new concepts for plant regulation in horticulture. *Sci. Hortic.* **31**, 31–43.

Osborne, D.J. (1984) Concepts of target cells in plant differentiation. *Cell Diff.* **14**, 161–169.

Paparozzi, E.T., and Tukey, H.B., Jr (1983) Developmental and anatomical changes in leaves of yellow birch and Red Kidney bean exposed to simulated acid precipitation. *J. Am. Soc. Hort. Sci.* **108**, 890–898.

Parameswaran, N., Fink, S., and Liese, W. (1985) Feinstrukturelle Untersuchungen an Nadeln geschadigter Tannen und Fichten aus Waldschadensgebieten im Schwarzwald. *Eur. J. For. Pathol.* **15**, 168–182.

Patterson, W.A., and Olson, J.J. (1983) Effects of heavy metals on radicle growth of selected woody species germinated on filter paper, mineral and organic substrates. *Can. J. For. Res.* **13**, 233–238.

Pell, E.J., and Weissberger, W.C. (1976) Histopathological characterization of ozone injury to soybean foliage. *Phytopathology* **66**, 856–861.

Percy, K.E., and Baker, E.A. (1990) Effects of simulated acid rain on epicuticular wax production, morphology, chemical composition and on cuticular membrane thickness in two clones of Sitka spruce [*Picea sitchensis* (Bong.) Carr.]. *New Phytol.* **116**, 79–87.

Percy, K.E., and Riding, R.T. (1981) Histology and histochemistry of elongating needles of *Pinus strobus* subjected to a long-duration, low-concentration exposure to sulfur dioxide. *Can. J. Bot.* **59**, 2558–2567.

Percy, K.E., Jensen, K.F., and McQuattie, C.J. (1992) Effects of ozone and acidic fog on red spruce needle epicuticular wax production, chemical composition, cuticular membrane ultrastructure and needle wettability. *New Phytol.* **122**, 71–80.

Platt-Aloia, K.A., Thomson, W.W., and Terry, N. (1983) Changes in plastid ultrastructure during iron nutrition-mediated chloroplast development. *Protoplasma* **114**, 85–92.

Polle, A., and Rennenberg, H. (1992) Field studies on Norway spruce at high altitudes: II. Defense systems against oxidative stress in needles. *New Phytol.* **121**, 635–642.

Polle, A., and Rennenberg, H. (1993a) The significance of antioxidants in plant adaptation to environmental stress. In: Mansfield, T., Fowden, L., and Stoddard, F. (Eds.) *Plant Adaptation to Environmental Stress.* James & James, London.

Polle, A., and Rennenberg, H. (1993b) Photooxidative stress in trees. In: Foyer, C., and Mullineaux, P. (Eds.) *Photooxidative Stresses on Plants: Causes and Amelioration.* CRC Press, Boca Raton, Florida.

Polle, A., Chakrabarti, K., and Rennenberg, H. (1991) Extracellular and intracellular peroxidase activities in needles of Norway spruce (*Picea abies* L.) under high elevation stress. In: Lobarzewski, J., Greppin, H., Penel, C., and Gaspar, T. (Eds.) *Molecular, Biochemical and Physiological Aspects of Plant Peroxidases*, pp. 447–453. University Press, Geneva.

Price, A., Lucas, P.W., and Lea, P.J. (1990) Age dependent damage and glutathione metabolism in ozone fumigated barley: a leaf section approach. *J. Exp. Bot.* **41**, 1309–1317.

Ranger, J., Discours, D., Mohamed, A.D., Moares, C., Dambrine, E., Merlet, D., and Rouiller, J. (1993) Comparaison des eaux liees et des eaux libres etudides dans les sols de trois pessieres vosgiennes. Application a l'etude du fonctionnement actuel de ces sols et consequences pour l'etat sanitaire des peuplements. *Ann. Sci. For.* **50**.

Reichl, A., Reissinger, M., and Hutzinger, O. (1987) Accumulation of organic air constituents by plant surfaces. III. Occurrence and distribution of atmospheric organic micropollutants in conifer needles. *Chemosphere* **16**, 2647–2652.

Rennenberg, H. (1984) The fate of excess sulfur in higher plants. *Annu. Rev. Plant Physiol.* **35**, 121–153.

Rennenberg, H. (1991) The significance of higher plants in the emission of sulfur compounds from terrestrial ecosystems. In: Sharkey, T.D., Holland, E.A., and Mooney, H.A. (Eds.) *Trace Gas Emissions by Plants*, pp. 217–265. Academic Press, San Diego, California.

Rennenberg, H., and Polle, A. (1992) Metabolic consequences of atmospheric sulphur influx into plants. In: Wellburn, A., and Alscher, R. (Eds.) *Proceedings of the Third International Symposium on Gaseous Pollutants and Plant Metabolism.* Elsevier Science, Barking, England.

Riding, R.T., and Percy, K.E. (1985) Effects of SO_2 and other air pollutants on the morphology of epicuticular waxes on needles of *Pinus strobus* and *P. banksiana*. *New Phytol.* **99**, 555–563.

Rinallo, C., Raddi, P., Gellini, R., and di Lonardo, V. (1986) Effects of simulated acid deposition on the surface structure of Norway spruce and silver fir needles. *Eur. J. For. Pathol.* **16**, 440–446.

Roberts, T.M., Skeffington, R.A., and Blank, L.W. (1989) Causes of type I spruce decline in Europe. *Forestry* **62**, 179–222.

Roelofs, J.G.M., and Van Dijk, H.F.G. (1986) The effect of airborne ammonium deposition on canopy ion-exchange in coniferous trees. In: Madiy, P. (Ed.) *Direct Effects of Dry and Wet Deposition on Forest Ecosystems, in Particular Canopy Interaction.* Air Pollution Research Report No. 4 (EEC Workshop, Lokeberg, Sweden, 1986/10/19–23), pp. 34–39.

Rosemann, D., Heller, W., and Sandermann, H. (1991) Biochemical plant responses to ozone. II. Induction of stilbene biosynthesis in Scots pine *(Pinus sylvestris* L.) seedlings. *Plant Physiol.* **97**, 1280–1286.

Roth-Holzapfel, M., Funke, W., and Rittner, P. (1992) Multielementanalysen an Invertebraten im Okosystem "Fichtenforst." 7 Statuskolloguium des PEF, Karlsruhe, Band 1, pp. 171–183.

Ruetze, M., Schmitt, U., Liese, W., and Kuppers, K. (1988) Histologische Untersuchungen an Fichtennadeln *(Picea abies* L. Karst.) nach Begasung mit SO_2, O_3 und NO_2. *Allg. Forstjagdztg.* **159**, 195–203.

Russo, F., and Brennan, E. (1980) Evidence for a cadmium and ozone interaction on *Populus tremuloides. J. Arboric.* **6**, 103–134.

Sauter, J.J., and Voss, J.-U. (1986) SEM—observations on the structural degradation of epistomatal waxes in *Picea abies* (L.) Karst.—and its possible role in the "Fichtensterben." *Eur. J. For. Pathol.* **16**, 408–423.

Schierl, R. (1989) Bestimmung des Komplexierungsgrades von Al, Fe, und Mn in Bodenlosungen durch Kationenaustausch. *Vom Wasser* **73**, 161–165.

Schierl, R., and Kreutzer, K. (1991) Einflusse von saurer Beregnung und Kalkung auf die Schwermetalldynamik im Hoglwaldexperiment. In: Kreutzer, K., and Gottlein, A. (Eds.) *Okosystemforschung Hoglwald. Beitrage zur Auswirkung von saurer Beregnung und jakkung in einem Fichtenalthbstand*, pp. 204–211. P. Parey, Hamburg.

Schmidt, E. (1936) Baumgrenzenstudien am Feldberg im Schwarzwald. *Thar. Forstl. Jahrb.* **87**, 1–43.

Schmitt, U., Liese, W., and Ruetze, M. (1986) Ultrastruktyrekke Veranderungen in gruen Nadeln geschadigter Fichten. *Angew. Bot.* **60**, 441–450.

Schmitt, U., Ruetze, M., and Liese, W. (1987) Rasterelektronenmikroskopische Untersuchungen an Stomata von Fichten-und Tannennadeln nach Begasung und saurer Beregnung. *Eur. J. For. Pathol.* **17**, 118–124.

Schraudner, M., Ernst, D., Langebartels, Ch., and Sandermann, H. (1992) Biochemical plant responses to ozone. III. Activation of the defense–related proteins b–1,3–glucanase and chitinase in tobacco leaves. *Plant Physiol.* **99**, 1321–1328.

Schroder, P., and Rennenberg, H. (1992) Characterization of glutathione S–transferase from dwarf pine needles *(Pinus mugo* Turra). *Tree Physiol.* **11**, 151–160.

Schroder, P., Lamoureux, G.L., Rusness, D.G., and Rennenberg, H. (1990a) Glutathione S-transferase activity in spruce needles. *Pest. Biochem. Physiol.* **37**, 211–218.

Schroder, P., Rusness, D.G., and Lamoureux, G.L. (1990b) Detoxification of xenobiotics in spruce trees is mediated by glutathione S-transferases. In Rennenberg, H., Brunold, Ch., De Kok, L.J., and Stulen, I. (Eds.) *Sulfur Nutrition and Sulfur Assimilation in Higher Plants. Fundamental, Environmental and Agricultural Aspects*, pp. 245–248. SPB Acad. Publ., The Hague.

Schulze, E.D. (1987) Tree responses to acid depositions into the soil. A summary of the COST workshop at Julich 1985. In: Mathy. P. (Ed.) *Air Pollution and Ecosystems. International EEC Symposium, 18–22 May, 1987, Grenoble, France*, pp. 225–241. Reidel, Dordrecht.

Schupp, R., Schatten, T., Willenbrink, J., and Rennenberg, H. (1992) Long-distance transport of reduced sulphur in spruce *(Picea abies* L.). *J. Exp. Bot.* **43**, 1243–1250.

Seiler, J.R., and Paganelli, D.J. (1987) Photosynthesis and growth response of red spruce and loblolly pine to soil applied lead and simulated acid rain. *For. Sci.* **33**, 668–675.

Smith, G.C., and Brennan, E. (1984) Response of silver maple seedlings to an acute dose of root-applied cadmium. *For. Sci.* **30**, 582–586.

Smith, I.K., Polle, A., and Rennenberg, H. (1990) Glutathione. In: Alscher, R.G., and Cumming, J.R. (Eds.) *Stress Responses in Plants: Adaptation and Acclimation Mechanisms*, pp. 201–215. Wiley & Sons, New York.

Soikkeli, S. (1981a) A review of the structural effects of air pollution on mesophyll tissue of plants at light and transmission electron microscope level. *Savonia* **4**, 11–34.

Soikkeli, S. (1981b) The types of ultrastructural injuries in conifer needles of northern industrial environments. *Silva Fennica* **15**, 399–404.

Soikkeli, S. (1981c) Comparison of cytological injuries in conifer needles from several polluted industrial environments in Finland. *Ann. Bot. Fenn.* **18**, 47–61.

Soikkeli, S., and Karenlampi, L. (1984) Cellular and ultrastrural effects. In: Treshow, M. (Ed.) *Air Pollution and Plant Life*, pp. 159–174. Wiley & Sons, Chichester, New York.

Steinnes, E., Solberg, W., Petersen, H.M., and Wrer, C.D. (1989) Heavy metal pollution by long range atmospheric transport in natural soils of southern Norway. *Water Air Soil Pollut.* **45**, 207–218.

Stewart, D., Treshow, M., and Harner, F.M. (1973) Pathological anatomy of conifer needle necrosis. *Can. J. Bot.* **51**, 983–988.

Sutinen, S. (1987a) Cytology of Norway spruce needles. I. Changes during ageing. *Eur. J. For. Pathol.* **17**, 65–73.

Sutinen, S. (1987b) Cytology of Norway spruce needles. II. Changes in yellowing spruces from the Taunus Mountains, West Germany. *Eur. J. For. Pathol.* **17**, 74–85.

Sutinen, S. (1987c) Ultrastructure of mesophyll cells of spruce needles exposed to O_3 alone or together with SO_2. *Eur. J. For. Pathol.* **17**, 362–368.

Sutinen, S., Skarby, L., Wallin, G., and Seliden, G. (1990) Long-term exposure of Norway spruce, *Picea abies* (L.) Karst., to ozone in open–top chambers. II. Effects on the ultrastructure of needles. *New Phytol.* **115**, 345–355.

Sutinen, S. (1986) Ultrastructure of mesophyll cells in and near necrotic spots on otherwise green needles of Norway spruce. *Eur. J. For. Pathol.* **16**, 379–384.

Sverdrup, H.U., and Warfvinge, P. (1988) Weathering of primary silicate minerals in the natural soil environment in relation to a chemical weathering model. *Water Air Soil Pollut.* **38**, 387–408.

Swiecki, T.J., Endress, A.G., and Taylor, O.C. (1982) The role of surface wax unsusceptibility of plants to air pollutant injury. *Can. J. Bot.* **60**, 316–319.

Takahama, U., Veljovic-Iovanovic, S., and Heber, U. (1992) Effects of the air pollutant SO_2 on leaves. Inhibition of sulfite oxidase in the apoplast by ascorbate and of apoplastic peroxidase by sulfite. *Plant Physiol.* **100**, 261–266.

Thimonier, A., Dupouey, J.L., and Timbal, J. (1992) Floristic changes in the herb-layer vegetation of a deciduous forest in the Lorraine Plain under the influence of atmospheric deposition. *For. Ecol. Manage.* **55**, 149–167.

Thoene, B., Schreder, P., Papen, H., Egger, A., and Rennenberg, H. (1991) Absorption of atmospheric NO_2 by spruce *(Picea abies* L. Karst.) trees. 1. NO_2 influx and its correlation with nitrate reduction. *New Phytol.* **117**, 575–585.

Thomson, W.W. (1975) Effects of air pollutants on plant ultrastructure. In: Mudd, J.B., and Kozlowski, T.T. (Eds.) *Responses of Plants to Air Pollution*, pp. 179–194.

Thomson, W.W., Dugger, W.M., Jr, and Palmer, R.L. (1966) Effects of ozone on the fine structure of the palisade parenchyma cells of bean leaves. *Can. J. Bot.* **44**, 1677–1682 + 6 plates.

Thomson, W.W., Nagashi, J., and Platt, K. (1974) Further observation on the effects of ozone on the ultrastructure of leaf tissue. In: Dugger, M. (Ed.) *Air Pollution Effects on Plant Growth*, pp. 83–93. ACS Symposium Series 3. American Chemical Society, Washington, D.C.

Tuomisto, H. (1988) Use of *Picea abies* needles as indicators of air pollution: Epicuticular wax morphology. *Ann. Bot. Fenn.* **25**, 351–364.

Turner, R.R., Bogle, M.A., and Baes, C.F., III. (1985) *Survey of Lead in Vegetation, Forest Floor and Soils of the Great Smoky Mountains National Park.* Report No. ORNL/TM 9416, Publication No. 2425. Oak Ridge National Laboratory, Environmental Sciences Division, Oak Ridge, Tennessee.

Turunen, M., and Huttunen, S. (1989) A review of the response of epicuticular wax of conifer needles to air pollution. *J. Environ. Qual.* **19**, 35–45.

Turunen, M., and Huttunen, S. (1991) Effect of simulated acid rain on the epicuticular wax of Scots pine needles under northerly conditions. *Can. J. Bot.* **69**, 412–419.

Ulrich, B. (1984) Interaction of indirect and direct effects of our pollutants in forests. In: Troyanowsky, C. (Ed.) *Air Pollution and Plants. Second European Conference on Chemistry and the Environment, 21–24 May, 1984, Lindau, FRG*, pp. 148–181. Federation of European Chemical Societies/VCH Verlagsgesellschaft.

Van Breemen, N., Driscoll, C.T., and Mulder, J. (1984) Acidic deposition and internal proton sources in acidification of soils and waters. *Nature* **307**, 599–604.

Van Gardeningen, P.R., Grace, J., and Jeffree, C.E. (1991) Abrasive damage by wind to the needle surfaces of *Picea sitchensis* (Bong.) Carr. and *Pinus sylvestris* L. *Plant Cell Environ.* **14**, 185–193.

Vapaavuori, E.M, Korpilathi, E., and Nurmi, A.H. (1984) Photosynthetic rate in willow leaves during water stress and changes in the chloroplast ultrastructure, with special reference to crystal inclusions. *J. Exp. Biol.* **35**, 306–321.

Wellburn, A.R. (1985) SO_2 effects on stromal and thylakoid function. In: Winner, W.E., Mooney, H.A., and Goldstein, R.A. (Eds.) *Sulfur Dioxide and Vegetation. Physiology, Ecology and Policy Issues*, pp. 133–147. Stanford University Press, Palo Alto, California.

Wellburn, A.R. (1990) Why are atmospheric oxides of nitrogen usually phytotoxic and not alternative fertilizers? *New Phytol.* **115**, 395–429.

Wellburn, A.R., Majernik, O., and Wellburn, F.A.M. (1972) Effects of SO_2 and NO_2 polluted air upon the ultrastructure of chloroplasts. *Environ. Pollut.* **3**, 37–49.

Whatley, J.M. (1971) Ultrastructural changes in chloroplasts of *Phaseolus vulgaris* during development under conditions of nutrient deficiency. *New Phytol.* **70**, 725–742.

Whitney, R.D., and Ip, D.W. (1991) Necrotic spots induced by simulated acid rain on needles of *Abies balsamea* saplings. *Eur. J. For. Pathol.* **21**, 36–48.

Winner, W.E., Mooney, H.A., Williams, K., and von Caemmerer, S. (1985) Measuring and assessing SO_2 effects on photosynthesis and plant growth. In: Winner, W.E., Mooney, H.A., Goldstein, R.A. (Eds.) *Sulfur Dioxide and Vegetation. Physiology, Ecology and*

Policy Issues, pp. 118–132. Stanford University Press, Palo Alto, California.

Wittig, R., and Neite, H. (1989) Distribution of lead in the soils of *Fagus sylvatica* forests in Europe. In: Ozturk, M.A. (Ed.) *Plant and Pollutants in Developed and Developing Countries*, pp. 199–206. Izmir, Turkey.

Wolfenden, J., Pearson, M., and Francis, J. (1991) Effects of over-winter fumigation with sulphur and nitrogen oxides on biochemical parameters and spring growth in red spruce (*Pices rubens* Sarg.). *Plant Cell Environ.* **14**, 35–45.

Ylimartino, A., Paakonen, E., and Holopainen, T. (1992) Foliar nutrient concentrations and epicuticular wax formation of Scots pine needles. (Abstr.) In: *Proceedings of the 15th International Meeting on Air Pollution and Interactions between Organisms in Forest Ecosystems, 9–11 September 1992*, p. 81. Tharandt, Dresden.

Zellnik, G., and Gailhofer, M. (1989) Feinstruktur der Chloroplasten von *Picea abies* verschiedener Standorte im Hohenprofil "Zillertal." *Phyton* **29**, 147–161.

Zobel, A., and Nighswander, J.E. (1991) Accumulation of phenolic compounds in the necrotic areas of Austrian and red pine needles after spraying with sulphuric acid: a possible bioindicator of air pollution. *New Phytol.* **117**, 565–574.

Zottl, H.N. (1985) Heavy metal levels and cycling in forest ecosystems. *Experientia* **41**, 1104–1113.

Zottl, H.W., and Lamparski, F. (1981) Schwermetalle (Pb, Cd) in der Bodenfauna des Sudschwarzwaldes. *Mitt. dt. Bodenkdl. Ges.* **32**, 509–518.

10 Methods to Assess the Effects of Chemicals on Arid and Semi-Arid Ecosystems

David Mouat, Amos Banin, and Bruce Jones
Desert Research Institute, USA

10.1 INTRODUCTION

Arid ecosystems possess a number of characteristics that form the basis for their response to natural and anthropogenic stress. Of these characteristics, none is more dominant than precipitation and water availability (MacMahon, 1981; Smith and Morton, 1990). Low annual precipitation (generally less than 25 cm/year) results in plant communities typically dominated by shrubs and grasses. Furthermore, spatial and temporal variability in precipitation and soil moisture result in heterogeneity in plant and animal community structures. Desert soils are generally infertile and nutrients tend to accumulate under shrubs rather than between them (Smith and Morton, 1990; McAuliffe, 1988).

Arid ecosystems are remarkably stable when compared to other ecosystems. Succession often appears non-existent or is difficult to document. Although little species turnover may exist over large spatial scales, considerable changes can be observed in species dominance within a local area. Furthermore, the rate and degree of natural and anthropogenic disturbance can have an effect on species turnover and succession in arid ecosystems (McAuliffe, 1988; Turner, 1990; Webb *et al.*, 1987).

Overexploitation of arid land resources, combined with low and erratic precipitation and relatively infertile soils, have led to severe land degradation and reduced productivity. Loss or decline in potential or actual biological productivity (desertification) is one of the most significant global environmental issues facing mankind (Speth, 1988). Because of low and erratic precipitation and relatively infertile soils, arid ecosystems recover very slowly following disturbance (Webb *et al.*, 1987).

Several human activities have affected, and continue to affect, the condition of arid ecosystems. These include grazing, land conversion (e.g., to agriculture), water development, salinization, recreation (off-road vehicle use), urbanization, and mining. Grazing by domestic livestock has had a dramatic effect on plant species

Methods to Assess the Effects of Chemicals on Ecosystems
Edited by R. A. Linthurst, P. Bourdeau, and R. G. Tardiff

composition; overgrazing often reduces species richness and increases aridity (e.g., decreases soil moisture), soil erosion, and invasion of exotic plant species (UCAR, 1990). Land conversion typically reduces the amount of native habitats and often results in development of limited water supplies. Agricultural conversions lead to the development of a monoculture, increases in soil bulk density, and decreases in infiltration. These changes may lead to increased soil salinity that can eventually preclude plant growth. Additionally, increased salinity combined with mobilized toxic trace elements from agricultural pesticides and fertilizers may have a dramatic effect on plant species composition (Kepner and Fox, 1991).

Rapid urbanization and development of arid regions has raised a concern about the declining air quality and its associated effects on arid environments. Of greatest concern are nitrogen oxide and sulphur dioxide emissions from automobiles and industry, and ozone. Unfortunately, existing air monitoring networks provide poor coverage of air pollutants within arid lands; therefore, concentrations and deposition rates of airborne pollutants is largely unknown. Additionally, few studies have addressed the impacts of airborne chemicals on arid ecosystems. Thompson *et al.* (1984) showed that certain desert annuals were extremely sensitive to ozone at concentrations below current standards. Hill *et al.* (1974) found that many native desert plants were highly resistant to injury from SO_2 and NO_2, although at least one plant (Indian ricegrass) showed injury at lower concentrations of SO_2 (1 ppm vs. 4 ppm) when soil moisture was increased. This observation suggests that a concentration that is safe during relatively dry years may cause injury in years with high rainfall. Dawson and Nash (1980) studied the impacts of a smelter on a desert shrub community in Arizona, and found that large shrubs were unaffected by smelter emissions but that shallow rooted plants were. They attributed lack of impact on large shrubs to their reduced number of stomata and the short length of time stomata are open (as compared to plants in more mesic environments). Impact on shallow-rooted species was attributed to chemical changes in the upper layer of soils.

Arid zones are spread over wide ranges of the world, between the latitudes 10–50°N and 10–40°S. Arid zones are particularly prominent at latitude 30°N and 30°S on account of global weather patterns. A "convection cell" in the atmosphere results from the excessive heating at the equator which causes hot air to rise. The rotation of the earth causes a downward convection of the air at around 30°N and 30°S. The rising air at the equator expands and cools, thus losing its water vapour content and forming the typical cloud cover of the equatorial regions. When the air is convected downwards near latitudes 30°N and 30°S, it becomes drier and land areas in this latitude comprise the primary desert belt of the planet.

In addition to this anticyclonic activity, arid zones may be caused by continentality (often either exacerbating or exacerbated by anticyclonic activity), topography, and cold ocean current. The one factor that they have in common is a lack of moisture in normal climate conditions (UNEP, 1992). Taken together, these areas comprise approximately 37 percent of the land area of the earth.

Numerous methods exist to determine the effects of natural and anthropogenic chemicals in arid ecosystems. This chapter is mot meant as an exhaustive treatise encompassing all chemical stressors and all methods, but rather to present an overview of the basic considerations involved both from the standpoint of the stressors and from the perspective of the techniques. Specifically, the methods will be briefly outlined for the assessment of the effects of chemicals that are largely unique to arid environments on soils, and will discuss techniques to assess land degradation as a function of climate and anthropogenic stressors.

10.2 METHODS TO ASSESS THE EFFECTS OF CHEMICALS ON SOILS

Of the desert and semi-desert areas, approximately 15×10^{-6} km^2 is characterized by a lack of leaching resulting in an accumulation of soluble salts, gypsum, carbonates, and sodification in the soils (UNFAO, 1971–1981).

The major chemical stress on arid-zone ecosystems is increased salinization in soil and water which may also lead to sodification of soils. The diminished leaching of arid zone soils combined with increased inputs of salts through urbanization and irrigation, and the limited removal of salts, results in a slowly (tens to hundreds of years) increasing salinity of soil and local water resources.

Increased salinity adversely affects plants through increased osmotic pressure and decreased water availability. Increased concentration of exchangeable sodium in soils, caused by the increased relative concentration of Na^+ in saline soil solutions, is causing soil-structure destabilization, decreased hydraulic conductivity, crust formation, and increased susceptibility to water erosion.

A well-established system of methods for salinity assessment in soils and water has been developed by the USDA Salinity Laboratory in Riverside, California. Originally, it consisted of analysing the concentration of the major soluble ions (Ca, Mg, Na, K, Cl, SO_4, HCO_3/CO_3) in soil extracts and in water. In recent years, with the advent of modern, rapid, multielemental analytical techniques such as AA and ICP, and the increased demand for the assessment of a wider range of environmental contaminants, more comprehensive elemental analysis schemes are conducted including nitrate, essential trace elements (Fe, Mn, Zn, Cu, B), and potentially toxic trace elements (e.g., Cd, Cr, Ni, As, Se, Hg).

The availability of comprehensive elemental concentration data in soil solutions and waters supplies the needed data for chemical speciation modelling in solution and prediction of precipitation–dissolution, sorption–desorption and partitioning–transport processes of elements in the soil system. Advanced computer codes using thermodynamic equilibria have been employed to fully characterize the chemistry of soils and water (Sposito and Mattigod, 1979). These are still in the developmental stage.

10.2.1 ANALYTICAL PROCEDURES FOR SOIL SOLUTIONS AND WATERS

10.2.1.1 Soil solution extraction

Since the extraction of a soil solution at the field-moisture range is a difficult task, the customary procedure is to conduct the extraction at a higher (water to soil) ratio which is determined by the soil itself to ensure appropriate representation. The saturated soil-paste (SP) is the preferred condition (Rhoades, 1982). In this method, the soil is brought to saturation with distilled water, the paste is permitted to reach chemical equilibrium and the soil solution is extracted by applying a vacuum.

10.2.1.2 Total salinity

A simple method to assess total salinity in solution has been the gravimetric method of total dissolved solids (TDSs). More recently, the measurement of electrical conductivity (EC) at standardized conditions has replaced the gravimetric method. The EC method is widely accepted for routine work due to its rapidity and good correlation with total concentration of charged ions in soil extracts and water.

10.2.1.3 Electrical conductivity of the solution

The electrical conductivity of the extract is used to estimate the following general solution concentration parameters:

1. total cation (or anion) concentration, meq/L = 10 × EC (ds/m, 25°C);
2. total salt concentration, mg/L = 640 × EC (ds/m, 25°C);
3. osmotic pressure, bars at 25°C = 0.39 × EC (ds/m, 25°C).

10.2.1.4 Analysis of ionic constituents

Several methods are available for the analysis of the ionic constituents in soil extracts. Where available, the use of ICP-AES or ICP-MS enables the rapid, simultaneous analysis of many of the major, minor, and trace components in the soil solution. Single-column ion chromatography has also been tested and optimized for the sequential simultaneous analysis of the major and minor ionic constituents in soil extracts (Nieto and Frankenberger, 1985a, 1985b).

The following compilation represents those methods routinely used in advanced field laboratories. Details of the technical steps can be obtained from the respective instrument manuals.

10.2.1.5 Soluble major cations

Atomic absorption is customarily used to analyse Ca^{2+} (0–0.4 meq/L concentration range). Mg^{2+} (0–0.1 meq/L) and K^{2+} (0–0.1 meq/L) (Rhoades, 1982). Soluble major anions include:

1. Chloride: automatic chloridometer.
2. Carbonate/bicarbonate: automatic potentiometric titrator (Rhoades, 1982).
3. Sulphate: sulphate analysis has not yet reached the level of instrumental sophistication as routinely available for most other soluble constituents in arid waters and soils. Available methods include turbidimetric analyses (Rhoades, 1982), and ion chromatography (e.g., Nieto and Frankenberger, 1985a).
4. Nitrate: a selective-ion electrode is used to measure nitrate activity potentiometrically using an expanded scale potentiometer (Keeney and Nelson, 1982).
5. Boron: spectrometric analysis at 420 nm.

10.2.1.6 Analytical procedures to assess solid phase composition

Soils of arid lands are modified by anthropogenic inputs of chemicals in much the same way as in other ecosystems. A process, somewhat unique to arid regions, is that of sodification or the increase in the proportion of exchangeable Na^+ above about 15 percent of the cation exchange capacity (CEC) of the soil. Pollution by heavy metals (e.g., Cd, Pb, Cr, Ni) is observed due to atmospheric fallout, municipal waste disposal, use of fertilizers (e.g., Cd in P-containing fertilizers), and prolonged use of reclaimed sewage effluents. Due to a general lack of leaching, low organic matter content, and high pH of arid soils, the retention of potentially toxic trace elements in the top layer of the soil is enhanced. This situation increases the probability of heavy metal introduction into the food chain or their direct ingestion by animals or humans.

10.2.1.7 Exchangeable cation composition of soils

Exchangeable ions are electrostatically bound to charged sites in the soil mineral and organic components. They are readily exchanged and displaced by excess of any neutral salt. In arid zone soils, the major contributor to the soil CEC are the mineral components, mostly clay minerals such as montmorillonite, illite, palygorskite, and kaolinite. The exchangeable ions are Ca^{2+}, Mg^{2+}, Na^{2+}, and K^+, usually appearing in that order of abundance. Increased proportion of Na^+ causes destabilization of the soil structure and adverse effects on its permeability to water.

The composition of exchangeable ions is determined after their displacement by a relatively concentrated salt solution. The displacing solution of choice is 1N ammonium acetate ($NH_4C_2H_3O_2$) (Thomas, 1982), since its cation is not part of the exchangeable ions in nature and it presents only limited interferences in the determination of the four exchangeable ions by flame photometry and/or atomic absorption spectrometry. Some dissolution of $CaCO_3$ or $CaSO_4 \cdot 2H_2O$ when present in the soil by the displacing solution, may cause erroneous estimation of exchangeable Ca^{2+}, and the method should be avoided in such soils, or calcium content be determined by the difference between the total cation exchange capacity and the sum of exchangeable (Mg + Na + K).

10.2.1.8 Extractable ("available") trace elements by DTPA

The DTPA (diethylenetriaminepentaacetic acid) method was developed to measure the available concentration of essential trace elements—Fe, Mn, Zn, and Cu. It may also be appropriate for the estimation of available Ni and Cd in polluted soils (Baker and Amacher, 1982). This method attempts to extract the labile pool of the metals in the soil which is readily available to plants growing in it. Further calibration of this method is required, since it is not an equilibrium extraction procedure and highly dependent on the conditions of extraction.

10.3 REMOTE SENSING TECHNIQUES

Arid ecosystems provide an interesting set of problems and issues with which to study the effects of environmental chemistry. A unique geochemistry exists over much of arid regions related primarily to the low rainfall and high evaporation. With clear skies and low cloud cover, remote sensing techniques might be used to identify, assess, and monitor the effects of chemical exposure in arid environments. Finally, with moisture in short supply, many arid ecosystems are near the edges of their ecological tolerances; as such, subtle changes in the environmental chemistry may bring about highly significant changes in attributes of the ecosystems.

Remote sensing systems can be used to identify, assess, characterize, map, and monitor contaminant exposure. This section is not meant to serve as a complete guide to all remote sensing systems or capabilities, but rather those that are felt to have direct use or potential use to analyze environmental stress in arid and semi-arid ecosystems. Remote sensing technologies that can be used for environmental assessment may be readily available ("off-the-shelf") or may be developmental. Clearly, the experience and knowledge of the person doing the assessment is a major factor in determining the use of a particular system.

10.4 BASIC REMOTE SENSING CONCEPTS AND PRINCIPLES

The use of remote sensing for the characterization and assessment of environmental stress is predicated on some basic assumptions and principles. The spectral properties of materials can be used not only for identification as to type, property, or composition, but also for characterization, that is an assessment of some sort of condition or quantification of properties of the material. Spectral remote sensing can be used to derive information about surface materials based on their reflectance and/or emittance behaviour. The technology can also be used to discriminate subsurface (metres to 10s of metres) phenomena provided the phenomena have affected (through capillary action, for example) the near surface environment.

The spectral reflectance and spectral emittance characteristics of vegetation, soils (and rocks), and water in different wavelength regions are a result of the chemical and physical properties of these materials. Thus, a strategy employing remote sensing for discriminating surface or near-surface phenomena must take advantage of the unique spectral manifestations of the materials, within specified wavelength regions, within certain temporal guidelines, analyzed by a specific set of techniques, and perhaps integrated with other methods or techniques.

The use of sensors to discriminate and characterize surface and subsurface materials is not restricted solely to systems operating within the electromagnetic spectrum. Sensors (e.g., radiation sniffers) that may be considered to be "direct sensors" can be considered in this context. Other considerations for a remote sensing-based strategy of the assessment of environmental chemistry in arid ecosystems include GIS (geographic information systems) integration, techniques for data analysis, and vegetation response.

10.4.1 SENSORS AND CAPABILITIES

10.4.1.1 Aerial photography

This process is used to characterize the site setting (including vegetation, drainage pattern, soils, etc.) for change detection, and for direct visual examination of contaminants (directly as in oil and chemical spills, and indirectly as in discarded barrels and soil and vegetation response). Aerial photography has the advantage of having very high spatial resolution, but limited spectral resolution.

10.4.1.2 Multispectral imagery (less than 2.4μm)

Examples include the lower spectral resolution Landsat satellite sensors (multispectral scanner, MSS, and thematic mapper, TM), the French SPOT satellite, and the Daedalus (Landsat TM simulator) airborne scanner, and the higher spectral resolution (24 channel) (hyperspectral) Geoscan airborne scanner. These available systems are useful for site characterization (the satellite systems are primarily useful for larger area site characterization) and for specific (but fairly coarse scale) soils and vegetation responses to the geochemistry. While the Landsat sensors have fairly coarse resolution (30 m and 80 m), their high temporal resolution characteristics and digital data format allows them to be quite useful for change detection studies. Thematic mapper (TM) simulators typically have the same band passes as are on the TM but are housed on aircraft. One advantage of these systems is that they allow for imaging convenient to a mission need, and that spatial resolution can be improved (sometimes considerably) over the coarser scale TM. The French SPOT (systems probatoire de l'observation de la Terre) satellite provides a critical link between the poorer spatial resolution but more appropriate spectral configuration of the Landsat MSS and TM and the very high spatial resolution of aerial photography.

10.4.1.3 Hyperspectral sensors

Hyperspectral remote sensing is predicated on the need for narrower band sensors to discriminate and characterize materials with greater precision than the coarser Landsat bandpasses. The Geoscan sensor is a 46 channel sensor (with 24 channels available for a given mission) operating in the visible, near-infrared, and thermal infrared portions of the spectrum. Bandpasses, while narrower than those of Landsat or SPOT, are nevertheless fairly coarse (20 to 40nm). Recent research describes the need for 10–20 nm for characterization of minerals in the 2.0 to 2.4μm spectral region and for 10 nm or less for discriminating vegetation spectral responses to geochemical conditions. The CASI (compact airborne spectrographic imager) is a commercially available system with very fine spectral resolution (2 nm) but unfortunately operates only in the 0.4–1.0 μm region.

10.4.1.4 Thermal infrared sensors

Thermal sensors directly measure the emitted thermal energy from objects (including the earth's surface). A variety of these sensors exist and range from broad-band temperature measuring devices to narrower-band multispectral instrumentation (e.g., the thermal capabilities of Geoscan).

10.4.1.5 Microwave systems

Passive systems can measure surface and subsurface temperature and soil moisture. Active microwave (RADAR) can differentiate surface and subsurface disturbances.

10.4.1.6 Lasers

The principle of laser remote sensing involves the projection of a narrow beam of coherent visible or near-infrared light and then measurement of the reflected radiation. Laser-induced fluorescence involves the measurement of emission spectra at specific wavelengths. The measurement of emission spectra near 690 nm and 740 nm can be diagnostic for monitoring plant stress.

10.4.1.7 Vegetation spectral response

Numerous researchers have made use of the response of vegetation to geochemical conditions to analyze the near-surface environment. Vegetation is known to respond to anomalous environmental geochemistry in three principal ways: taxonomically, structurally, and spectrally. The taxonomic response involves species of community differences in areas having anomalous environmental geochemistry. The structural response involves growth pattern (or "phenologic") and physiognomic indicators. These might include stunting, more open vegetation, or affected flowers and fruits. The spectral response may involve the spectral manner in which the previous two are manifest. Researchers have described a movement of the red edge of plant reflectance to shorter or longer wavelengths as being indicative of stress. One critical issue involves the separation of moisture stress from chemical stress. Research conducted at the Desert Research Institute, the USGS, the University of California at Davis, and elsewhere suggests that this separation is feasible.

Several other systems have been developed, which, although not currently operational, have considerable potential to assess environmental contamination. Chief among these are airborne and satellite hosted hyperspectral imaging Spectrometers. These systems (the AVIRIS is a current prototype) allow for very fine spectral resolution (2–10nm) over the 0.4–2.4μm range and allow for highly diagnostic information on surface materials.

Ground techniques are frequently used to examine vegetation affected by environmental chemistry. The use of these techniques includes various types of spectroradiometers that may either duplicate the bandpasses of the airborne and satellite sensors or, in some cases, provide considerably greater spectral detail. These spectra were acquired by a Personal Spectrometer-2 (PS-2), an instrument operating in the visible and near-infrared which can acquire spectra having a resolution of 2 nm. The top curves illustrate the first derivative of the spectra and clearly show a shift of one spectrum towards the blue from the red edge. That spectrum comes from a plant with a concentration of 1377 ppb MITC (Table 2), while the control plant (with the longer wavelength of the maximum inflection point) has a concentration of only 6 ppb MITC.

10.5 DESERTIFICATION/LAND DEGRADATION

An issue of considerable importance in arid and semi-arid environments is that involving land degradation. The stressors bringing about desertification and the response of the ecosystem to these stressors have profound implications in the natural and cultural landscape of areas affected. Methods to assess land degradation are equally diverse, ranging from ground-based soils and vegetation measurements, census and other socioeconomic analyses, and synoptic (including remote sensing) techniques.

Land degradation, or desertification, is a phenomenon involving climate, soils, flora, fauna, and humans. It is the process of change in these ecosystems which can be measured by reduced productivity of desirable plants, alterations in the biomass, and diversity of the micro and macro fauna and flora, accelerated soil deterioration, and increased hazards for human occupancy (Dregne, 1977). It may be regarded as a form of dryland ecosystem degradation because of human use as well as natural factors. The significance of the deterioration derives from its magnitude in the amount of land and numbers of people affected, the rate at which it occurs, and its implications for the future well-being of mankind. Dryland ecosystems under excessive pressure of human use or changes in land use may undergo a loss in productivity resulting in a possible inability to recover (Reining, 1978). It destroys the food-producing capacity of vast tracts of dryland areas in every continent of the world. Many millions of people living in almost 100 countries suffer its effects (UNEP, 1992).

A serious difficulty in examining land degradation in arid regions is their inherent climatic variability. For example, the more extreme arid regions may receive all of their average annual precipitation in one rainfall event (UNEP, 1992). One, two, or several years of below average rainfall may occur in a given region. The identification, monitoring, and combating of land degradation must take this inherent variability into account.

10.6 INDICATORS OF DESERTIFICATION

Changes in indicators of desertification are guides as to whether desertification is becoming more or less of a problem. An "indicator" is defined as a statistic or the presence of a phenomenon judged to carry a specific informative value. An indicator must be diagnostic of an interrelated set of phenomena. Desertification is a dynamic process. Its indicators must, therefore, also be dynamic in order to show the progress of desertification. A number of indicators of desertification are necessary in order to evaluate its effect on the land. Climatic conditions (precipitation, temperature, wind, etc.) are not included among the indicators of desertification even though they may play a considerable role. Indicators of desertification have been grouped into these categories: physical, biological, and social.

10.6.1 PHYSICAL INDICATORS

Physical indicators of land degradation include soil erosion, salinization, depth to ground water, extent and distribution of surface water, numbers and duration of dust and sand storms, presence or absence of soil crusts, amount of soil organic matter, and quality of surface runoff. A potentially valuable physical indicator of desertification is surface reflectance, or albedo. Albedo, or the degree to which light is reflected, can provide integrated information on plant cover and density, soil erosion, salinization, waterlogging, and soil moisture.

Both soil and ground water are subject to salinization, particularly where irrigation has been used on arid and subhumid land. A high evapotranspiration rate in an arid climate can result in the buildup of soil minerals. Even when water used for irrigation is not of particularly poor quality, if the evapotranspiration rate is very high the water evaporates, rather than draining off or leaching through the soil, leaving whatever minerals the water contained and causing a gradual buildup of salts or alkali. Salts and alkali within the soil are also raised to the surface through capillary action. Methods to assess salinization are provided in a separate section within this chapter.

Degradation of soil, involving a number of soil physical and chemical properties, accompanies these changes (these include a decrease in soil permeability, porosity, depth of penetration of water, and amount of available water). At an advanced stage, saline and alkali deposits are visible as white, grey, or black patches. These surface manifestations are clearly visible from airborne and satellite-hosted sensors and, in fact, may be discriminated at an early stage through the use of these types of techniques.

10.6.1.1 Biological indicators

Biological indicators of land degradation include changes in structural, functional, and compositional diversity and impoverishment at multiple levels of biological organization (i.e. genetic, population, community, ecosystem, and regional landscape). Specifically, these include species diversity, productivity, cover (including leaf area index), above-ground biomass, absorbed photosynthetically active radiation, yield, and other measurements.

Schlesinger *et al.* (1990) have examined desertification through changes in ecosystem function and within the context of spatial and temporal distribution of soil resources relative to vegetation. They have found, through research conducted in the Jornada Experimental Range in southern New Mexico, that when net, long-term desertification of productive grasslands occurs, a relatively uniform distribution of water, nitrogen, and other soil resources is replaced by an increase in their spatial and temporal heterogeneity. This heterogeneity leads to the invasion of grasslands by shrubs. Invasion by shrubs must be considered in part due to the absence of wildfire. In these new plant communities, the soil resources are concentrated under shrubs, while wind and water remove materials from intershrub spaces and transport soil materials to new positions in the landscape.

Changes in vegetation cover (and leaf area index) are also indicative of desertification. Vegetation cover is strongly influenced, in rangelands, by the location of surface water sources. A tendency to overgraze around water sources can itself become a focus of desertification.

10.6.1.2 Social indicators

Social indicators of desertification are those that are related to human occupancy of areas subject to desertification. They include changes in land use and water use (including irrigation, dryland agriculture, and pastoralism), settlement patterns (including diversification of settlement and abandonment), human biological parameters (such as health indices), and social process parameters (such as migration, redistribution, and marginalization). Socioeconomic indicators applicable to people living in areas undergoing desertification may not be uniquely related to desertification. These indicators are general, and are typically used to examine the behaviour of people whose actions may lead to desertification or may be responding to its effects. To the extent that desertification is anthropogenic, i.e., caused by the impact of man on the semi-arid environments, social indicators may also serve as early warning signals (adapted from Reining, 1978).

10.7 MEASUREMENT OF DESERTIFICATION

Several written accounts and photographic records of the western United States illustrate a pattern that is nearly ubiquitous for the region: the vegetation and other land cover of the region existing in the late nineteenth century is considerably different from that which has developed since (Bahre, 1991; Hastings and Turner, 1965; Humphrey, 1956). A quantitative baseline, however, is absent. Various resource inventories may be used to establish such a baseline under present conditions.

The US National Cooperative Soil Survey (NCSS) provides a mechanism to assess certain issues related to desertification. The revised universal soil loss equation, water erosion prediction project, and the wind erosion prediction models have been developed to predict the amount of soil erosion that will occur on a given soil under various climatic environments and management practices. These models use a soil survey database to help pinpoint soil degradation through erosion.

Soil surveys and ecological site inventories provide excellent baseline data for site specific and management unit analyses. The information, however, is more difficult to aggregate for more regional analyses of desertification and climate change processes, and replications of the inventories to detect changes over time are expensive. Statistical grid samples represent one solution, but are still time consuming and costly. The use of remote sensing and other measurement techniques in conjunction with initial baseline inventories represents a strong possibility for desertification assessments and other analyses.

A most prominent manifestation of desertification is the land surface albedo. The albedo (reflectance integrated over the upward hemisphere of directions; sometimes referred to as "brightness") of land surfaces is an indication of degradation. Increasing albedos are thought to indicate erosion, salinization, overgrazing, and other deleterious land surface effects (Mouat *et al.*, 1990). Measurements of albedo may be made through the use of satellite observations. Surface albedo helps determine how much solar energy is absorbed and hence the surface temperature and evapotranspiration.

Various investigators (Justice *et al.*, 1985; Malo and Nicholson, 1990) have made use of remote sensing for deriving vegetation indices for the purpose of assessing vegetation parameters associated with desertification. One such index is the Normalized Difference Vegetation Index (NDVI). The NDVI is typically given as

$$NDVI = \frac{(Ch2 - Ch1)}{(Ch2 + Ch1)} \quad (1)$$

where Ch1 represents data from a visible channel (typically a red band — e.g., 0.63–0.69 μm) and Ch2 represents data from a near infrared channel (e.g., 0.76–0.90 μm). Two satellite sensors that are often used for the assessment of NDVI are the Advanced Very High Resolution Radiometer (AVHRR) and the Landsat Thematic Mapper (TM). The AVHRR with its high temporal resolution (images are acquired at every location twice daily) has enjoyed increased use in desertification (through the use of NDVI) studies. NDVI has been used in Africa and elsewhere to demonstrate the effects of overgrazing, especially during times of drought (Justice *et al.*, 1985). The NDVI and other vegetation indices might allow us to obtain an understanding of trends of vegetation phenology in the context of climatic and cultural events. Malo and Nicholson (1990) used AVHRR-derived NDVI to study the dynamics of vegetation and rainfall in the Sahel of West Africa. They found that the temporal and spatial patterns of monthly NDVI closely replicate those of rainfall. They also found that the ratio of NDVI to rainfall provides a rough quantitative measure of the efficiency of water use, with the highest efficiencies found in the plant formations of the driest environments.

While many consider the AVHRR to have too coarse a spatial resolution to be of use for most detailed studies, the Landsat multispectral scanner (MSS) and TM as well as the French SPOT satellite system provide considerably more spatial and spectral detail, and may provide information on both soil salinity and erosion as indicators of desertification.

While remote sensing techniques provide a spatial picture of land processes, other techniques must be employed if the recently acquired (since 1972) satellite images are to be placed into a longer term perspective to assess longer term climate changes. Changes in climate can be examined through the use of the Palmer Drought Severity Index (PDSI). The PDSI is a useful climatic integration that can link modern climatic data to biological processes. It is a useful measure of recent (approximately 100 years) climatic variations, and is derived from a combination of monthly precipitation, temperature, and soil moisture retention information. It offers an integrated measure of moisture availability, i.e., effective precipitation, (Wharton *et al.*, 1990).

If PDSI is combined with other data layers, including those that can be considered to be anthropogenic, an environmental index that perhaps more closely mirrors the pattern of land degradation can be developed. The Desert Research Institute in conjunction with the Environmental Protection Agency (under the auspices of the Environmental Monitoring and Assessment Program, or EMAP) has developed a preliminary Drylands Risk Index to characterize land degradation. The data layers used to develop this index include PDSI, vegetation type, vegetation greenness, total herbivory (measured in terms of total animals vs. carrying capacity), and demographics. Other data layers that could be added include exotics or native species and soil erosion.

The Drylands Risk Index currently provides equal weighting to each of the data layers within the context of a geographic information system. The method of developing ecological risk assessments for the analysis of ecosystem degradation in arid zones has a considerable benefit in allowing scientists and land managers the opportunity to change values of one or more data layers. This could be done to model ecosystem response given either a change in management status or in natural stressors (e.g., climate change).

10.8 REFERENCES

Bahre, C.J. (1991) *A Legacy of Change*. University of Arizona Press, Tucson, 231 pp.

Baker, D.E., and Amacher, M.C. (1982) Nickel, copper, zinc, and cadmium. In: Page, A.L. (Ed.) *Methods of Soil Analysis*, Part 2, 2nd edn., pp. 323-336. Agronomy Monograph No. 9. American Society of Agronomy, Madison, Wisconsin.

Dawson, J.L., and Nash, T.H. (1980) Effects of air pollution from copper smelters on a desert grassland community. *Environ. Exp. Bot.* **20**, 61.

Dregne, H.E. (1977) Desertification of arid lands. *Econ. Geogr.* **53**, 322–331.

Hastings, J.R., and Turner, R.M. (1965) *The Changing Mile*. University of Arizona Press, Tucson, 317 pp.

Hill, A.C., Hill, S., Lamb, C., and Barrett, T.W. (1974) Sensitivity of native desert vegetation to SO^2 and NO^2 combined. *J. Air Pollut. Control Assoc.* **24**, 153–157.

Humphrey, R.R. (1956) History of vegetational changes in Arizona. *Ariz. Catalog* **11**, 32–35.

Keeney, D.R., and Nelson, D.W. (1982) Nitrogen-inorganic forms. In: Page, A.L. (Ed.) *Methods of Soil Analysis*, Part 2, 2nd edn., pp. 663–679. Agronomy Monograph No.9. American Society of Agronomy, Madison, Wisconsin.

Justice, C.O., Townshend, R.G., Holben, B.N., and Tucker, C.J. (1985) Analysis of the phenology of global vegetation using meteorological satellite data. *Int. J. Remote Sens.* **6**, 1271–1318.

Kepner, W.G., and Fox, C.A. (Eds.) (1991) *Environmental Monitoring and Assessment Program: Strategic Monitoring Plan for Arid Ecosystems.* US Environmental Protection Agency, Las Vegas, Nevada.

McAuliffe, J.R. (1988) Markovian dynamics of simple and complex desert plant communities. *Am. Nat.* **131**, 459–490.

MacMahon, J.A. (1981) Introduction. In: Goodall and Perry (Eds.) *Arid Land Ecosystems: Structure, Functioning, and Management,* Vol. 2, pp. 263–268. International Biol. Prog. No. 17, Cambridge University Press, Cambridge.

Malo, A.R., and Nicholson, S.E. (1990) A study of rainfall and vegetation dynamics in the African Sahel using normalized difference vegetation index. *J. Arid Environ.* **19**, 1–24.

Mouat, D.A., Fox, C.A., and Rose, M.R. (1990) Ecological indicator strategy for monitoring arid ecosystems. In: *Proceedings of the International Symposium on Ecological Indicators,* 16–19 October 1990, Ft Lauderdale, Florida. Elsevier Science, New York.

Nieto, K.F., and Frankenberger, W.T., Jr. (1985a) Single column ion chromatography: I. Analysis of inorganic anions in soils. *J. Soil Sci. Soc. Am.* **49**, 87–592.

Nieto, K.F., and Frankenberger, W.T., Jr (1985b) Single column ion chromatography. II. Analysis of ammonium, alkali metals and alkali earth cations in soils. *J. Soil Sci. Soc. Am.* **49**, 592–596.

Reining, P. (Ed.) (1978) *Handbook on Desertification Indicators.* American Association for the Advancement of Science, Washington, D.C., 141 pp.

Rhoades, J.D. (1982) Soluble salts. In: Page, A.L. (Ed.) *Methods of Soil Analysis,* Part 2, 2nd edn., pp. 167–179. Agronomy Monograph No. 9. American Society of Agronomy, Madison, Wisconsin.

Schlesinger, W.H., Reynolds, J.F., Cunningham, G.L., Huenneke, L.F., Jarrell, W.M., Virginia, R.A., and W.G. Whitford. (1990) Biological feedbacks in global desertification. *Science* **247**, 1043–1048.

Smith, D.M.S., and Morton, S.R. (1990) A framework for the ecology of arid Australia. *J. Arid Environ.* **18**, 255–278.

Speth, G. (1988) Introduction. In: Sears, P.B. (Ed.) *Deserts on the March.* Island Press, Washington, D.C.

Sposito, C., and Mattigod, S.V. (1979) *GEOCHEM: A Computer Program for the Calculation of Chemical Equilibria in Soil Solutions and Other Natural Water Systems.* Kearney Foundation of Soil Science, University of California, Riverside.

Thomas, G.W. (1982) Exchangeable cations. In: Page, A.L. (Ed.) *Methods of Soil Analysis,* Part 2, 2*nd* edn., pp. 159–165. Agronomy Monograph No. 9. American Society of Agronomy, Madison, Wisconsin.

Thompson, C.R., Olszyk, D.W., Kats, G., Bytnerowicz, A., Dawson, P.J., and Wolf, J.W. (1984) Effects of ozone and sulfur dioxide on annual plants of the Mojave Desert. *J. Air Pollut. Control Assoc.* **34**, 1017–1022.

Turner, R.M. (1990) Long-term vegetation change at a fully protected Sonoran Desert site. *Ecology* **7**, 464–477.

UCAR (1990) Arid Ecosystem Interactions: Lessons from North American Experiences. Draft Report, Office of Interdisciplinary Earth Sciences, Boulder, Colorado.

UNEP (United Nations Environment Program) (1992) *World Atlas of Desertification.* Edward Arnold, London.

UNFAO (United Nations Food and Agricultural Organization) (1971–1981) *FAO/UNESCO Soil Map of the World, 1:5,000,000*, Vols. 1-10. UNESCO, Paris.

Webb, R.H., Steiger, J.W., and Turner, R.M. (1987) Dynamics of Mojave Desert shrub assemblages in the Panamint Mountains, California. *Ecology* **68**, 478–490.

Wharton, R.A., Wigand, P.E., Rose, M.R., Reinhardt, R.L., Mouat, D.A., Klieforth, H.E., Ingraham, N.L., Davis, J.O., Fox, C.A., and Ball, J.T. (1990) The North American Great Basin: a sensitive indicator of climatic change. In: Osmond, C.B., Pitelka, L.F., and Hidy, G.M. (Eds.) *Plant Biology of the Great Basin*, pp. 323–359. Ecological Studies Vol. 80. Springer-Verlag, Berlin, 375 pp.

11 Methods to Assess the Effects of Chemicals on Soils

H. A. Verhoef and C. A. M. van Gestel
Vrije Universiteit, The Netherlands

11.1 INTRODUCTION

The industrialization of our society has led to an increased production and emission of both xenobiotic and natural chemical substances. Many of these chemicals will end up in the soil. Various soil constituents have a great capacity to retain chemicals, especially those with apolar molecules or positively charged divalent and trivalent ions. Consequently, the soil is a net sink for all kinds of chemicals, and concentrations are often considerably higher than in any other environmental compartment. This situation may lead to smaller or larger impacts on the functioning of soil ecosystems. Important ecological functions of the soil are those associated with organic matter decomposition, mineralization of nutrients, and synthesis of humic substances. For that reason, an increasing need exists for methods to assess the side effects of these chemicals on soil ecosystems (OECD, 1989).

In this chapter, an overview will be provided of methods to determine the effects of chemicals on soil ecosystems. An overload of chemicals will affect both abiotic soil properties and, directly and indirectly, soil biota. This paper will, therefore, start with a description of the possible sources and consequences of chemical pollution for abiotic soil properties. Subsequently, methods are described for the determination of the effects of chemicals on soil organisms.

When considering methods to assess the effects of chemicals on soil biota, two types of tests can be distinguished. The first contributes to the prediction of the potential effects of single chemicals on soil ecosystems. For that purpose, mainly single-species laboratory tests are conducted; at times, more complex micro-ecosystem, mesocosm, or field studies are carried out. This type of testing, which aims at establishing dose-response relationships and the estimation of LC_{50}, EC_{50}, or NOEC values, may be called "prognosis."

The second type of method is aimed at assessing the potential ecological risk of a certain case of soil pollution. In such a situation, several chemicals may be involved. To determine whether a specified case of soil pollution poses a real hazard for soil biota, both laboratory and field studies may be performed. This

Methods to Assess the Effects of Chemicals on Ecosystems
Edited by R. A. Linthurst, P. Bourdeau, and R. G. Tardiff

type of testing may be called "diagnosis."

In this chapter the focus will be on terrestrial ecosystems; studies on groundwater ecosystems will not be taken into account.

11.2 QUANTIFICATION OF INPUT OF CHEMICALS IN THE SOIL

Before methods to assess the effects of chemicals on soil components can be presented, information must be provided about the various ways chemicals enter the soil and how these inputs can be quantified.

11.2.1 MAJOR ELEMENTS

Focusing on major elements, Table 11.1 summarizes the most important processes contributing to the input of chemicals to a soil ecosystem. The relative importance of the input differs greatly among elements as well as among areas of low versus high pollution. The precision in the measurement of input quantities is mainly a question of equipment, methods, and experimental design.

Table 11.1. Major inputs of elements to soil ecosystems

	N	S	P	K	Na	Ca	Mg	Cl
Wet deposition	X	X		X	X	X	X	X
Dry deposition								
Gaseous input	X	X						
Particle input	X	X	?	?	X	?	X	X
N_2-fixation	X							
Mineral weathering		X	X	X	X	X	X	

11.2.1.1 Wet deposition

This is the major process for the input of nitrogen, sulphur, and chloride, and is significant for other elements. Excluding problems such as the definition of input by mist and fog, the measurement of wet deposition is essentially the estimation of rainfall and its elemental content.

Type and design of rain gauges can be found in *Tropical Soil Biology and Fertility: A Handbook of Methods* (Anderson and Ingram, 1989, p. 8), and their position relative to the ground surface is usually the reason for an underestimation in rainfall, depending on wind and evaporation losses in exposed areas (Eriksson, 1980). To avoid chemical changes in the sample, frequent sampling and rapid analysis are preferred to preservation.

11.2.1.2 Dry deposition

This is defined as the direct transfer of gases and particles to different ecosystem surfaces (= receptors). Two methods are noted here:

1. Micrometeorological methods are indirect, and based on the assumption that transmission of chemical compounds is a process similar to transmission of heat and momentum. The method has been proven to produce useful results for SO_2 deposition to grass-covered areas, but does not fully satisfy to measure deposition to forests.
2. Receptor- (or ecosystem-) oriented mass balance methods are more direct methods, but with problems and limitations. For chloride, sodium, and sulphur at high deposition, throughfall measurements have been useful to estimate dry deposition, when combined with wet deposition measurements (Grennfelt *et al.*, 1985).

11.2.1.3 Nitrogen fixation

This method is well known as the acetylene-reduction (AR) method for measuring nitrogenase activity.

Isotope techniques, such as incubation in $^{15}N_2$-containing atmosphere are attractive; but for long-running experiments, practical problems arise. Further methods are the ^{15}N-isotope dilution technique and the classic total nitrogen difference method, based on a comparison of total N-yield in a N-fixing crop and that of a non-N fixing reference crop.

11.2.1.4 Mineral weathering

Methods to estimate current weathering rates include mass balance of watersheds, radiometric methods, and mineral bag technique.

Mass balance of watersheds or lysimeters is widely used and gives the best estimates of current weathering (e.g. Likens *et al.*, 1977). The approach is indirect, leaving weathering as the residue in the mass balance equation:

$$E(rw) = [E(efflux) - E(influx)] + E(ps) \qquad (1)$$

where E(rw) is element release by weathering, *E* (efflux) is elemental losses mainly from leaching, E (influx) is input of elements mainly by dry and wet deposition, and E(ps) is the change in elemental storage in plants and soil.

The results of mass balance studies depend on the accuracy of estimation of all possible sources and sinks for elements in the soil. Main sinks are elemental storage in biomass and humus, but also accumulation in the soil by microbial activity, redox reactions, surface exchange processes, and formation of secondary minerals. These sinks may easily turn into sources if biological or chemical conditions are altered. For instance, present-day acid deposition depletes base cations from the exchange sites in the soil profile.

Radiometric methods using $^{87}Sr/^{86}Sr$ ratios are used to estimate calcium weathering.

For the mineral bag technique, selected soil or mineral fraction is put into a non-biodegradable mesh bag, and placed in the field for a specific period of time. The bag is then returned to the laboratory for analysis.

11.2.2 ORGANIC CHEMICALS AND METALS

Major sources for the input of organic chemicals and heavy metals in soils are agriculture, industries, and traffic. Both diffuse and point sources can be identified, and some chemicals (e.g., pesticides) are applied in quite a controlled way enabling a proper prediction of the input in the soil.

To assess the deposition of pesticides that are generally applied under controlled conditions, sheets of aluminium foil or other inert substances can be placed on the soil and analysed after spraying. For other chemicals that may be released in a less controlled manner, chemical analysis of soil samples is needed to quantify the input. In all cases, rather specific analytical techniques are required to determine chemical concentrations in soils. Before analysis, complicated extraction and purification steps are often required. Generally, extraction with an organic solvent (e.g., hexane, acetonitrile, toluene or acetone) is applied, followed by analysis by HPLC, GC, or GC-MS.

To determine the soil content of heavy metals, digestion of soil samples with strong acids (e.g., $HNO_3/HClO_4$) is required. After that, the destruate can be analysed by atomic absorption spectrophotometry.

11.3 METHODS TO QUANTIFY EFFECTS OF CHEMICAL INPUT ON ABIOTIC SOIL CHARACTERISTICS

11.3.1 NITROGEN

Recent concerns over nitrogen deficiencies have led to others concerning excess nitrogen availability and the potential for forest decline and surface water pollution. High input of N can lead to N-saturation with serious environmental impacts on soil chemistry and water quality and on fluxes of radioactively active (or "greenhouse") gases.

In Table 11.2 the characteristics of N-saturated forest soil are listed. The values for the characteristics are endpoints. To get information about the deposition level upon which these characteristics start to change, the "critical load" concept has been introduced. The definition is "the maximum deposition of elements that will not cause chemical changes leading to long-term harmful effects on ecosystem structure and function" (after Nilsson and Grennfelt, 1988). A regional assessment of critical loads is very important to formulate optimal policies for emission reductions. The generic approach to map critical loads is presented in Figure 11.1. In Table 11.3, an illustration is given of some critical chemical amounts for forest soil (water).

Table 11.2. Characteristics of *N*-saturated forest soils

Characteristic	Value	Method
N cycled	25–50% NO_3^- 50–75% NH_4^+	Anderson and Ingram, 1989; Faber and Verhoef, 1991
DOC concentration	Low	Complete (Nelson-Sommers) or partial (Walkley-Black) oxidation (Anderson and Ingram, 1989)
C/N ratio	Low	Kirsten, 1979
Ca, Mg concentration	Low	Flame atomic absorbance spectrophotometry
H^-, Al_i concentration	High	pH based on free protons, on the exchangeable fraction extracted with KCl, or on the fraction titrated with a base; and ICP
N_2O production	High	Lloyd, 1985; Harrison *et al.*, 1990
CH_4 production	Low	Lloyd, 1985; Harrison *et al.*, 1990

11.3.2 SULPHUR

Sulphur is transformed in soils by processes similar to those occurring in the nitrogen cycle. Like nitrogen, sulphur can be oxidized, reduced, assimilated, or mineralised from organic matter. The major differences between the two cycles are (1) no process is equivalent to N-fixation and (2) losses of gaseous sulphur from soils are not equivalent to N-losses due to denitrification. Recent interest in soil sulphur transformations results from an increased awareness of the fertilizer value of the element and the recognition of the importance of the sulphate ion, which reaches the soil in acid rain, as the major counter-anion involved in cation leaching from soils. Sulphate adsorption by soils is an important property affecting the availability of sulphate to plants and the leaching of sulphate and associated cations. Sulphate adsorption is particularly important in soils subjected to acid precipitation, since it determines the impact of acid rain on cation mobility and leaching. Soil temperature and moisture influence sulphate adsorption, whereas desorption on waterlogging may also be an important reaction in soils exposed to atmospheric pollution.

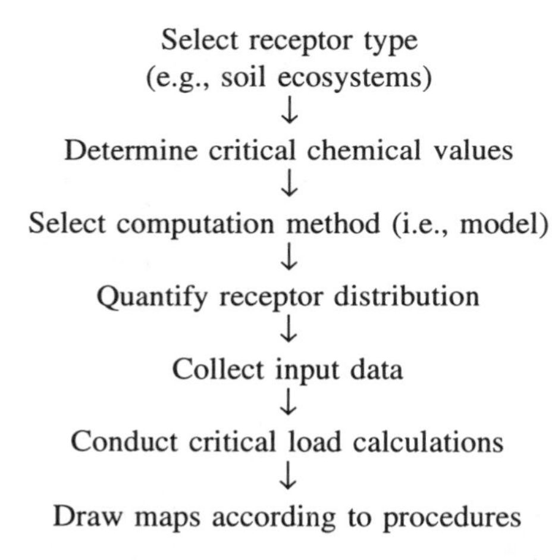

Figure 11.1. Flowchart to map critical loads and areas where they have been exceeded (De Vries, 1991)

Many agents have been used to extract sulphate and other sulphur ions from the soil (see Wainwright, in Harrison *et al.*, 1990). Important is 0.01 M $Ca(H_2PO_4)_2$ which appears to remove sulphate from the same pool of soil sulphur that is available for plants. Sulphate can be measured by methods including gravimetrically, turbidimetrically with barium chloride, spectrophotometrically using methylene blue, by titrimetric methods, adsorption chromatography, ion exchange chromatography, ion-selective electrodes, and thin-layer or gas chromatography (Williams, 1979).

Recent studies on the damage to ecosystems caused by oxides of sulphur and acid rain involve the use of lysimeters, and collectors to measure through-fall, stem flow, and litter deposition (Ulrich and Mayer, 1980). Unfortunately, most of these studies have omitted the microbial transformations of the element.

Recent laboratory studies have been concerned with the microbial cycling of sulphur in soils exposed to heavy atmospheric pollution from point sources. The first approach was to remove soils from the field at intervals throughout the season, and to have them analysed in the laboratory for S-ions and sulphur oxidizing micro-organisms. In the latter approach, soils were sampled from sites exposed to point-source pollution and from relatively unpolluted sites that had essentially the same soil type, vegetation, and climate as the polluted sites. They were packed into plastic tubes. Several soil columns were placed in the polluted site and several in the non-polluted site. In this way, an assessment of effects of pollution on relatively unpolluted soil could be determined as well as the time taken for heavily polluted soils to regain characteristics more typical of relatively unpolluted soil. Exposure duration was about 1.5–2 years. Similar exposure periods were found in a reciprocal transplant experiment with soil cores over a gradient of N and S input over Europe.

Table 11.3. Critical levels of chemicals for forest soils

Criteria	Unit	Soil
[Al]	mol per m^3	0.2
Al/Ca	mol per mol	1
pH	–	4.0[a]
[Alk][c]	mol per m^3	-0.3[a]
NO_3	mol per m^3	0.1[b]
NH_4/K	mol per mol	5

[a]For forest top soils, pH of 3.7 and alkalinity of -0.4 are suggested.
[b]Related to vegetation changes (from De Vries, 1991).
[c][HCO_3] + [RCOO] - [H] – [Al]

11.3.3 PHOSPHORUS

Both forests and grasslands are frequently phosphorus deficient to a variable degree, and this deficiency limits their productivity. Fertilizer application is, therefore, a principal means of increasing timber and grass production.

In forestry, tree needle analysis has been used for many years as the main guide in the assessment of phosphorus fertilizer requirements. Recent publications, however, indicate that this type of analysis is unrealistic as a predictor of fertilizer responses in commercial forest trees (Axelsson, 1984; McIntosh, 1984). Analysis of the forest or grassland soils has been proposed as an alternative (Hunter *et al.*, 1985). Extraction methods for P_i rely on three different principles:

1. Anion exchange resin acts as a sink for solution P_i, and thereby offsets the equilibrium between dissolved and soluble P_i. "Exchangeable" P_i as well as

some of the more soluble precipitated P forms will enter the solution, bind to the resin, and can then be measured.

2. Changes in pH cause changes in the solubility of P_i. Acid will extract calcium P. Alkaline solutions will solubilize Al and Fe bound P_i. Different P_i compounds have different solubilities at various pH values, and this can be used to characterize soil P_i composition or to evaluate labile P_i.
3. Specific anions can bring P_i into solution by competing for adsorption sites and/or lowering the solubilities of cations that bind P_i. Fluoride has for instance been used under conditions of controlled pH, to release P from Al-bound forms, by forming insoluble aluminium fluoride. Organic anions have also been used to bind or chelate cations and release P_i into solution.

Methods to extract P_o have employed alkaline solutions (e.g., $NaHCO_3$ or NaOH) or various organic solvents (e.g., acetylacetone which dissolves organic matter). Little progress has been made towards characterising P_o extracts in terms of the mechanisms for bringing P_o into solution, binding modes in the soil and its availability to plants (Anderson and Ingram, 1989, p. 113).

A new, physiologically based root bioassay has been developed, that appears to be sensitive in assessing P-deficiency in plants. The bioassay relies on the negative relationship between the rate of metabolic uptake of ^{32}P-labelled phosphorus by roots from a standardized solution in the laboratory and the amount of phosphorus supply in the original rooting environment. This method has been successfully applied to forest stands and grasslands (Harrison *et al.*, 1985, 1986).

11.3.4 CARBON

The CO_2 concentration in the atmosphere has increased by 25 percent over the past 100 years, and a consensus exists that a doubling of the concentration may occur by the middle of the next century. The largest terrestrial carbon sources and sinks influencing CO_2 fluxes are the forests, which account for approximately two-thirds of the photosynthesis. The effects of a doubling of the CO_2 concentration on the growth and development of trees is known for a few species. A general finding is an increase of the tissue density of the leaves, a change in leaf structure, and an increase in the C/N ratio of the tissues. This changed C/N ratio may reduce the decomposition rates of plant material and modify the nutrient availability (Coûteaux *et al.*, 1991). On the other hand, increased atmospheric concentrations of CO_2, together with trace gases such as methane (CH_4), nitrous oxide (N_2O), and chlorofluorohydrocarbons (CFCs), are effecting changes in the global heat balance, resulting in significant changes in climate over the next century. Twice as much carbon is found in the top metre of soil compared to the amount in the atmosphere, and CO_2 emissions from soils will increase as organic decomposition is enhanced at higher temperatures. CO_2 emissions are particularly sensitive to temperatures between 0°C and 5°C. Based on a recent model predicting global emissions from soil organic matter (Jenkinson *et al.*, 1991), a world temperature rise of 0.3°C per decade has been estimated to result in an additional release of CO_2 from soil organic matter over the next 60 years, equivalent to about 19 percent of that released by combustion of fossil fuels if present use of fuel were to continue unabated. These calculations suggest that increased decomposition of soil organic carbon could make an important contribution to the greenhouse effect.

Methods to measure C/N ratios have already been given. Measurements of CO_2 evolution under field conditions are described and standardized (Anderson and Ingram, 1989). Methods to measure soil organic carbon are given by Anderson and Ingram (1989).

11.3.5 ORGANIC CHEMICALS AND METALS

Effects of organic chemicals on the abiotic soil properties are rarely mentioned in the literature. Some chemicals, such as the herbicide paraquat, are incorporated into clay particles, but the extent to which this may influence the swelling and shrinking behaviour of the clay is unknown. Other chemicals or their degradation products are incorporated into the soil organic matter. This is a physical or a biological process, which may result, in case of chlorinated organics, in chlorination of the soil organic matter. The extent to which this process occurs and the consequences for the soil characteristics are unknown. Presently, no methods exist to determine the potential impact of organic chemicals on soil abiotic properties.

Metals generally occur in the soil solution as positively charged cations, competing for negatively charged adsorption places on the soil particles. An overload of metals will affect the ionic balance of the soil; it may also lead to a release of other, less strongly bound, metals or cations from the soil. Often this process is slow, not affecting soil abiotic properties to a great extent. Only in the case of flooding a soil with salt water, containing an excess of cations, was the swelling and shrinking properties of clays shown to be severely affected. No methods can be given to measure the impact of metals on soil abiotic properties.

11.4 METHODS TO ASSESS THE POTENTIAL RISK OF CHEMICALS FOR SOIL ORGANISMS (PROGNOSIS)

A brief description is presented of single-species laboratory tests, microcosm tests, and field tests. For an extended overview of these tests, the reader is referred to Van Straalen and Van Gestel (1992a, 1992b).

11.4.1 SINGLE-SPECIES LABORATORY TOXICITY TESTS

Among soil invertebrates, only earthworms have seriously been considered as test organisms during the past decade, and some standardized test methods are available. For microfauna and mesofauna, only few tests are available, although these animals are among the most numerous and species-rich groups of soil animals. Many species, however, are promising test animals, because they are easy to culture and their size allows for small-scale experimental set-ups with many replications.

Besides these soil animals, higher plants have also been considered for testing, and standardized tests with some plant species are available.

In several tests, artificial substrates (nutrient solution, agar, silica gel, filter paper) are used, the composition of which greatly affects toxicity. However, extrapolation of these test results to the field remains problematic. If concentrations in test solutions can be equated with pore water concentrations, sorption data may be used to express the toxicity per unit of soil. The validity of this extrapolation, however, still has to be investigated. The same extrapolation problems may arise for tests in which the main route of exposure is via the food. For such tests, a conversion of food concentrations to soil concentrations may be needed.

11.4.1.1 Higher plants

A test with higher plants has been described in an international test guideline (OECD, 1984b), while some others are under discussion. The OECD guideline 208 on higher plant toxicity testing uses several plant species, representing different agricultural crops and both monocotylodoneous and dycotylodoneous species.

11.4.1.2 Protozoans and nematodes

Protozoans and nematodes live in the soil pore water, and the best way to test them is to use methods similar to those used in aquatic toxicology. Among the protozoans the ciliates *Tetrahymena pyriformis*, *Colpoda cucullus*, and *Paramecium aurelia* have been considered (Berhin *et al.*, 1984; Nyberg and Bishop, 1983) for test animals, as have the nematode species *Caenorhabditis elegans*, *Panagrellus silusiae*, and *Plectus parietinus* (Sturhan, 1986; Haight *et al.*, 1982; Van Kessel *et al.*, 1989); however, an accepted test procedure is unavailable.

11.4.1.3 Isopods and millipedes

Isopods are an interesting group of animals in heavy metal research, because of their unique ability to concentrate extreme amounts of metals in their bodies (Hopkin, 1989). Their use as a test animal for soil toxicity studies, however, is restricted to a few cases, and no attempts have yet been made to arrive at standardization (Dallinger and Wieser, 1977; Van Capelleveen, 1987; Hopkin, 1990; Van Straalen and Verweij, 1991; Eijsackers, 1978b). *Porcellio scaber*, *Oniscus asellus*, and *Trichoniscus pusillus* are three species frequently investigated. Among these, *T. pusillus* seems to be the most suitable as a test species, as it has a somewhat shorter life-cycle compared to *P. scaber* and *O. asellus*. All three species are very easy to culture, and do not require special conditions.

Usually isopods are kept on a plaster substrate, and are fed with partly decomposed leaves, either intact or ground, to which chemicals can be added. Increase in growth over several weeks is observed, but is rather variable, even for one individual. Reproduction is difficult to assess, because, after mating, females may retain the sperm for a long period before producing eggs, which are carried in a brood pouch. Tests require a minimum period of four weeks.

Millipedes (Diplopoda) are another important group of saprotrophic soil invertebrates, but they have never been considered seriously as test animals. The most widely investigated species is *Glomeris marginata* (Hopkin *et al.*, 1985). The species *Cylindroiulus britannica* is also well suited as a test animal. Test conditions for millipedes are similar to those for isopods.

11.4.1.4 Oribatid mites

A reproduction toxicity test using the parthenogenetic oribatid mite *Platynothrus peltifer* has been described by Denneman and Van Straalen (1991). This seems to be the only oribatid used so far in soil toxicity experiments, although oribatids comprise hundreds of species, and are usually the most numerous group of arthropods in forest soils.

In the test with *P. peltifer*, the animals are exposed to contaminated algae, and the number of eggs are counted. The test is very laborious, as the animals hide their eggs in small crevices; it is also a rather lengthy test (9 to 12 weeks) because of the low rate of egg production in this species and its long life-cycle (1 year), which are remarkable features for such a small animal (± 1 mm).

P. peltifer appeared to be rather resistant to cadmium, copper, and lead in terms of lethality, but very susceptible in terms of egg production. Due to their peculiar habits, species such as *P. peltifer* tend to be forgotten in the development of toxicity tests. It is, however, the most sensitive soil invertebrate tested so far for cadmium, while it is more sensitive than springtails for copper and lead (Denneman and Van Straalen, 1991; Van Straalen *et al.*, 1989).

11.4.1.5 Collembola

Collembola are a relatively well investigated group of soil animals. Several species have been used frequently in toxicity experiments: *Onychiurus* spp. (Eijsackers, 1978a; Bengtsson *et al.*, 1985; Mola *et al.*, 1987), *Folsomia candida* (Thompson and Gore, 1972; Tomlin, 1977; Iglisch, 1986), *Tullbergia granulata* (Subagja and Snider, 1981), and *Orchesella cincta* (Van Straalen *et al.*, 1989). The first three species are parthenogenetic (thelytokous); *O. cincta* is sexual, sperm being transferred indirectly through spermatophores deposited on the substrate by the male. Three different exposure systems have been described: (1) through feeding on fungi grown on contaminated agar, (2) through feeding on directly contaminated food, and (3) residual exposure (treated substrate, e.g. sand, leaves, soil).

When testing Collembola with contaminated fungi, the animals are kept on a plaster of Paris substrate in a Petri dish, and fed on a piece of agar, overgrown with hyphae (e.g., *Verticillium bulbillosum*) (Bengtsson *et al.*, 1983, 1985). Egg production, growth and survival are recorded regularly throughout a period of several weeks. The advantage of this system is that substances are offered in a natural way, i.e., after being taken up and possibly transformed to naturally occurring complexes. Concentration levels in the fungus, however, are difficult to maintain or to set to specific values.

When testing Collembola with directly contaminated food, chemicals are added in water or acetone solution to the food (algae, yeast, ground leaf material). Food can be offered as droplets on filter paper discs, while the animals are kept on a plaster or sand substrate. In this manner, concentrations can be manipulated easily, while growth, egg production, and survival are monitored over a period of several weeks (Van Straalen *et al.*, 1989).

The third system of testing Collembola is to use the artificial soil medium developed for earthworm toxicity tests (Jancke, 1989; Wohlgemuth *et al.*, 1990; ISO, 1991). Juvenile Collembola (*Folsomia candida*) are placed in artificial soil with dry yeast provided for food. After 28 days, the number of remaining animals and their offspring are counted after flotation extraction. The *Folsomia* test is very easy to carry out; it requires little attention during the test; and it gives reproducible results. Another advantage is the use of artificial soil similar to the earthworm test; thus, experimental results can be compared between earthworms and springtails. The only disadvantage of the test is that reproduction cannot be observed directly, and cannot be separated from juvenile mortality and hatching success. The *Folsomia* test is now undergoing the process of international standardization (ISO, 1991).

11.4.1.6 Enchytraeids

Enchytraeids can be cultured easily on agar and on (artificial) soil substrates, when fed rolled oats (Römbke and Knacker, 1989; Römbke, 1989). The toxicity tests described in the literature all use species of the genus *Enchytraeus*. The well-known species *Cognettia sphagnetorum* can also be bred easily in the laboratory, but its tendency to fragmentate upon handling makes it less suitable for toxicity tests.

Westheide *et al.* (1991) described a test in which the test chemical is incorporated in 1.5 percent nutrient agar. Two species are used, *Enchytraeus* cf. *globuliferus* and *E. minutus*. Reproduction, measured as the number of cocoons and juveniles produced, is the endpoint studied in this test. This method seems to provide an easy and reproducible test, but, because of the use of an agar substrate, results cannot be translated to real soil.

Römbke (1989) described a test with *Enchytraeus albidus*, using the OECD artificial soil prescribed for earthworm toxicity tests. Adult *E. albidus* are exposed for 28 days to different concentrations of the test chemical, mixed homogeneously through the artificial soil. Survival of adult worms and the number of juveniles produced are the endpoints studied. In this way, both acute and sublethal effects are combined in one test. The reproducibility of the method cannot be judged and, because only one chemical was tested, no conclusions can be drawn with respect to the sensitivity of the sublethal endpoint. An important positive aspect of this test is that the substrate used is the same artificial soil used in the internationally accepted earthworm toxicity tests (OECD, 1984a; EEC, 1985).

11.4.1.7 Lumbricids

In the guidelines of OECD (1984a) and EEC (1985), *Eisenia fetida* and its sibling species *E. andrei* are recommended. Both species are commonly found in compost and dung heaps, and can be cultured easily in the laboratory on a substrate of horse dung or cow dung (OECD, 1984a). According to the existing guidelines on acute toxicity testing with earthworms (OECD, 1984a; EEC, 1985), other real soil dwelling species may also be used. Such species are, however, hard to culture in the laboratory, because they have long generation times and need large volumes of soil. So, for practical reasons, the use of the two *Eisenia* species is recommended.

Three acute toxicity tests exist. In the filter paper contact test (OECD, 1984a), adult earthworms of the species *Eisenia* spp. are exposed to filter paper wetted with a solution of the test substance. Mortality is assessed, and the 48-hour LC_{50} value is expressed as μg per cm^2. The method has been shown to be easy, fast, and highly reproducible (Edwards, 1983). Several authors including Heimbach (1984) have demonstrated, however, that this test has no predictive value for the effect of chemicals on earthworms in the soil; it can only be used to rank chemicals.

In the artificial soil test (OECD, 1984a), adult earthworms of the species *Eisenia* spp. are exposed to the test chemical, which is mixed through an artificial soil substrate for 14 days. This artificial soil is made up by mixing (dry weight) 10 percent sphagnum peat, 20 percent kaolin clay, 70 percent quartz sand, while some $CaCO_3$ is added to adjust the pH to 6.0±0.5. The moisture content of the substrate is adjusted to about 55 percent (w/w) or to 40–60 percent of the water holding capacity. Mortality is the only test parameter, and LC_{50} values are expressed as mg per kg dry soil. Van Gestel and Ma (1990) have demonstrated that results obtained in this artificial soil can easily be translated to natural soils by using sorption data. For this reason, the use of the artificial soil is acceptable, and the test can be concluded to have enough predictive value with respect to effects that occur in the field (Heimbach, 1992; Van Gestel, 1992). The test has been shown to be reproducible.

In the Artisol test (Ferrière *et al.*, 1981), adult earthworms of the species *Eisenia* spp. are exposed to chemicals mixed through a substrate of amorphous silica gel (Artisol) for fourteen days. Survival is the only test parameter, and results are expressed in terms of LC_{50} values. The silica gel substrate does not bear any resemblance to natural soil; thus for reasons of ecological realism and extrapolation towards natural soil, this test cannot be recommended.

Recently, two sublethal toxicity tests have been described. In both tests, the OECD artificial soil and the earthworm species *Eisenia* spp. are used. In the first test (Van Gestel *et al.*, 1989), chemicals are mixed homogeneously through the artificial soil, and after three weeks of exposure effects on the growth and cocoon production by adult earthworms are determined. The worms are fed by supplying a small amount of (untreated) cow dung in a small hole in the middle of the soil. By incubating cocoons produced for five weeks in untreated artificial soil, effects on hatchability (percent fertile cocoons, number of juveniles per cocoon) and the total number of offspring per adult worm can be determined.

The second method (Kokta, 1992) was developed to determine the sublethal effects of pesticides on earthworms. The pesticide is sprayed onto the soil surface, and earthworms are fed by applying about 0.5 g cow dung per animal to the soil surface once a week. Pesticides are applied in two treatment levels, corresponding with the recommended dose and a five fold dose. After six weeks incubation, adult worms are removed from the substrate and weighed. The test substrate containing cocoons and juveniles is incubated for another four weeks. Food is given when required. After ten weeks, the juveniles are extracted from the substrate by hand-sorting or heat extraction and counted. Effects on earthworm growth and on the total number of offspring produced per tray are determined. The method has been subjected to a (German) ring test, and will be revised on the basis of the results of it. The method is only applicable for pesticides, and the recovery of all juveniles from the artificial soil by hand-sorting is difficult. This hinders comparison of this method with that of Van Gestel *et al.* (1989). Furthermore, the method of pesticide application is not standardized, which may be the reason for the variability observed in the first ring test (Kokta, 1992).

11.4.1.8 Molluscs

In the limited number of toxicity tests using terrestrial molluscs, exposure was via the food. Russel *et al.* (1981) described a method for toxicity experiments with the garden snail *Helix aspersa*. The snails were kept in polyethylene boxes, filled with a substrate of moist quartz sand covered by a piece of woven glass towel. The snails are fed a diet of ground Purina Lab-Chow (formulation for rats, mice, and hamsters) supplemented with $CaCO_3$. Parameters affected include survival, reproductive behaviour, dormant state, new shell growth, and food consumption. Similar test methods using the snail *Helix pomatia* or the slug *Arion ater* have been described by other authors (Marigomez *et al.*, 1986; Meincke and Schaller, 1974; Moser and Wieser, 1979).

11.4.1.9 Beneficial arthropods

Arthropods that may improve the production of agricultural products are designated as "beneficials," and commercial interest exists in designing and applying pesticides in such a way that beneficials are least affected. The working group on "Pesticides and Beneficial Organisms" of the International Organization for Biological and Integrated Control of Noxious Animals and Plants (IOBC) has contributed significantly to designing ecotoxicological test methods and decision schemes to evaluate the hazard of pesticides. These methods have been reviewed by various authors (IOBC, 1988; Croft, 1990; Samsøe-Petersen, 1990).

The hymenopteran groups Ichneumonidae, Braconidae, and Chalcidoidea contain a large number of parasitoid species. The female insect deposits an egg in or on a host (usually an insect egg or larva), which is then gradually eaten as the offspring develop. The host selection process and the life-cycle of the parasitoid are finely tuned to the host, and many species will attack only a single or a few host species. Furthermore, other hymenopteran species used in toxicity tests include *Diaeretiella rapae*, an internal parasite of aphids such as *Myzus persicae*, *Phygadeuon trichops*, a parasite of *Delia* species (bulb flies), *Coccygomimus* (= *Pimpla*) *turionellae*, a polyphagous parasite of Lepidoptera (Tortricidae, Geometridae, Noctuidae), and *Opius* sp., a parasite of leaf mining insects. The methods used for these species are similar to those described for *Trichogramma* and *Encarsia*.

Within the order of the Coleoptera, the families Carabidae (ground beetles), Staphylinidae (rove beetles), and Coccinellidae (lady birds) contain representatives that are common in agricultural fields and are recognized for their predation of pests.

Among the various arthropod groups, spiders seem to be particularly sensitive. This phenomenon often appears in field tests with pesticides, where catches of surface active spiders are reduced in a manner similar to that of predatory mites following pesticide application. The families Erigonidae and Linyphiidae (money spiders) are important groups with a great species richness.

The recommendations made by the International Commission for Plant Bee Relations (ICPBR) have been included in a guideline of the European and Mediterranean Plant Protection Organization (EPPO) to evaluate the hazards of pesticides to the honey bee *Apis mellifera* (OEPP/EPPO, 1991). Several countries have slightly different national guidelines to test pesticides on honey bees.

Several other beneficial arthropods have been proposed as test species (IOBC, 1988; Hassan *et al.*, 1985): *Chrysoperla carnea* (Neuroptera, Chrysopidae), *Anthocoris nemorum* (Heteroptera, Anthocoridae), *Syrphus corollae* and *Syrphus vitripennis* (Diptera, Syrphida), and *Drino inconspicua* (Diptera, Tachinidae).

11.4.2 MICROCOSM TESTS INCLUDING THOSE ON SOIL MICROFLORA

11.4.2.1 Microcosm tests

Single-species tests are carried out under rather artificial conditions, and disregard ecological interactions between different species. To evaluate effects of chemicals under more natural conditions, model ecosystems, microcosms or micro-ecosystems have been designed that simulate certain aspects of real ecosystems, and are yet simple enough for experimental use. Decomposing invertebrates have been considered for such systems, because their activities can be assessed conveniently in terms of system functions such as leaf litter fragmentation and nutrient conversions (Teuben and Roelofsma, 1990; Eijsackers, 1991).

Several terrestrial model ecosystems have been described (Giesy, 1980; Anderson, 1978b; Verhoef and de Goede, 1985; Hågvar, 1988; Teuben and Roelofsma, 1990; Teuben and Verhoef, 1992; Mothes-Wagner *et al.*, 1992), without attempt to arrive at standardization. The system may either be closed or open to the ambient air, and contains intact core samples from a natural habitat (e.g., Chaney *et al.*, 1978; Ausmus *et al.*, 1978) or a more or less standardized soil (e.g., Bond *et al.*, 1976). For ecotoxicological tests, the use of standardized soils seems to be most appropriate, since it allows the chemical to be mixed homogeneously through the soil, and it minimizes experimental variation between replicate units. The effects of various pretreatments, such as drying, sterilizing, inoculation, litter type, age of the litter, however, have a significant impact on the behaviour of the system and need to be investigated thoroughly (Van Wensem *et al.*, in press).

Natural rainfall may be simulated, and leachate can be collected (e.g., Verhoef and Meintser, 1991; Bengtsson *et al.*, 1988). Various chemical analyses of the leachate solution may indicate aspects of decomposer activity: dissolved organic carbon, NH_4, NO_3, pH, and Ca. The advantage of this procedure is that repeated sampling in time from the same soil column is possible. A disadvantage of the leaching procedure is that the humidity of soil and litter is unstable, and is difficult to standardize. Moreover, toxicants added to the system may be displaced through the column or leached out. Microbial respiration can be estimated by measuring CO_2 production in the microcosms. For that purpose, CO_2-free air (20.8 percent O_2; 79.2 percent N_2) is guided through the soil column, and the resulting CO_2 is subsequently measured by infrared gas analysis (Teuben and Roelofsma, 1990). Verhoef and Dorel (1988), Verhoef *et al.* (1989), and Verhoef and Meintser (1991) have used these types of microcosms, filled with pine litter, to study the effects of gaseous (NH_3) and wet ($(NH_4)_2SO_4$) atmospheric deposition. N-deposition eliminates the stimulation of mineral leaching by the collembolan *Tomocerus minor*. Neither survival nor growth of the animals are affected. Reproduction, however, is negatively influenced. In pine litter, which has been confronted with high N input for several decades, *T. minor* slows down mineral leaching by stimulating microbial growth.

Van Wensem (1989) and Van Wensem *et al.* (1991) added chemicals to poplar leaf litter, which is incubated for 4 weeks, after which some replicates are terminated to determine DOC, NH_4, NO_3, and pH. The remaining replicates are incubated for another four weeks, with eight isopods (*Porcellio scaber*) added to each system. Survival and growth of the isopods can be assessed, as well as particle size distribution and concentrations of minerals of the remaining litter. The organotin fungicide triphenyltin hydroxide increased the concentration of soluble ammonium in the litter, due partly to excretion by isopods and partly to stimulating effects on the microflora. In systems with isopods, the organotin decreased ammonation in treatment levels higher than 10 µg per g, but in systems without isopods the organotin had no significant effect. The addition of isopods in this case, therefore, made the system quite sensitive, which was unexpected (triphenyltin is a fungicide), and would not have been noticed in a single-species test using isopods.

Mothes-Wagner *et al.* (1992) described a more complex microcosm system, consisting of 25 litres of natural or standardized soil that are inoculated with nematodes (*Pelodera strongyloides*) and enchytraeides (*Enchytraeus coronatus*) and sown with bush beans (*Phaseolus vulgaris*). After emergence of the beans, spider mites (*Tetranychus urticae*) are introduced. After a preincubation period of about six months in the laboratory, in a greenhouse, or in the field, the systems can be treated with the test chemical. Test parameters are survival, reproduction, and population growth of the introduced organisms. Furthermore, measurement of several histological and enzymatic parameters in these organisms is recommended. As no substantial test results are available, the predictive value and sensitivity of this system cannot be evaluated.

11.4.2.2 Tests on soil microbial processes

Tests on single species of isolated microorganisms in artificial substrates are not regarded as representative for the soil ecosystem, and will, therefore, not be considered here.

Based on a series of workshops held during the 1970s and 1980s, Somerville and Greaves (1987) formulated several recommended tests to assess the side effects of pesticides on the soil microflora. For all microbial tests in soil, the use of freshly sampled soil containing an active microflora was considered essential. Prolonged storage and drying of the soil should be avoided. For a proper assessment of the effect of chemicals, at least two different soil types should be used (Somerville and Greaves, 1987). A short description is given here of several tests on microbial processes related to the conversion of nutrients in soil. Unless stated otherwise, all tests are carried out in the dark at a temperature of 20±2°C, and the test chemicals are mixed homogeneously through the soil. Generally, soils are tested at a moisture content corresponding to field capacity or to 40–60 percent of the water holding capacity.

Test for soil respiration and mineralization of substrates

In these tests the production of CO_2 from small soil samples (≤100 g) treated with the test chemical is measured continuously or semi-continuously. The tests should run for a minimum of 30 days. Tests may be performed in either unamended soil or in soil amended with a substrate. For this purpose mostly 0.5 percent (w/w) lucerne or horn meal is used (Somerville and Greaves, 1987). The disadvantage of this soil respiration test is that the activity of the total soil microflora is determined. When certain species are affected by the test chemical, this will often not be noticed, as other (less sensitive) species may take over the activity of the sensitive ones.

During the past decade some new test methods have been developed which aim to determine chemical effects on more specific groups of soil micro-organisms. One is the addition of a readily degradable substrate and the determination of the short-term respiration rate. Such a test was described by Haanstra and Doelman (1984) using glutamic acid as a substrate. Soils are amended with glutamic acid and the CO_2-production is measured. Glucose may be used as a substrate. The duration of the test is no longer than 100–120 hours. The test appeared to be quite sensitive to heavy metals. These short-term respiration tests may be combined with a biomass determination, and seem to be more sensitive than the traditional respiration tests. Another alternative may be found in the addition of more persistent substrates such as lignin or cellulose (Ljungdahl and Eriksson, 1985). Only a few soil microorganisms are capable of degrading these substrates, and, when they are affected by the test chemical, no others can take over their activity.

The disadvantage of the previously described soil respiration or substrate degradation methods is that less sensitive species of micro-organisms may grow on the substrate during the test. This results in a shift among the microflora towards more resistant species, masking the possible elimination of sensitive species (Van Beelen *et al.*, 1991). For this reason, Van Beelen *et al.* (1990) developed test methods using the mineralization of low concentrations of ^{14}C-acetate, ^{14}C-chloroform or other labelled substrates. The amount of substrate applied is very low (1 μg per L) to ensure that no growth of the microflora will occur. This amount of substrate is added to a slurry of the test soil, prepared by mixing the homogenized soil with an equal weight of ground water. The test chemicals are added to the slurry in the desired concentration levels. Samples are incubated at 10°C. The test duration depends on the capacity of the microflora in the soil sample to mineralize the test substrate, and is chosen depending on the half-life of the acetate mineralization. Acetate mineralization is measured by determining the amount of $^{14}CO_2$ released from the sample and by determining the amount of ^{14}C remaining in the suspension at the end of the test.

Test for ammoniation and nitrification

In ammoniation tests, the release of inorganic nitrogen from soil organic matter or a substrate (e.g., plant material or horn meal) is studied in a way comparable to soil respiration tests. The influence of nitrification, i.e., the conversion of ammonia into nitrate, may also be studied in these tests. Ammoniation is performed by a wide variety of soil micro-organisms, and is, therefore, relatively insensitive to perturbation. The advantages of nitrification are (1) that fewer species of micro-organisms are involved in this process and (2) that the process is considered to be of ecological and agricultural importance. Therefore, either combining these parameters in one test or running a separate test on nitrification is recommended (Somerville and Greaves, 1987). Nitrification tests can be performed in soil amended with either $(NH_4)_2SO_4$ or with organic substrates such as lucerne or horn meal. For this purpose, substrate equivalent to approximately 100 mg N per kg soil is added, and the disappearance of NH_4^+ and the appearance of NO_3^- is monitored. In case the rate of NO_3^- formation does not follow the disappearance rate of the NH_4^+, the soil should also be checked for the formation of NO_2^-. To check whether the test soil is capable of nitrification and whether the organic matter amendment is suitable for ammoniation and nitrification, studies are also recommended (Somerville and Greaves, 1987).

Test for nitrogen fixation

Tests on both symbiotic and asymbiotic nitrogen fixation can be identified. Tests on symbiotic nitrogen fixation in fact consider the unique relationship between the host plant and *Rhizobium*, and, therefore, include in one test effects on both the plant and the bacteria. These experiments are conducted in a soil suitable for growth of the plant. The plant, seeds, or soil can be inoculated with *Rhizobium* if no suitable bacteria are present in the soil (Somerville and Greaves, 1987). Effects on plant growth and the degree of nodulation should be included. In tests on asymbiotic nitrogen fixation, the degree of acetylene reduction (or formation of ethylene from acetylene) by soil samples is determined in relation to the addition of the test chemical.

Test for denitrification

Denitrification is the conversion of nitrate to atmospheric nitrogen, and will especially take place under anaerobic conditions. This process may be relevant in soil, as microsites may become anaerobic. In this test, soils are generally flooded with a layer of water, and nitrate is supplied as a substrate. Additionally, an organic substrate such as glucose is added to the soil. Since the process cannot be quantified, the formation of nitrogen gas, the disappearance of nitrate, and the formation of nitrite are measured as test parameters (Anderson, 1978a).

11.4.2.3 Tests on enzyme activity in soil

As in the tests on microbial processes, chemicals are also mixed homogeneously through the soil in tests on enzyme activity. Moisture content is adjusted to field capacity or to 40–60 percent of water holding capacity, and all incubations are done in the dark at 20°C. Several soil enzymes, relevant to microbial processes in soil, can be used as test parameters, as noted below.

Soil enzyme: urease

At several intervals, small soil samples (6–7 g) are taken, and incubated with 5 ml of demineralized water and 1.0 ml of a solution containing 60 mM urea. Incubation is at 35°C for 5 hours on a shaking water bath. A phenylmercury acetate solution in 2 M KCl is added to the soil samples to stop the urease reaction. After 10 minutes of shaking, the soil suspensions are filtered. The filtrates are analysed photometrically at 525 nm for urea concentrations (NEN 5796, 1989).

Soil enzyme: dehydrogenase

Soil samples (5–10 g) are incubated with a solution of TTC (2,3,5-triphenyl tetrazolium chloride) in 0.1 M tris buffer solution (pH 7.6), and incubated for 24 hours at 30 or 37°C. The reduced triphenyl formazan formed is extracted with methanol and quantified by measuring the absorbance at 485 nm (Casida *et al.*, 1964; Thalmann, 1968). Dehydrogenase reflects a broad range of microbial oxidative activities, and does not consistently correlate to microbial numbers, CO_2 evolution or O_2-consumption. Additionally, dehydrogenase activity may depend upon the nature and concentration of amended C-substrates and alternative electron acceptors (Somerville and Greaves, 1987). Rossell and Tarradellas (1991) concluded that short-term (substrate-induced) dehydrogenase activity may reflect the impact of chemicals on the physiologically active biomass of the soil microflora.

Soil enzyme: phosphatase

At several intervals, 0.5 g soil samples are taken, and incubated with 5 mM *p*-nitrophenylphosphate (*p*-NPP) for 1 hour in a shaker at 20°C. Phosphatase activity is measured as the amount of *p*-nitrophenol formed using a spectrophotometer (Tabatabai and Bremner, 1969). Phosphatase is said to bear little relation to total phosphate availability in soils (Somerville and Greaves, 1987). Its relevance for microbial activity in soil may, therefore, be questionable.

Somerville and Greaves (1987) stated that soil enzyme activities would be of little value to monitor side effects of pesticides on microflora. The main reasons for this were:

1. The total enzymatic activity of the soil is made up of various fractions, and quantifying the contribution of each to the catalysis of a particular substrate is extremely difficult; furthermore, many enzymes are formed extracellularly, and will still be active when the micro-organisms responsible for their production have been eliminated.
2. There is no universally agreed methodology, and almost any result can be achieved by varying assay conditions (temperature, pH, substrate). Although tests on enzyme activity have been described by many authors, few data are available to judge the reproducibility of these methods. Also soil animals, such as collembola and isopods, significantly influence the activity of several enzymes, such as urease (Verhoef and Brussaard, 1990), dehydrogenase, and cellulase (Teuben and Roelofsma, 1990). Therefore, discriminating between direct and indirect effects of the tested chemicals on microorganisms is difficult. Other enzymes that are more or less frequently used as test parameters for microbial activity in soil are: arylsulphatase, β-glucosidase, β-acetylglucosaminidase, saccharase, galactosidase, protease, and phosphodiesterase.

11.4.3 FIELD TESTS

The reliability of microcosm studies in the laboratory to interpret field conditions is much debated. Microcosms differ from the field situation concerning the influence of temperature and moisture dynamics, the influence of root presence, and the composition of the soil biota community. A recent study compared microcosm studies in the laboratory with mesocosm studies and direct field measurements concerning microbial respiration, enzyme activities, and availability of macronutrients in interaction with soil animals; these soil process variables appeared to be of the same order of magnitude (Teuben and Verhoef, 1992).

The tests described in the preceding sections provide only a rough estimate of the possible hazard imposed by a chemical in the environment. In many cases, this degree of precision is sufficient. Usually the laboratory test is considered to be a "worst case" situation, since test animals are exposed to a constant concentration that is relatively available, because the test substrate is prepared freshly. Under field conditions, exposure may be lower since the chemical is not distributed uniformly over the habitat, and bioavailability will often be lower due to various sorption processes. By contrast, the laboratory test considers the test organism under optimal conditions, without secondary stresses, such as those of food shortage, drought, and cold. The uncertainties attached to the laboratory-to-field extrapolation can be avoided by conducting experiments under semi-field or field conditions.

Various organizations have recommended test protocols for field investigations (IOBC, 1988; OEPP/EPPO, 1991). In several cases, guidelines for field tests are part of the national registration procedures for pesticides. Furthermore, considerable scientific research has been done in which side-effects of pesticides on non-target arthropods have been documented, and several reviews on this aspect have been published (Edwards and Thompson, 1973; Eijsackers and Van de Bund, 1980; Inglesfield, 1988; Jepson, 1989). Some attempts have also been made to develop standardized procedures for field tests to assess the effects of pesticides on earthworms (Kula, 1992).

11.4.3.1 Cage tests using selected arthropod species

Some of the arthropods used as laboratory test species can also be exposed to chemicals under semi-field conditions, while exposed in cages. Hassan *et al.* (1985) lists protocols developed for *Trichogramma cacoeciae, Phygadeuon trichops, Coccygomimus turionellae, Phytoseiulus persimilis, Aleochara bilineata, Chrysoperla carnea*, and *Drino inconspicua*. The usual procedure is to treat a group of plants with a spray of the chemical. A cage is then put over the plants after which test animals, with hosts or food, are introduced. The cage is installed either in a greenhouse or outdoors under a cover to provide shelter from rain and excessive sunshine. After an adequate duration of exposure, the performance of beneficials is compared with water-treated controls.

Cage tests have been described in detail for testing with pollinators (Felton *et al.*, 1986; OEPP/EPPO, 1991). Bees (*Apis mellifera*) from small colonies are made to forage on a flowering crop in cages measuring minimally 2 × 2 × 3 m, with a 3 mm mesh netting. The product is applied to the plants, and not to the cage walls, by spraying. The EPPO-guideline does not require replication of the treatment. Effects are recorded at several intervals, preferably 0, 1, 2, 4, and 7 days after treatment. Observations are made on the number of dead bees, on foraging activity, and on behaviour. The results are compared with a blank control (usually water-sprayed) and a positive control (a reference product known to be hazardous to bees, e.g., parathion).

11.4.3.2 Honey bee field test

OEPP/EPPO (1991) also provides a guideline for field tests on honey bees. A chemical to be tested is applied to a plot of at least 1500 m^2, with the crop for which the chemical is intended, or another crop attractive to bees (rape, *Phacelia*), in full flower. Per treatment, three colonies of honey bees are placed in or on the edge of the plot. Test plots should be separated by at least 500 to 1000 m^2 to avoid bees foraging on the wrong plot. Replication of the treatment is considered desirable, but is not required in the EPPO guideline. A blank control (untreated, or treated with a reference product known to present a low hazard to bees), as well as a positive control (e.g., parathion, dimethoate) are applied to separate test plots.

After treatment, observations are made at several intervals, preferably after 0, 1, 2, 4, 7, and 14 days. Meteorological data are recorded during the entire period of the trial. Several parameters are estimated such as the number of foraging bees in the crop, behaviour of bees on the crop and around hives, mortality of bees (using dead bee traps), pollen collection (using pollen traps), pollen in collected honey, number of bees on frames, brood status in frames, and residues in dead bees, pollen wax, and honey. For the test to be valid, mortality in the negative control should not exceed 15 percent, while mortality in the positive control should be statistically significant.

11.4.3.3 Arthropod fauna in arable crops

Hassan *et al.* (1985) summarize the recommendations made by the IOBC Working Group Pesticides and Beneficial Organisms for full-scale field tests. The methods are suitable for a variety of crops, but have been applied mostly to winter wheat.

The trial is laid out in a replicated block design: three large fields with similar agronomic history are each divided into three treatment areas, where each treatment area covers at least 3 ha. The treatments are a spray with the product to be tested, a blank treatment (water spray), and a positive control (e.g., dimethoate).

Sampling is planned on seven occasions: 10 and 5 days before treatment, 2, 5, 10, and 20 days after treatment, and just before harvest. Sampling activities are concentrated in the central parts of each treatment area. Crop foliage fauna is collected with a suction net sampler (e.g., the Dietrich vacuum sampler). Soil surface fauna are sampled using pitfall traps, left in the field for 5 days. Visual inspection, water traps, and sticky traps can provide additional information on those arthropods sampled inefficiently by pitfalls or suction samplers.

The fauna collected are identified to at least the family level. Some groups where species can be recognized easily can be further subdivided (e.g., carabid beetles).

In addition to the large-scale experiments suggested for arthropods in winter wheat, smaller set-ups have been suggested by Edwards and Thompson (1973) and Eijsackers and Van de Bund (1980). Treatment plots of 3 × 3 m are recommended for microarthropods (Collembola, mites), while 10 × 10 m plots are suitable for studies on beetles and spiders. In all cases, however, the plots should be fenced, preferably using a polythene sheet, protruding 15 cm below ground and 40 cm above. The barrier should limit the immigration from neighbouring plots by surface-active arthropods such as beetles.

The statistical treatment of data from field experiments is not harmonized. Yet, such harmonization may be important since the probability of finding effects of the treatment will depend on the power of the statistical analysis. Stewart-Oaten *et al.* (1986) suggested the use of "pseudoreplication in time," to allow for a detailed evaluation of the effects of a treatment in relation to a control. In this design, also called BACI (Before After Control Impact comparison), the correlation between the observations from control plots and treatment plots before treatment is used to assess the effects in the treatment plots as deviations from the expectations made on the basis of the control plots (Everts *et al.*, 1989; Jagers op Akkerhuis and Van der Voet, 1992).

11.4.3.4 Arthropod fauna in orchards

Hassan *et al.* (1985) summarize the standardized methods developed by the IOBC Working Group "Integrated Protection in Orchards." The methods consist of catching the fauna in collectors placed under the trees to which a chemical has just been applied. The collectors can be trays, canvas sheets, or funnels ("Steiner funnels"), with at least 0.5 m^2 of collecting area. In each trial, both the control (water spray) and the test treatment are followed by a "cleaning" treatment, 48 hours after the initial treatment. The "cleaning" treatment consists of dichlorvos at double the recommended dose, which will remove all beneficials present. The effectiveness of the treatment can thus be expressed in relation to the total population of beneficials present in the treated tree. The use of a reference chemical with each treatment, e.g., phosalone, is also recommended.

The design of the trial is a complete randomized block design, or a balanced incomplete block design. Each replicate is represented by one tree with one or more fauna collectors; six to eight replicates per treatment are recommended. The trees should be separated by at least one untreated tree.

The fauna are gathered from the collectors 24 hours and 48 hours after the treatment, as well as 24 hours after the "cleaning" treatment. Only the arthropods, of which there is at least an average of ten individuals per collector, are considered. The fauna are identified to the species or the family level.

11.4.3.5 Earthworm field tests

Although no standardized guidelines exist yet for the study of pesticide side-effects on earthworms, recommendations for the performance of such a test have been formulated by Kula (1992).

Field studies with earthworms can be performed on arable land or on permanent grass; in both cases, a minimum number (100 individuals per m^2) of earthworms is required, and the relevant species (*Aporrectodea caliginosa* and *Lumbricus terrestris*) must be present. Minimum plot size should be 100 m^2, and at least four replicate plots should be used per treatment. A study should include a control, the highest recommended dose, and a toxic standard (benomyl). In many cases also a manyfold (e.g., fivefold) of the recommended dose should be studied. At each sampling time, at least two samples (sampling area 0.25 m^2) should be taken from each replicate plot. Preferably treatment should take place in the spring, and samples should be taken 1, 4 to 6, and 12 months after application. For the sampling of earthworms, the formaldehyde method and electrical sampling methods seem to be the most useful.

11.5 METHODS TO ASSESS THE IMPACT OF SOIL CONTAMINATION ON SOIL ORGANISMS (DIAGNOSIS)

11.5.1 LABORATORY AND FIELD BIOASSAYS

The main characteristic of laboratory bioassays is that the potential toxicity of samples of field soil is studied in the laboratory using laboratory-bred test organisms. So, in such a test, well known organisms having similar characteristics are used, and the test can be performed under controlled conditions ruling out other possible disturbing influences. The advantage of bioassays is that they provide a direct indication of the toxicity of a specific soil, and integrate the effect of all substances present. The disadvantage is that the specific chemicals causing the observed effects cannot be identified. Bioassays can also be applied to determine the bioavailability of pollutants in soils as an indication of the potential risk for higher trophic levels.

Laboratory bioassays with earthworms (*Eisenia fetida*) and higher plants (*Cyperus esculentus*) have been described by Marquenie and Simmers (1988) and Van Gestel *et al.* (1988); the latter authors used these bioassays to study the influence of soil clean-up on the bioavailability of metals. In these methods, cylinders with a diameter of about 18 cm are filled with a 20 cm layer of soil. The cylinders have a perforated bottom and are placed in a dish filled with water. After one week of preincubation, test organisms are introduced. After 4 weeks, earthworms are sorted out of the soil; after incubation on wet filter paper for 24–48 hours to void the gut, they are analysed for metal content. Plants are harvested, and shoots are analysed. Van Gestel *et al.* (1993) described similar bioassays, using lettuce (*Lactuca sativa*) and radish (*Raphanus sativus*), to determine metal bioavailability in soils. For these organisms, smaller amounts of soil are needed (about 400 mg), and as in the bioassays using *C. esculentus*, some nutrient solution was added to the soils to stimulate plant growth.

Bioassays using the earthworm species *Lumbricus terrestris* and *E. fetida* have been described by Menzie *et al.* (1992). They studied survival of the earthworms after exposure to contaminated soil for 28 and 14 days, respectively, and also determined the uptake of some selected chemicals in the earthworms.

Menzie *et al.* (1992) and Callahan *et al.* (1991) studied the potential risk of contaminated soils using *in situ* bioassays with earthworms. Contaminated soil was placed in plastic buckets placed in the ground from which the soil was taken. The buckets were constructed to allow for free exchange of air and water. Adult earthworms of the species *Lumbricus terrestris* were placed in the buckets, and observations were made after 1 and 7 days. Survival and morbidity (burrowing, coiling, shortening, swelling, lesions) were the test parameters, and earthworms (including gut content) were analysed to determine bioaccumulation of the main pollutants. This bioassay proved to give a good indication of the possible risk of polluted soil.

Two observations merit comment. First, the bioassays last only 7 days, which might be too short to allow for an equilibrium in the uptake of highly lipophilic chemicals. Thus, uptake of these chemicals by the earthworms may be underestimated. Second, uptake might be misjudged, because of the presence of contaminated soil in the earthworm gut. From the authors' experience, the gut content of an earthworm may account for about 50 percent of its dry weight.

Kopezski (1992) used small enclosures (3 cm long; 4.8 cm in diameter) to study the impact of acidification on population growth and decomposition activity of the collembola species *Folsomia candida* and *Heteromurus nitidus* in a forest soil. The enclosures were filled with 1 g of hazel leaf litter and 1 g of wafers, and buried into the soil. By using cheese cloth for the bottom and top ends of the enclosures, free contact with the surrounding soil was ensured. After 6 months, samples were analysed. A significant correlation appeared between animal numbers and soil pH, with *H. nitidus* being most sensitive. Decomposition showed a somewhat weaker correlation with soil pH.

To study the effects of N-deposition on the interactions between soil fauna and microflora and the effects on mineralization in coniferous forest soils, lysimeters have been used by Berg and Verhoef (1992). Intact soil cores with total soil fauna or absence of mesofauna are treated with three $(NH_4)_2SO_4$ concentrations (10, 50, and 200 kg *N* per ha per g). The lysimeters are defaunated by means of microwave treatment. Before installation, they are incubated with a soil-spore suspension. Faunal groups extracted from the soil are added to the different treatments. Migration of fauna between the lysimeter and the surrounding soil can occur through holes in the lysimeter. The lysimeters are covered by a gauze lid and covered just above the top by a plexiglass roof preventing rain input. The lysimeters are watered every two weeks by hand.

11.5.2 MUTAGENICITY TESTS

Besides the bioassays using invertebrates or higher plants, special methods may be necessary to assess the potential risk of contaminated soils. Some compounds present in soil as contaminants (e.g., PAHs) are known mutagens, and their effects can be assessed by genotoxicity tests developed for drinking water, surface water, and sediments. These tests may be applicable for industrially contaminated sites, waste disposals, or sewage sludge-amended soils.

The usual approach to assess mutagenicity is to record the number of revertants in a *Salmonella thyphimurium* strain, plated with the test substance on a histidine deficient medium. To include those mutagens that require metabolic activation, a rat liver microsome suspension is added (Ames *et al.*, 1975). In addition to this test, several other procedures have been proposed, which are often more sensitive than this test (Van der Gaag, 1989).

To apply mutagenicity tests to contaminated soils, an extract must be obtained. The extraction may be crucial to the validity of the results: extractions with solvents such as methanol or dimethylsulphoxide often induce a stronger mutagenic response in the *Salmonella* test, compared to water leachates; this difference in potency is especially true for superficial soil horizons (Kool *et al.*, 1989; Donnelly *et al.*, 1991).

The ecological relevance of mutagenicity test results for soils is difficult to evaluate, as this field of research is still underdeveloped. Mutagenicity is detected not only at contaminated sites but also in uncontaminated soil (Kool *et al.*, 1989; Brown *et al.*, 1985), and often bears no clear relationship with the levels of known chemical mutagens in soil (Donnelly *et al.*, 1991).

11.5.3 FIELD STUDIES (BIOMONITORING)

Biomonitoring studies in soil often deal with the study of the effect of chemical stress on the structure and function of entire ecosystems, or at least at the community level. For that purpose, studies are performed to determine the impact on litter decomposition in forest ecosystems or on communities of soil mesofauna. Foodweb models for soil organisms have been constructed for agro-ecosystems (Hunt *et al.*, 1987), and are developed for a coniferous forest soil, based on stratified litterbag experiments (Berg and Verhoef, 1992). With these models, effects of excessive N-input on the soil ecosystem can be estimated. No standardized guidelines exist for such studies.

Also certain species of organisms can be selected, and followed in time to detect certain deviations that may be due to the impact of chemical contamination, a process called biomonitoring. Isopods are, for instance, excellent bioindicators for metal contamination, because of their ability to concentrate metals in their body tissues (Hopkin, 1989).

Tolsma *et al.* (1991) recommended inclusion of primary producers (the plant species *Urtica dioica* and *Holcus lanatus*), detritivores (earthworms and isopods), and carnivores (mice or moles) in a biomonitoring system for the terrestrial environment. These organisms were selected because of their capability to accumulate metals, PAH, and organochlorine pesticides.

11.6 CONCLUSIONS

The array of methods used in testing terrestrial invertebrates is diverse, mainly because different tests have been developed with different aims. Many methods are still poorly described, especially in relation to the medium to which the chemical is applied, and the consequences for bioavailability.

The potential for standardization of a test system is important when one strives for international use of test methods. For some species, a standardized method may not be possible; this holds for the oribatid mite *Platynothrus peltifer*, that has a very long life cycle and a low reproduction rate, not allowing for a proper determination of effects on reproduction within a reasonable test duration. For others, such as the earthworms and the collembola *Folsomia candida*, tests have already been standardized at an international level, as is the case for many tests on beneficial arthropods.

The usefulness of a test system to derive environmental quality criteria also comprises the substrate used. When real or artificial soils are used as a test substrate, test results may be applied directly to derive soil quality criteria; however, this is not the case for tests on soil organisms using other substrates or exposure routes, such as water, agar, or nutrient solutions used in tests with protozoans and nematodes. Results of such tests cannot be translated directly to soil quality criteria. This conclusion holds also for exposure routes used in tests on beneficial arthropods; such tests can only be useful for the risk assessment of pesticides when test results can be related to natural exposure routes.

By one approach, each species should be tested under its optimal conditions, and inter-species harmonization of conditions (e.g., by using the same test substrate) is not to be recommended. The usefulness of more than one test on the same chemical is, however, very limited when the results for two species cannot be compared to each other. Standardization of the test substrate should, therefore, be considered in the further development of test methods.

Only a few methods are available to determine the potential risk of contaminated soils. Because of the complex nature of contaminated soil, in which often a mixture of chemicals is involved, such methods are urgently needed. Also more knowledge is needed on the toxicity of combinations of toxicants.

For a suitable risk assessment, a battery of tests should be available. Such a test battery should contain organisms representing different taxonomic groups as well as the soil community. For that purpose, the Health Council of the Netherlands (1991) selected 24 parameters to be applied for both diagnosis and prognosis of chemical effects in terrestrial ecosystems and sediments. A conclusion was that, especially for higher levels of organization, tests are lacking. For an initial screening of possible effects, a test system should consider higher plants, decomposition capacity of the soil, invertebrates, and vertebrates (the latter being outside the scope of this paper).

11.7 REFERENCES

Ames, B.N., McCann, J., and Yamasaki, E. (1975) Methods for detecting carcinogens and mutagens with the *Salmonella*/mammalian microsome mutagenicity test. *Mutat. Res.* **113**, 173–215.

Anderson, J. (1978a) Some methods for assessing pesticide effects on non-target soil microorganisms and their activities. In: Hill, I.R., and Wright, S.J.L. (Eds.) *Pesticide Microbiology*. Academic Press, London.

Anderson, J.M. (1978b) Competition between two unrelated species of soil Cryptostigmata (Acari) in experimental microcosms. *J. Anim. Ecol.* **47**, 787–803.

Anderson, J.M., and Ingram, J.S.I. (1989) *Tropical Soil Biology and Fertility: A Handbook*

of Methods. C.A.B. International.
Ausmus, B.S., Dodson, G.J., and Jackson, D.R. (1978) Behavior of heavy metals in forest microcosms. III. Effects on litter-soil carbon metabolism. *Water Air Soil Pollut.* **10**, 19–26.
Axelsson, B. (1984) Ultimate forest productivity: what is possible? In: *Proceedings of the International Union of Forest Research Organisations Symposium on Forest Site and Continuous Productivity*, pp. 61–69.
Bengtsson, G., Gunnarsson, T., and Rundgren, S. (1983) Growth changes caused by metal uptake in a population of *Onychiurus armatus* (Collembola) feeding on metal polluted fungi. *Oikos* **40**, 216–225.
Bengtsson, G., Gunnarsson, T., and Rundgren, S. (1985) Influence of metals on reproduction, mortality and population growth in *Onychiurus armatus* (Collembola). *J. Appl. Ecol.* **22**, 967–978.
Bengtsson, G., Berden, M., and Rundgren, S. (1988) Influence of soil animals and metals on decomposition processes: a microcosm experiment. *J. Environ. Qual.* **17**, 113–119.
Berg, M.P., and Verhoef, H.A. (1992) Effects of nitrogen deposition on nutrient cycling and soil biota in coniferous forest soils in The Netherlands; A field approach. In: Teller, A., Mathy, P., and Jeffers, J.N.R. (Eds.) *Responses of Forest Ecosystems to Environmental Changes*, pp. 707–708. Elsevier, Amsterdam.
Berhin, F., Houba, C., and Remade, J. (1984) Cadmium toxicity and accumulation by *Tetrahymena pyriformis* in contaminated river waters. *Environ. Pollut.* (Series A) **35**, 315–329.
Bond, H., Lighthart, B., Shimabuku, R., and Russell, L. (1976) Some effects of cadmium on coniferous forest soil and litter microcosms. *Soil Science* **121**, 278–287.
Brown, K.W., Donnelly, K.C., Thomas, J.C., Davel, P., and Scott, B.R. (1985) Mutagenicity of three agricultural soils. *Sci. Total Environ.* **41**, 173–186.
Callahan, C.A., Menzie, C.A., Burmaster, D.E., Wilborn, D.C., and Ernst, T. (1991) On-site methods for assessing chemical impact on the soil environment using earthworms: a case study at the Baird and McGuire superfund site, Holbrook, Massachusetts. *Environ. Toxicol. Chem.* **10**, 817–826.
Casida, L.E., Jr, Klein, D.A., and Satoro, T. (1964) Soil dehydrogenase activity. *Soil Sci.* **98**, 371–376.
Chaney, W.R., Kelly, J.M., and Strickland, R.C. (1978) Influence of cadmium and zinc on carbon dioxide evolution from litter and soil from a black oak forest. *J. Environ. Qual.* **7**, 115-119.
Coûteaux, M.M., Morisseau, M., Célérier, M.L., and Bottner, P. (1991) Increased atmospheric CO_2 and litter quality: decomposition of sweet chestnut leaf litter with animal food webs of different complexities. *Oikos* **61**, 54–64.
Croft, B.A. (1990) *Arthropod Biological Control Agents and Pesticides*. John Wiley & Sons, New York.
Dallinger, R., and Wieser, W. (1977) The flow of copper through a terrestrial food chain. I. Copper and nutrition in isopods. *Oecologia* **30**, 253–264.
Denneman, C.A.J., and Van Straalen, N.M. (1991) The toxicity of lead and copper in reproduction toxicity tests using the oribatid mite *Platynothrus peltifer*. *Pedobiologia* **35**, 305–311.
De Vries, W. (1991) *Methodologies for the Assessment and Mapping of Critical Loads and of the Impact of Abatement Strategies on Forest Soils*. Report 46. DLO The Winand Staring Centre, Wageningen. 109 pp.

Donnelly, K.C., Brown, K.W., Anderson, C.S., Thomas, J.C., and Scott, B.R. (1991) Bacterial mutagenicity and acute toxicity of solvent and aqueous extracts of soil samples from an abandoned chemical manufacturing site. *Environ. Tox. Chem.* **10**, 1123–1131.

Edwards, C.A. (1983) *Report of the Second Stage in Development of a Standardized Laboratory Method for Assessing the Toxicity of Chemical Substances for Earthworms.* Commission of the European Communities, DOC XI/83/700, Brussels.

Edwards, C.A., and Thompson, A.R. (1973) Pesticides and the soil fauna. *Res. Rev.* **45**, 1–79.

EEC (1985). EEC Directive 79/831. Annex V, Part C. Methods for determination of ecotoxicity. Level I. C(II)4: Toxicity for earthworms. Artificial soil test. DG XI/128/82.

Eijsackers, H. (1978a) Side effects of the herbicide 2,4,5-T affecting mobility and mortality of the springtail *Onychiurus quadriocellatus* Gisin (Collembola). *Zeitschr. angew. Entomol.* **86**, 349–372.

Eijsackers, H. (1978b) Side effects of the herbicide 2,4,5-T affecting the isopod *Philoscia muscorum* Scopoli. *Zeitschr. angew. Entomol.* **87**, 28–52.

Eijsackers, H. (1991) Litter fragmentation by isopods as affected by herbicide application. *Neth. J. Zool.* **41**, 277–303.

Eijsackers, H., and Van de Bund, C.F. (1980) Effects on soil fauna. In: Hance, R.J. (Ed.) *Interactions Between Herbicides and the Soil*, pp. 255–305. Academic Press, London.

Eriksson, B. (1980) Statistical analysis of precipitation data. Part II: frequency analysis of monthly precipitation data. *Swedish Meteorol. Hydrol. Inst. Reporter*, No. RMK **17**, 1–152.

Everts, J.W., Aukema, B., Hengeveld, R., and Koeman, J.H. (1989) Side-effects of pesticides on ground-dwelling predatory arthropods in arable ecosystems. *Environ. Pollut.* **59**, 203–225.

Faber, J.H., and Verhoef, H.A. (1991) Functional differences between closely related soil arthropods with respect to decomposition processes in the presence and absence of pine tree roots. *Soil Biol. Biochem.* **23**, 15–23.

Felton, J.C., Oomen, P.A., and Stevenson, J.H. (1986) Toxicity and hazard of pesticides to honey bees: harmonization of test methods. *Bee World* **67**, 114–124.

Ferrière, G., Fayolle, L., and Bouché, M.B. (1981) Un nouvel outil, essentiel pour l'écophysiologie et l'écotoxicologie: l'élevage des lombriciens en sol artificiel. *Pedobiologia* **22**, 196–201.

Giesy, J.P. Jr (1980) *Microcosms in Ecological Research.* Technical Information Center, US Department of Energy, Springfield.

Grennfelt, P., Larsson, S., Leyton, P., and Olsson, B. (1985) Atmospheric deposition in the Lake Gårdsjön area, S.W. Sweden. In: Anderson, F., and Olsson, B. (Eds.) *Lake Gårdsjön: An Acid Forest Lake and Its Catchment. Ecological Bulletin* **37**, 101–108.

Haanstra, L., and Doelman, P. (1984) Glutamic acid decomposition as a sensitive measure of heavy metal pollution in soil. *Soil Biol. Biochem.* **16**, 595–600.

Hågvar, S. (1988) Decomposition studies in an easily-constructed microcosms: effects of microarthropods and varying soil pH. *Pedobiologia* **31**, 293–303.

Haight, M., Mudry, T., and Pasternak, J. (1982) Toxicity of seven heavy metals on *Panagrellus silusiae*: the efficacy of the free-living nematode as an *in vivo* toxicological bioassay. *Nematologica* **28**, 1–11.

Harrison, A.F., Dighton, J., Hatton, J.C., and Smith, M.R. (1985) A phosphorus-deficiency bioassay for trees and grasses growing in low nutrient status soils. In: *Proceedings of the VIth International Colloquium for the Optimisation of Plant Nutrition 3*, pp. 957–963. AIONP, Montpellier.

Harrison, A.F., Hatton, J.C., and Taylor, K. (1986) Application of a root bioassay for determination of P-deficiency in high-altitude grasslands. *J. Sci. Food Agric.* **37**, 10–11.

Harrison, A.F., Ineson, P., and Heal, O.W. (1990) *Nutrient Cycling in Terrestrial Ecosystems; Field Methods, Application and Interpretation.* Elsevier Science, London. 454 pp.

Hassan, S.A., Bigler, F., Blaisinger, P., Bogenschütz, H., Brun, J., Chiverton, P., Dickler, E., Easterbrook, M.A., Edwards, P.J., Englert, W.D., Firth, S.I., Huang, P., Inglesfield, C., Klingauf, F., Kühner, C., Ledieu, M.S., Naton, E., Oomen, P.A., Overmeer, W.P.J., Plevoets, P., Reboulet, J.N., Rieckmann, W., Samsøe-Petersen, L., Shires, S.W., Stäubli, A., Stevenson, J., Tuset, J.J., Vanwetswinkel, G., and Van Zon, A.Q. (1985) Standard methods to test the side-effects of pesticides on natural ennemies of insects and mites developed by the IOBC/WRPS Working Group "Pesticides and Beneficial Organisms." *OEPP/EPPO Bull.* **15**, 214–255.

Health Council of The Netherlands (1991) *Quality Parameters for Terrestrial Ecosystems and Sediments. A Selection of Practical Exotoxicological Assessment Methods.* Publication No. 91/17E. Health Council of the Netherlands, The Hague.

Heimbach, F. (1984) Correlation between three methods for determining the toxicity of chemicals for earthworms. *Pest. Sci.* **15**, 605–611.

Heimbach, F. (1992) In: Becker, H., Edwards, P.J., Greig-Smith, P.W., and Heimbach, F. (Eds.) *Ecotoxicology of Earthworms.* Intercept Press, Andover, Hants., UK.

Hopkin, S.P. (1989) *Ecophysiology of Metals in Terrestrial Invertebrates.* Elsevier Applied Science, London.

Hopkin, S.P. (1990) Species-specific differences in the net assimilation of zinc, cadmium, lead, copper and iron by the terrestrial isopods *Oniscus asellus* and *Porcellio scaber*. *J. Appl. Ecol.* **27**, 460–474.

Hopkin, S.P., Watson, K., Martin, M.H., and Mould, M.L. (1985) The assimilation of metals by *Lithobius variegatus* and *Glomeris marginata* (Chilopoda; Diplopoda). *Bijdr. Dierk.* **55**, 88–94.

Hunt, H.W., Coleman, D.C., Ingham, E.R., Ingham, R.E., Elliott, E.T., Moore, J.C., Rose, S.L., Reid, C.P.P., and Morley, C.R. (1987) The detrital food web in a shortgrass prairie. *Biol. Fertil. Soils* **3**, 57–68.

Hunter, I.R., Nicholson, G., and Thorn, A.J. (1985) Chemical analysis of pine litter: an alternative to foliage analysis? *New Zeal. J. Forest Sci.* **15**, 101–109.

Iglisch, I. (1986) Hinweise zur Entwicklung von Testverfahren zum Nachweis subakuter Wirkungen von Chemikalien. *Angew. Zool.* **2**, 199–218.

Inglesfield, C. (1988) Pyrethroids and terrestrial non-target organisms. *Pestic. Sci.* **27**, 387–428.

IOBC (1988) Guidelines for testing the effects of pesticides on beneficials: short description of test methods. *WPRS Bull* **11**, 1–143.

ISO (1991) *Soil Quality-Effects of Soil Pollutants on Collembola: Determination of the Inhibition of Reproduction.* ISO/TC 190/SC4/WG2, N26.

Jagers op Akkerhuis, G.A.J.M., and Van der Voet, H. (1992) A dose-effect relationship for the effect of deltamethrin on a linyphiid spider population in winter wheat. *Arch. Environ. Contam. Toxicol.* **22**, 114–121.

Jancke, G. (1989) Modelversuche zur subakuten und subletalen Wirkung von Herbiziden auf Collembolen im Hinblick auf ein Testsystem für Umweltchemikalien. *Zool. Beitr. N.F.* **32**, 261–299.

Jenkinson, D.S., Adams, D.E., and Wild, A. (1991) Model estimates of CO_2 emissions from soil in response to global warming. *Nature* **351**, 304–306.

Jepson, P.C. (1989) The temporal and spatial dynamics of pesticide side-effects on non-target invertebrates. In: Jepson, P.C. (Ed.) *Pesticides and Non-target Invertebrates*, pp. 95-127. Intercept Ltd, Wimborne.

Kirsten, W.J. (1979) Automatic methods for the simultaneous determination of carbon, hydrogen, nitrogen, and sulphur, and for phosphor alone in organic and inorganic materials. *Anal. Chem.* **51**, 1173–1175.

Kokta, C. (1992) A laboratory test on sublethal effects of pesticides on *Eisenia fetida*. In: Becker, H., Edwards, P.J., Greig-Smith, P.W., and Heimbach, F. (Eds.) *Ecotoxicology of Earthworms*, pp. 55–62. Intercept Press, Andover, Hants.

Kool, H.J., van Kreyl, C.F., and Persad, S. (1989) Mutagenic activity in groundwater in relation to mobilization of organic mutagens in soil. *Sci. Total Environ.* **84**, 185–199.

Kopezski, H. (1992) Versuch einer aktiven Bioindikation mit den bodenlebenden Collembolen-Arten *Folsomia candida* (Willem) und *Heteromurus nitidus* (Templeton) in einem Buchemwald-Ökosystem. *Zool. Anz.* **228**, 82–90.

Kula, H. (1992) Measuring effects of pesticides on earthworms in the field: test design and sampling methods. In: Greig-Smith, P.W., Becker, H., Edwards, P.J., and Heimbach, F. (Eds.) *Ecotoxicology of Earthworms*, pp. 90–99. Intercept Press, Andover, Hants.

Likens, G.E., Bormann, F.H., Pierce, R.S., Eaton, J.S., and Johnson, N.M. (1977) *Biogeochemistry of a Forested Ecosystem*. Springer Verlag, New York.

Ljungdahl, L.G., and Eriksson, K.-E. (1985) Ecology of microbial cellulose degradation. In: Marshall, K.C. (Ed.) *Advances in Microbial Ecology*. Plenum Press, New York.

Lloyd, D. (1985) Simultaneous dissolved O_2 and redox measurements: use of polarographic, bioluminescence and mass spectrometric monitoring combined with dual wavelength spectrometry. In: Degn, H., Cox, R.P., and Toftlund, H. (Eds.), pp. 37–53. Reidel, Dordrecht, The Netherlands.

McIntosh, R. (1984) *Fertiliser Experiments in Established Conifer Stands*. Forestry Commission Forest Record 127. HMSO, London.

Marigomez, J.A., Angulo, E., and Saez, V. (1986) Feeding and growth responses to copper, zinc, mercury and lead in the terrestrial gastropod *Arion ater* (Linné). *J. Moll. Stud.* **52**, 68–78.

Marquenie, J.M., and Simmers, J.W. (1988) A method to assess potential bioavailability of contaminants. In: Edwards, C.A., and Neuhauser, E.F. (Eds.) *Earthworms in Waste and Environmental Management*, pp. 367-375, SPB Acad. Publ., The Hague.

Meincke, K.-F., and Schaller, K.-H. (1974) Uber die Brauchbarkeit der Weinbergschnecke (*Helix pomatia* L.) im Freiland als Indikator fur die Belastung der Umwelt durch die Elemente Eisen, Zink und Blei. *Oecologia* (Berl.) **15**, 393–398.

Menzie, C.A., Burmaster, D.E., Freshman, J.S., and Callahan, C.A. (1992) Assessment of methods for estimating ecological risk in the terrestrial component: a case study at the Baird and McGuire superfund site in Holbrook, Massachusetts. *Environ. Toxicol. Chem.* **11**, 245–260.

Mola, L., Sabatini, M.A., Fratello, B., and Bertolani, R. (1987) Effects of atrazine on two species of Collembola (Onychiuridae) in laboratory tests. *Pedobiologia* **30**, 145–149.

Moser, H., and Wieser, W. (1979) Copper and nutrition in *Helix pomatia* (L.). *Oecologia* (Berl.) **42**, 241–251.

Mothes-Wagner, U., Reitze, H.K., and Seitz, K.-A. (1992) Terrestrial multispecies toxicity testing. 1. Description of the multispecies assemblage. *Chemosphere* **24**, 1653–1667.

NEN 5796 (1989) Bodem–Bepaling van de urease-aktiviteit (Dutch standard for determining soil urease activity).

Nilsson, O., and Grennfelt, P. (1988). *Critical Loads for Sulphur and Nitrogen*. pp. 418. Miljøreport 1988: 15. Nordic Council of Ministers, København.

Nyberg, D., and Bishop, P. (1983) High levels of phenotypic variability of metal and temperature tolerance in *Paramecium*. *Evolution* **37**, 341–357.

OECD (1984a) Guideline for testing of chemicals no. 207. Earthworm acute toxicity tests. Adopted 4 April 1984. Organisation for Economic Co-operation and Development, Paris.

OECD (1984b) Guideline for testing of chemicals no. 208. Terrestrial plant growth test. Adopted 4 April 1984. Organisation for Economic Co-operation and Development, Paris.

OECD (1989) *Report of the OECD Workshop on Ecological Effects Assessment*. OECD Environment Monograph No. 26. Organisation for Economic Co-operation and Development, Paris.

OEPP/EPPO (1991) *Draft Guideline for Evaluating the Hazards of Pesticides to Honey Bees, Apis mellifera* L. European Plant Protection Organization.

Rossell, D., and Tarradellas, J. (1991) Dehydrogenase activity of soil microflora: significance in ecotoxicological tests. *Environ. Toxicol. Water Qual.* **6**, 17–33.

Römbke, J. (1989) *Enchytraeus albidus* (Enchytraeidae, Oligochaeta) as a test organism in terrestrial laboratory systems. *Arch. Toxicol.* **13**, 402–405.

Römbke, J., and Knacker, T. (1989) Aquatic toxicity test for enchytraeids. *Hydrobiol. J.* **180**, 235–242.

Russell, L.K., DeHaven, J.I., and Botts, R.P. (1981) Toxic effects of cadmium on the garden snail (*Helix aspersa*). *Bull. Environ. Contam. Toxicol.* **26**, 634–640.

Samsøe-Petersen, L. (1990) Sequences of standard methods to test effects of chemicals to terrestrial arthropods. *Ecotox. Environ. Safety* **19**, 310–319.

Somerville, L., and Greaves, M.P. (Eds.) (1987) *Pesticide Effects on Soil Microflora*. Taylor & Francis, London.

Stewart-Oaten, A., Murdoch, W.W., and Parker, K.R. (1986) Environmental impact assessment: "Pseudoreplication" in time? *Ecology* **67**: 929–940.

Sturhan, D. (1986) Influence of heavy metals and other elements on soil nematodes. *Rev. Nématol.* **9**, 311.

Subagja, J., and Snider, R.J. (1981) The side effects of the herbicides atrazine and paraquat upon *Folsomia candida* and *Tullbergia granulata* (Insecta, Collembola). *Pedobiologia* **22**, 141–152.

Tabatabai, M.A., and Bremner, J.M. (1969) Use of *p*-nitrophenyl phosphate for assay of soil phosphatase activity. *Soil Biol. Biochem.* **1**, 301–307.

Teuben, A., and Roelofsma, T.A.P.J. (1990) Dynamic interactions between functional groups of soil arthropods and microorganisms during decomposition of coniferous litter in microcosm experiments. *Biol. Fertil. Soils* **9**, 145–151.

Teuben, A., and Verhoef, H.A. (1992) Relevance of micro- and mesocosm experiments for studying soil ecosystem processes. *Soil Biol. Biochem.* **24**(11), 1179–1183.

Thalmann, A., (1968) Zur Methodik der Bestimmung der Dehydrogenaseaktivität im Boden mittels Triphenyltetrazoliumchlorid (TTC). *Landw. Forsch.* **21**, 249–258.

Thompson, A.R., and Gore, F.L. (1972) Toxicity of twenty-nine insecticides to *Folsomia candida*: laboratory studies. *J. Econ. Entomol.* **65**, 1255–1260.

Tolsma, D.J., Van Hattum, B., Denneman, W.D., Aiking, H., Wegener, J.W.M., and Cofino, W.P. (1991) *Aanbevelingen voor de Integratie Vanbiotische Parameters in het RIVM-Bodemkwaliteitsmeetnet*. Report No. R-91/16. Institute for Environmental Studies, Vrije Universiteit, Amsterdam.

Tomlin, A.D. (1977) Toxicity of soil applications of the fungicide benomyl, and two analogues to three species of Collembola. *Can. Ent.* **109**, 1619–1620.

Ulrich, B., and Mayer, R. (1980) Throughfall and stemflow: a review. In: Nicholson, I.A., Paterson, I.S., and Last, F.T. (Eds.) *Methods for Studying Acid Precipitation in Forest Ecosystems*, pp. 21-27. Cambridge, UK.

Van Beelen, P., Fleuren-Kemilä, A.K., Huys, M.P.A., van Mil, A.C.H.A.M., and van Vlaardingen, P.L.A. (1990) Toxic effects of pollutants on the mineralization of substrates at low experimental concentrations in soils, subsoils and sediments. In: Arendt F., Hinseveld, M., and van den Brink, W.J. (Eds.) *Contaminated Soil*. p. 431–438. Kluwer Academic, The Netherlands.

Van Beelen, P., Fleuren-Kemilä, A.K., Huys, M.P.A., van Montfort, A.C.P., and van Vlaardingen, P.L.A. (1991) The toxic effect of pollutants on the mineralization of acetate in subsoil microcosms. *Env. Tox. Chem.* **10**, 775–789.

Van Capelleveen, H.E. (1987) *Ecotoxicity of Heavy Metals for Terrestrial Isopods*. Ph.D. Thesis, Vrije Universiteit, Amsterdam.

Van der Gaag, M.A. (1989). Rapid detection of genotoxins in waste water: New perspectives with the sister-chromatid exchange assay "in vivo" with *Nothobranchius rachowi*. In: Løkke, H., Tyle, H., and Bro-Rasmussen, F. (Eds.) *Proceedings of the First European Conference on Ecotoxicology*, pp. 259–262. DIS Congress Service, Vanløse.

Van Gestel, C.A.M. (1992) The influence of soil characteristics on the toxicity of chemicals for earthworms: a review. In: Becker, H., Edwards, P.J., Greig-Smith, P.W., and Heimbach, F. (Eds.) *Ecotoxicology of Earthworms*, pp. 44–54. Intercept Press, Andover, Hants., UK.

Van Gestel, C.A.M., and Ma, W.-C. (1990) An approach to quantitative structure-activity relationships in terrestrial ecotoxicology: earthworm toxicity studies. *Chemosphere* **21**, 1023–1033.

Van Gestel, C.A.M., Adema, D.M.M., De Boer, J.L.M., and De Jong, P. (1988) The influence of soil clean-up on the bioavailability of metals. In: Wolf, K., Van den Brink, W.J., and Colon, F.J. (Eds.) *Contaminated Soil*, pp. 63–65. Kluwer Academic, Dordrecht, The Netherlands.

Van Gestel, C.A.M., van Dis, W.A., van Breemen, E.M., and Sparenburg, P.M. (1989) Development of a standardized reproduction toxicity test with the earthworm species *Eisenia fetida andrei* using copper, pentachlorophenol, and 2,4-dichloroaniline. *Ecotoxicol. Environ. Safety* **18**, 305–312.

Van Gestel, C.A.M., Dirven-van Breemen, E.M., and Kamerman, J.W. (1993) The influence of soil clean up on the bioavialability of heavy metals for earthworms and plants. In: *Proceedings of the EUROSOL Conference*, Maastricht, 6–12 September 1992.

Van Kessel, W.H.M., Brocades Zaalberg, R.W., and Seinen, W. (1989) Testing environmental pollutants on soil organisms: a simple assay to investigate the toxicity of environmental pollutants on soil organisms using cadmium chloride and nematodes. *Ecotox. Environ. Safety* **18**, 181–190.

Van Straalen, N.M., and Verweij, R.A. (1991) Effects of benzo(a)pyrene on food assimilation and growth efficiency in *Porcellio scaber* (Isopoda). *Bull. Environ. Contam. Toxicol.* **46**, 134–140.

Van Straalen, N.M., Schobben, J.H.M., and de Goede, R.G. (1989) Population consequences of cadmium toxicity in soil microarthropods. *Ecotoxicol. Environ. Safety* **17**, 190–204.

Van Straalen, N.M., and Van Gestel, C.A.M. (1992a) Ecotoxicological test methods using terrestrial arthropods. Detailed Review Paper for the OECD Test Guidelines Programme.

Van Straalen, N.M., and Van Gestel, C.A.M. (1992b) Soil, Chapter 15. In: Calow, P. (Ed.) *Handbook of Ecotoxicology*. Blackwell Scientific, Oxford.

Van Wensem, J. (1989) A terrestrial micro-ecosystem for measuring effects of pollutants on isopod-mediated litter decomposition. *Hydrobiologia* **188/189**, 507–516.

Van Wensem, J., Jagers op Akkerhuis, G.A.J.M., and van Straalen, N.M. (1991) Effects of the fungicide triphenyltin hydroxide on soil fauna mediated litter decomposition. *Pest. Sci.* **32**, 307–316.

Van Wensem, J., Verhoef, H.A., and van Straalen, N.M. (in press) Litter degradation stage as a prime factor for isopod interaction with mineralization processes. *Soil Biol. Biochem.*

Verhoef, H.A., and Brussaard, L. (1990) Decomposition and nitrogen mineralization in natural and agroecosystems: the contribution of soil animals. *Biogeochemistry* **11**, 175–211.

Verhoef, H.A., and de Goede, R.G.M. (1985) Effects of collembolan grazing on nitrogen dynamics in a coniferous forest. In: Fitter A.H., Atkinson, D., Read, D.J., and Usher, M.B. (Eds.) *Ecological Interactions in Soil*, pp. 367–377. Blackwell Scientific, Oxford.

Verhoef, H.A., and Meintser, S. (1991) The role of soil arthropods in nutrient flow and the impact of atmospheric deposition. In: Veeresh, G.K. (Ed.) *Advances in Management and Conservation of Soil Fauna*, pp. 497–506. Vedam Books, New Delhi.

Verhoef, H.A., and Dorel, F.G. (1988) Effects of ammonia deposition on animal-mediated nitrogen mineralization and acidity in coniferous forest soils in The Netherlands. In: Mathy, P. (Ed.) *Air Pollution and Ecosystems*, pp. 847–851. Reidel, Dordrecht, The Netherlands.

Verhoef, H.A., Dorel, F.G., and Zoomer, H.R. (1989) Effects of nitrogen deposition on animal-mediated nitrogen mobilization in coniferous litter. *Biol. Fertil. Soils* **8**, 255–259.

Westheide, W., Bethke-Beilfuss, D., and Gebbe, J. (1991) Effects of benomyl on reproduction and population structure of enchytraeid oligochaetes (Annelida)—sublethal tests on agar and soil. *Compar. Biochem. Physiol.* **100C**, 221–224.

Williams, W.J. (1979) *Handbook of Anion Determination*, pp. 500–609. Butterworths, London.

Wohlgemuth, D., Kratz, W., and Weigmann, G. (1990) The influence of soil characteristics on the toxicity of an environmental chemical (cadmium) on the newly developed mono-species test with the springtail *Folsomia candida* (Willem). In: Barcelo, J. (Ed.) *Environmental Contamination, Fourth International Conference, Barcelona, October 1990*, pp. 260–262. CEP Consultants Ltd, Edinburgh.

12 Methods to Assess the Effects of Chemicals in Cold Climates

W. L. Lockhart, D. C. G. Muir, R. Wagemann
Department of Fisheries and Oceans, Canada

G. Brunskill
Institute of Marine Sciences, Australia

T. Savinova
Academy of Sciences of Russia, Russia

12.1 INTRODUCTION

The concept of "cold" climate is a relative one, and it will be examined mainly by reference to methods being used to study contamination of aquatic ecosystems in arctic drainage of Canada. The most pervasive chemical contamination problems in this area are posed by a few stable contaminants (e.g., PCBs, heavy metals) found consistently in animal and human tissues. The presence of these materials has prompted questions about their sources, the pathways that supplied them, their geographic distribution, their future trends, and the biological implications associated with them. Few, if any, new chemical products are developed for exclusive use in the Arctic, and so the requirement is for a retrospective evaluation of contaminants already present. The source of these contaminants to the people is largely animal tissues consumed as a normal part of the diet (Kinloch *et al.*, 1992) hence value judgments must be made in which the risks of consuming such foods are weighed against their nutritional benefits.

Evaluating the biological meaning of known exposures requires knowledge of associations between exposures and responses. These typically rely on experimental laboratory toxicology, but few laboratories have worked experimentally with arctic species. Furthermore, the more highly valued arctic species like whales and walrus seem unlikely to ever be studied toxicologically in laboratory settings. Most experimental work has been done instead with temperate species convenient to maintain in laboratories; extension of results and conclusions to arctic species is by inference. An emerging effort is the examination of animals and plants for "bioindicators," biochemical or other changes that may be associated with exposure to contaminants, and this approach allows the direct examination of

Methods to Assess the Effects of Chemicals on Ecosystems
Edited by R. A. Linthurst, P. Bourdeau, and R. G. Tardiff

arctic species.

A significant difference between arctic and temperate settings is in the exposure to those contaminants that are subject to decomposition by processes dependent on heat or light. Rates of decomposition are slow under arctic winter conditions, so exposure to these materials can be more prolonged than might be expected from experience in temperate climates. The presence of ice cover for much of the year impedes the exchange of oxygen and other materials between water and atmosphere. This situation means that phenomena like the "weathering" (Payne and McNabb, 1984) of petroleum oils do not occur readily in arctic waters for much of the year.

The aquatic organisms of greatest concern in the Arctic are those that have economic and cultural values and those that are consumed by people, namely fish, marine mammals, and birds. These animals typically live a long time and reach large sizes. Ecosystem studies by Rapport *et al.* (1985) and toxicity studies by Neuhold (1987) indicate that these biological characteristics are associated with high sensitivity to stress. Most of these animals are long-lived predators, and so they integrate processes over long periods of time and several trophic levels, and often over wide geographic ranges. Given the economic, nutritional, cultural, and ecological importance of these animals, both residue and biological studies have been focused on them. This choice is partly a matter of sampling opportunities in the Arctic; these are the animals being taken in traditional subsistence hunting, hence tissue samples can be obtained without killing additional animals.

12.2 ORGANOCHLORINE COMPOUNDS IN ARCTIC FISHES, MARINE MAMMALS, AND SEA BIRDS

Interest in stable organochlorine compounds has been derived not only from their potential for exerting biological effects, but also for their power to illustrate hemispheric or global dispersal processes. These materials (polychlorinated camphenes (PCCs, toxaphene), DDT- and chlordane-related compounds, and PCBs, chlorinated dioxins and furans, hexachlorocyclohexanes, chlorobenzenes, dieldrin, and mirex) are reported consistently in tissues of arctic fishes and marine mammals (Muir *et al.*, 1992). Persistent chlorinated hydrocarbons and heavy metals have been detected in seabirds from different Arctic regions in Canada (Vermeer and Reynolds, 1970; Vermeer and Peakall, 1977; Nettleship and Peakall, 1987; Elliot *et al.*, 1992; Hart *et al.*, 1991) and in Norway (Fimreite and Bjerk, 1979; Holt *et al.*, 1979; Norheim and Kjos-Hansen, 1984; Ingebrigtsen *et al.*, 1984; Barrett *et al.*, 1985; Norheim, 1987; Norheim and Borch-Iohnsen, 1990; Daelemans *et al.*, 1992). Contamination levels in birds from Russian Arctic (Savinova, 1991, 1992), Greenland (Braestrup *et al.*, 1974), and Alaska regions (Ohlendort *et al.*, 1982) are less well known.

12.2.1 SPATIAL AND TEMPORAL TRENDS

Ringed seals (*Phoca hispida*) have been used for temporal and spatial trend studies, because they are a widely distributed and relatively sedentary species. Care must be taken to evaluate seals of similar sex, age, and blubber thickness, because these factors are known to influence levels of organochlorines in blubber (Addison and Smith, 1974; Muir *et al.*, 1988).

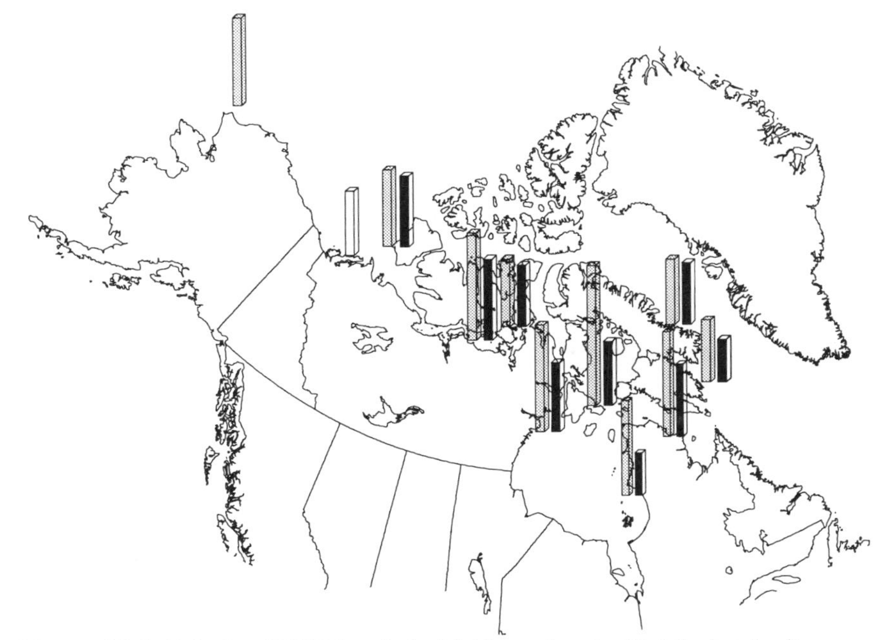

Figure 12.1. Relative ΣPCB levels in blubber of male (lightly hatched), female (black), and combined sexes (unshaded) of ringed seals from Canadian and Alaskan locations

Ringed seals from seven locations in the Canadian Arctic (Muir *et al.*, 1993) and the Chukchi Sea (Becker *et al.*, 1989) had very similar ΣPCB levels in blubber (Figure 12.1), with mean concentrations ranging from 0.51 to 1.2 μg per g. Levels of ΣPCB in ringed seals from Spitzbergen were about twice as high as concentrations in the Canadian Arctic, consistent with the relative proximity of Spitzbergen to agricultural and industrial areas of Europe (Oehme *et al.*, 1988). PCB levels in Canadian arctic ringed seals are 10 times lower than reported for land-locked ringed seals in Finland (Helle *et al.*, 1983), and up to 50 times lower than reported for the same species in the Baltic Sea during the mid-1970s (Bergman *et al.*, 1981). The elevated levels of PCBs, especially coplanar congeners (Olsson *et al.*, 1990), in Baltic ringed seals have been associated with the reproductive failure of this population (Helle *et al.*, 1976). In a 2-year laboratory study, common seals (*Phoca vitulina*) fed fish from the Wadden Sea, with high levels of PCBs, had significantly reduced retinol and thyroid hormones relative to controls and reduced reproductive success (Reijnders, 1986; Brouwer *et al.*, 1989).

An additional complication in evaluating possible effects of organochlorine compounds in marine mammals is the decline in concentrations due to bans on open uses of PCB oils and the application of many chlorinated pesticides that were implemented in the 1970s and 1980s. Addison *et al.* (1986) concluded that PCBs declined about threefold in ringed seals from Holman Island, in the western Canadian Arctic, between 1972 and 1981. For ΣDDT, however, much of the decline (30–40 percent) between 1972 and 1981 could be explained by thicker blubber of the animals collected in 1981. Little is known, however, about temporal trends of chlordane or PCCs in marine mammals, because measurement of these groups in arctic animals began only in the mid-1980s.

Levels of chlorinated hydrocarbons in Arctic seabirds vary widely depending for the most part on the species of bird and its ecology-feeding mode, migration route, sex, age, physiological, and biochemical parameters (rate of normal metabolic processes, quantity and composition on lipids, and hepatic microsomal mono-oxygenases activities). Levels of organochlorine substances have been monitored in sea birds eggs in different Arctic regions from the 1960s. In 1968–1969, concentrations of DDT in kittiwake (*Ryssa tridactyla*) eggs from the Atlantic Canadian coast were 2–13 ppm (Vermeer and Reynolds, 1970). In 1968–1973, the level of DDT in kittiwake eggs from Norway was, on average, 1.2 ppm (Holt *et al.*, 1979). Eggs from the same species collected in the mid-1970s at Prince Leopold Island in the Canadian Arctic archipelago had higher PCB levels (5.2 ppm) than eggs of the northern fulmar (*Fulmarus glacialis*) (1.93 ppm) or the thick-billed murre (*Uria lomvia*) (0.01 ppm). DDE/PCB ratios in eggs and livers of fulmars and murres were much lower than in kittiwakes, probably reflecting the lower levels of DDE in the latter. Compared with other seabirds, kittiwakes appear to have a greater capacity to metabolize and excrete organochlorine substances (Nettleship and Peakall, 1987). This capacity is probably related to the metabolic rates of kittiwakes, being higher that those of other species (Gabrielsen *et al.*, 1987). Values in the range 0.05–1.0 ppm DDT and 1.1–2.4 ppm PCB have been reported in the livers of kittiwakes from different Arctic regions (Bourne and Bogan, 1976; Nettleship and Peakall, 1987; Savinova, 1991). The concentrations of DDT and PCB found in kittiwakes from the east coast of the Kola peninsula (Savinova, 1991) were 3–5 times lower than in those from David Strait and Bear Island (Bourne, 1976).

A decrease of 55-60 percent in ΣDDT and 69–86 percent in PCBs was observed in thick-billed murre eggs collected at Prince Leopold Island, Lancaster Sound, between 1976 and 1987. Similar declines in DDT and PCB residues occurred in livers of kittiwakes and northern fulmars, but not in those of thick-billed murres from the same region (Nettleship and Peakall, 1987). In livers and muscles of herring gulls (*Larus argentatus*) from the Barents Sea ΣDDT content declined three- to sixfold during the last decade, but the decline was not observed for PCBs (Savinova, 1992).

In 1982–1983, puffin (*Fratercula arctica*) eggs from two colonies in Northern Norway contained, on average, about 20 percent higher residues of PCBs and *p,p'*-DDE than corresponding mean levels in Alaskan puffin eggs (Ingebrigtsen *et al.*, 1984; Ohlendort *et al.*, 1982). By contrast, lower PCB residues were found in eggs and adult brain tissues collected from puffins in eastern Canadian coastal waters (Pearce *et al.*, 1989).

From a toxicological viewpoint, some congeners are more significant than others, and may account for most of the toxic effects. Daelemans *et al.* (1992) report on PCB congeners in glaucous gulls (*Larus hyperboreus*) and black guillemots (*Cepphus grylle*) from the Svalbard area. Particularly high levels of organochlorine compounds have been recorded in glaucous gulls (Bourne, 1976). This species acts partly as a predator; during the breeding season gulls prey on other seabird's eggs and chicks. The average PCB concentration in the liver of the glaucous gull from Svalbard was 20.9 ppm, i.e., about 160 times higher than that found in the liver of the black guillemot (Daelemans *et al.*, 1992). The total concentration of three selected non-ortho-congeners represented only 0.18 percent of total PCB concentration. Congener 126 showed the highest average concentration (0.11 percent of total PCB) followed by 169 (0.04 percent) and 77 (0.03 percent).

The concentration levels in arctic seabirds, in the main, have been considerably below those expected to produce lethal effects. On the other hand, much evidence demonstrated that levels high enough to produce sublethal effects are frequently attained. As a group, birds are more resistant to acutely toxic effects of PCBs than mammals. LD_{50}s for various species of birds varied from 604 to more than 6000 mg Aroclor per kg diet (Eisler, 1986).

Predacious fish such as burbot and lake trout can also be used to assess spatial and temporal trends in organochlorine compounds in freshwater ecosystems. Temporal trend studies of PCBs in lake trout in the Great Lakes have demonstrated the need to compare fish of the same age class (Devault *et al.*, 1986). PCB levels in lake trout among inland lakes in Ontario, receiving similar inputs of contaminants from the atmosphere, appear to vary with the length of the pelagic food chain and the lipid content of the trout. Thus, lake trout from lakes with *Mysis* or smelt have higher PCB levels than those from lakes lacking *Mysis* or pelagic forage fish (Rasmussen *et al.*, 1990). A 16-year temporal trend study in Lake Storvindeln in northern Sweden showed that DDT and PCB levels in pike (*Esox lucius*) declined about three- to fourfold during the period from 1968 to 1984, coinciding with the ban on DDT and PCB use in Western Europe during that time. Concentrations of organochlorine substances were reported in burbot liver from seven locations on a northwesterly transect from northwestern Ontario to Fort McPherson in Northwest Territories (Muir *et al.*, 1990). Lipid normalized concentrations of ΣPCB and ΣDDT showed significant declines with increasing north latitude (ΣPCB in Figure 12.2). The decreasing concentrations were most apparent for hexa-, hepta- and octachlorobiphenyls, which were close to detection limits in the northern fish but readily detected in fish of similar size and sex in northwestern Ontario lakes. However, more volatile organochlorine compounds such as α-HCH, toxaphene, and tri- and tetrachlorobiphenyls showed no decline. The results were consistent with the hypothesis that inputs of semi-volatile organochlorines decrease with increasing north latitude and distance from North American sources.

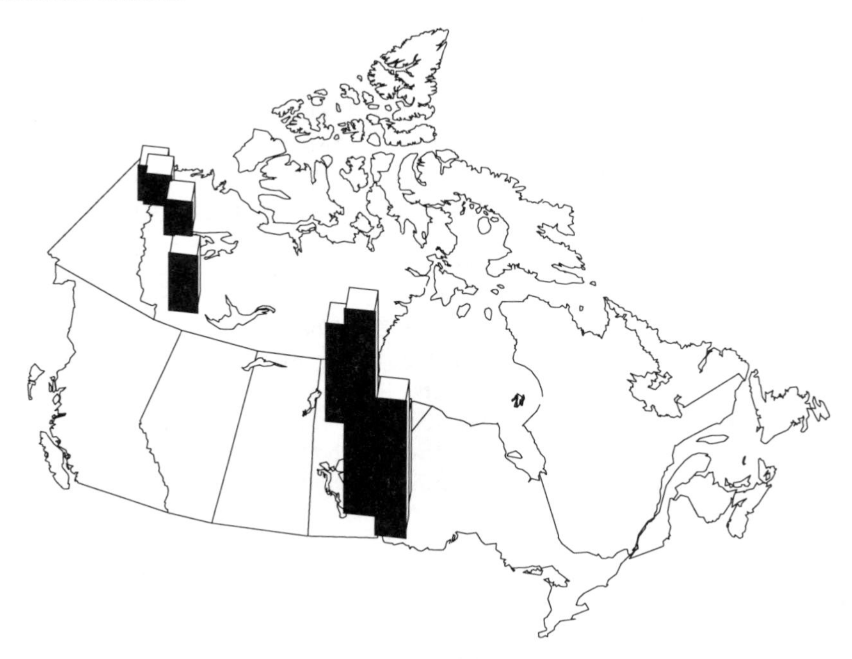

Figure 12.2. Lipid normalized relative total PCB concentrations in burbot liver from locations in western and northern Canada; maximum bar height is 1.9 μg/g (Muir *et al.*, 1990)

12.3 METALS

Metals are always detected in biological samples, and the major difficulty is in delineating "normal" from "abnormal." Several approaches have been used:

1. Comparing tissue levels in animals of the same species from many different geographic areas, taking into account their age and other biological variables.
2. Comparing tissue levels in animals from a given study area with those in animals from a "pristine" area. Unfortunately, no such areas may remain.
3. Comparing tissue levels in animals from the study area with previously established no-observed-effect levels in controlled studies with other animals.
4. Comparing tissue levels (hard tissues) in the study area with those in prehistoric

specimens of the same or similar species and assuming that the prehistoric levels, not having been subject to anthropogenic sources of contaminants, are "normal."

Comparing levels found in animals from different habitats is one of the most common designs in point-source pollution studies. Subtle variations in the concentrations of mercury, lead and cadmium, particularly the latter, in tissues of some arctic marine mammals have been reported, depending on the species and the location of capture. In the Canadian Arctic, some ringed seals (*Phoca hispida*) were reported to have high levels of mercury and lead (Smith and Armstrong, 1975; Wagemann, 1989). High cadmium levels were found in northern fur seals (*Callorhinus ursinus*) and narwhal (*Monodon monoceros*) (Goldblatt and Anthony, 1983; Wagemann *et al.*, 1983). Wagemann *et al.* (1990) found relatively higher levels of cadmium in liver and kidneys of beluga (*Delphinapterus leucas*) from some arctic locations compared to others (Figure 12.3).

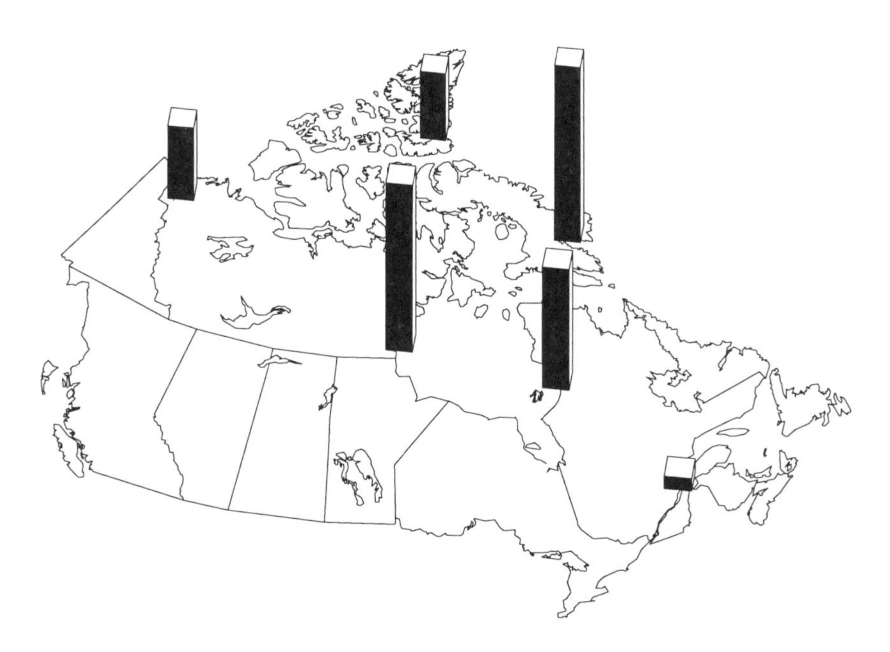

Figure 12.3. Relative Cd concentration (μg/g) in kidney of beluga whales from various locations in the Canadian Arctic and the St Lawrence estuary; the highest bar represents 106 μg/g.

To date, only belugas and polar bears (*Ursus maritimus*) have been systematically surveyed for heavy metals in tissues over a large geographic area of the Canadian Arctic (Norstrom *et al.*, 1986; Wagemann *et al.*, 1990). These surveys indicated some dependence of the cadmium, lead, and mercury concentrations in organs on geographic area. Cadmium in the kidney of belugas (both cortex and medulla homogenized) was higher in animals from the eastern Arctic than the western Arctic (Figure 12.3); lead in liver increased from northwest to southeast, as did mercury, with the exception of the group from the Mackenzie Delta, which had higher mercury levels than other arctic groups. Anomalously high mercury levels in animals from the Beaufort Sea have also been reported for other marine mammals (Smith and Armstrong, 1975). The trend of increasing cadmium from west to east in belugas was also reported for liver cadmium in polar bears (Norstrom *et al.*, 1986).

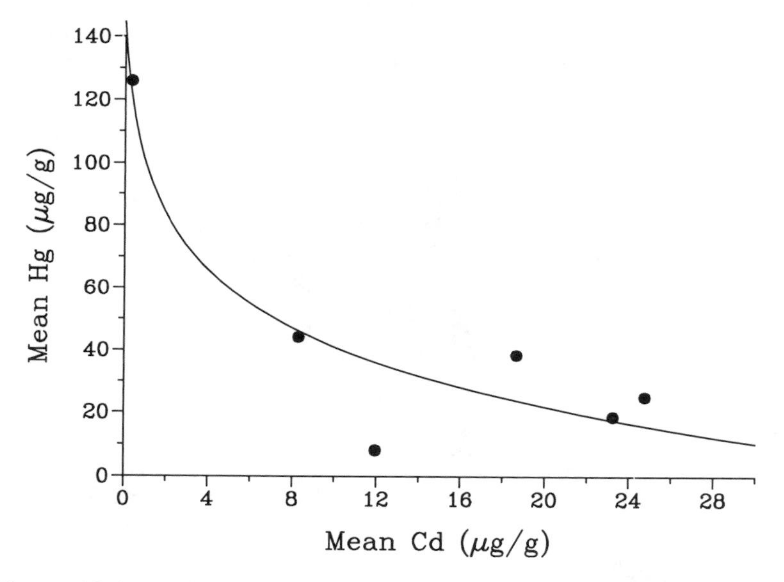

Figure 12.4. Mean Hg vs. mean Cd in beluga liver of groups from different locations in the Canadian Arctic

A compounding factor when comparing metal levels is the influence of other metals and the interrelationships among them. For example, a highly significant inverse relationship between mercury and cadmium was found for belugas from the various arctic locations; the higher the cadmium concentration, the lower the mercury and vice versa (Figure 12.4). The biochemical basis of this relationship remains unexplained. Additional correlations among metals in marine mammal tissues have been reported repeatedly, and can be accepted as having general validity, such as that between mercury and selenium in liver and between zinc and cadmium in liver and kidney. The former correlation apparently arises largely because of the formation of mercuric selenide particles in the liver, and the latter because both metals are constituents of metallothionein, and can replace each other in that metalloprotein (Webb, 1979).

The heavy metals content of arctic seabirds has not been as well investigated as chlorinated hydrocarbon levels. Heavy metals have been determined in different bird species collected in the 1980s off the west coast of Spitsbergen (Norheim and Kjos-Hansen, 1984; Norheim, 1987; Norheim and Borch-Iohnsen, 1990); in gulls, fulmar, long-tailed duck (*Clangula hyemalis*) from the eastern coast of Kola peninsula in 1989 (Savinova, 1992); in black guillemot, kittiwake, eider (*Somaterla mollissima*), glaucous gull, fulmar, and Brunnich's guillemot (*Uria aalge*) from Greenland (Nielsen and Dietz, 1989).

The concentrations of trace elements in Arctic seabirds examined in these studies are in a good agreement with values reported in the literature for Atlantic Canadian seabirds (Elliott *et al.* 1992), and generally represent normal physiological levels. By contrast, the content of Cu in liver of eider duck from Spitsbergen was about 40 times higher than in other species (Norheim and Kjos-Hansen, 1984; Norheim, 1987). The high level of Cu in either may reflect the fact that this species feeds mainly on mussels, snails, and crustaceans, which have haemocyanin as their blood pigment.

12.4 BIOCONCENTRATION AND BIOACCUMULATION

Given the emphasis on arctic animals as vectors of contaminants to people consuming them, the processes that brought the contaminants to the animals themselves are of interest. The accumulation of organic contaminants has been described using several types of models (Landrum *et al.*, 1992; Thomann *et al.*, 1992), and these work well for fish and other aquatic organisms if the required model input data are available. For arctic animals generally, and especially for the marine mammals, some parameters to model are not known, and so they have to be extrapolated from temperate species.

12.5 CONTAMINANT HISTORIES REPRESENTED IN LAKE SEDIMENTS

Sediments have been recognized increasingly as long-term sinks for several contaminants, especially in harbours and areas receiving industrial wastes (Smith and Levy, 1990; MacDonald *et al.*, 1992). Campbell *et al.* (1985) examined field samples of the yellow water lily *Nuphar variegatum*, and used correlations with sediment and water levels of metals to show that much of the copper in the plants was derived from the sediment, while the zinc was derived from the water column. Sediments accumulate over time, and analysis of sediment layers allows calculation of rates of change and hence predictions of future concentrations. Dated layers of lake sediments have been analyzed for trace metals (Johnson, 1987), polycyclic aromatic hydrocarbons (Gschwend and Hites, 1981), and organochlorine compounds

(Eisenreich *et al.*, 1989). These studies have allowed reconstruction of the histories of contaminant inputs to several areas. With some precautions, past and current fluxes can be extrapolated to predict future trends. This approach is particularly relevant to arctic cases where inputs are derived largely from atmospheric deposition.

Some sediment cores are mixed by natural or coring processes, and are not suitable for chronological contaminant analyses; processes of mixing and diffusion in sediments are not understood fully. Virtually all sediments are mixed to some extent, even annually laminated sediments, either by resuspension, *in situ* particle mixing, or diffusion. A general indication of the *in situ* integrity of the sediment column sample (core) is the appearance of simple profiles of percent water, loss on ignition (organic matter), and some conservative parameters of inorganic, non-biological, non-contaminant elements, such as Fe, Al, Si, or Ti. The depth integral of Pb^{210} and Cs^{137} can be used as an indication of the degree of success at recovering the sediment-water interface, if annual fluxes or depositional history of these nuclides is generally known (Crusius and Anderson, 1991; Anderson *et al.*, 1987). Interpretation of the chronology of the sediment core from Pb^{210} and other radiochemical tracers of sediment accumulation, sediment mixing, and diffusion can be done using models (Robbins, 1986). This approach requires some knowledge of the annual fluxes of Pb^{210} and other sedimentation tracers for the sampled region that can be obtained from soil or glacier unit area samples.

Cores are usually taken from the maximum depth of the lake, where fine, organic sediments are deposited after cycles of deposition and resuspension in shallower, more turbulent water. Lakes focus these sediments into smaller, deeper locations, and bathymetric information is required to find these locations and to judge the importance of the focusing. The annual flux and cumulative burden of contaminants in such deep water sediments need to be corrected for focusing. As an estimate of the focusing factor, the ratio of excess Pb^{210} integral has been used by calculating excess Pb^{210} flux from the profundal lake sediment cores, and comparing it to the excess Pb^{210} integral from several soil profiles (hopefully not focused) in the lake drainage basin.

An example of the application of these techniques is given in Figure 12.5 (in units of concentration) for a core taken in 1988 from Far Lake, Northwest Territories, a site with no local pollution sources within the drainage (63°38′N, 90°40′W). Mean ages for these slices have been assigned from simple models of Pb^{210} accumulation, and are considered reliable down to slice 11; values below those dates were estimated by extrapolation from the upper slices. The profile of mercury concentrations in the slices indicates an accelerating increase in mercury supply rate to the sediments over the time covered. During the current century, the concentration has approximately doubled, in keeping with changes indicated in air concentrations of mercury in the northern hemisphere (Slemr and Langer, 1992). Taking the sedimentation rate as indicated by the mass of sediment accumulated per unit area and time (77 g per m^2 per yr), the background and contaminant flux of mercury (Hg) to the site can be calculated (background flux = 77 g per m^2 per yr

× 40 ng Hg per g = 3080 ng per m^2 per yr; contaminant Hg flux = 77 g per m^2 per yr × [100 – 40] ng Hg per g = 4620 ng Hg per m^2 per yr). Over the 80 years of excess Hg inputs to the lake, approximately 360 micrograms of Hg have been deposited per m^2 (the integral contaminant burden), and are potentially available to lake biota from the top 20 cm of the lake sediments. Johnson (1987) made flux calculations for several lakes in Ontario, and found that the flux often correlated with Hg levels in fish from the same lakes.

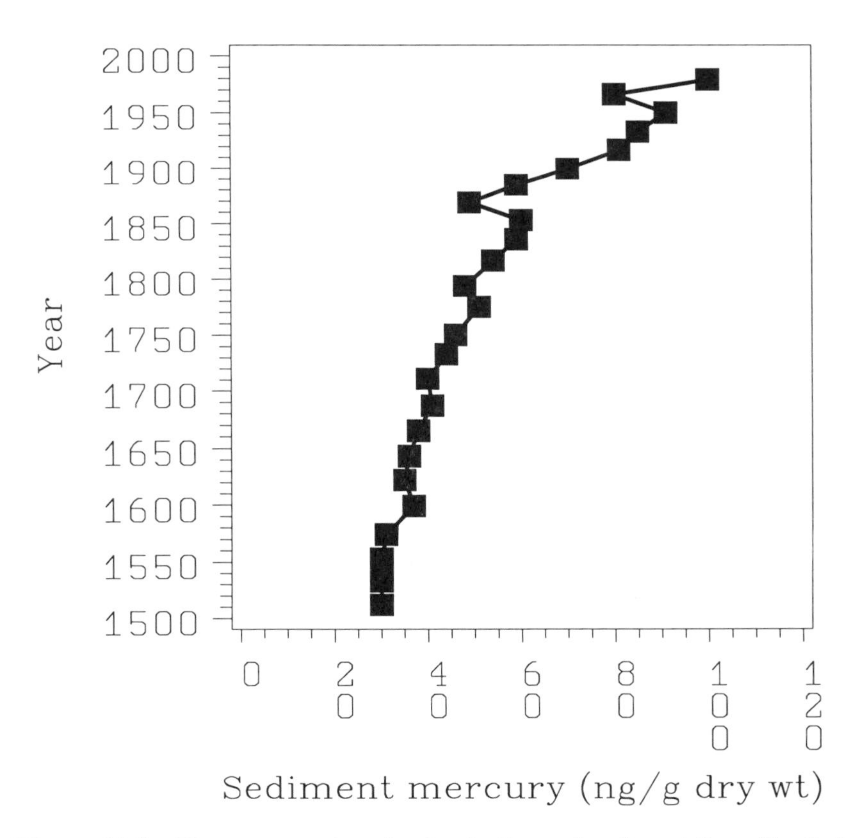

Figure 12.5. Hg concentrations in dated slices of sediment from Far Lake, Northwest Territories, Canada; dates are calculated using Pb^{210} for the top eleven slices, and extrapolated for deeper slices

12.6 BIOLOGICAL RESPONSES

Relatively little pollution response work has been done in the Arctic or with arctic species. The question for biologists is whether the increasingly well-documented inputs of contaminants are significant biologically. The response to chemicals by animals has sometimes been determined by measurements on populations rather than on individuals. Case histories like the disappearance of several species of fish from acidified lakes (Beamish *et al.*, 1975) or the loss of fish from an area receiving pulp mill effluent (Kelso, 1977) provide convincing examples of effects observed directly on the abundance and distribution of animals. Two problems exist with the use of population characteristics to indicate effects of chemicals. One is that the population must already have been affected before changes in its structure can be established. The second is that to establish causal relationships using population surveys by themselves is extraordinarily difficult. Nonetheless, the population, not the individual, must be maintained, and population responses must be used as the standard against which the usefulness of other approaches must ultimately be judged. Individual responses can help to test the hypothesis that the effect is, or is not, due to the suspected cause.

A useful tool to test for some causal relationships is the tissue level of the hypothesized causative agent. If dose-response relationships have been established linking biological responses with tissue levels, then the tissue levels themselves can be used as surrogates for biological responses that may be more difficult to measure. This analysis requires some knowledge of how these may change with other biological variables like species, age, gender, reproductive history, and the type of tissue. For all avian species tested, PCB residues of 310 mg per kg fresh weight or higher in brain were associated with an increased likelihood of death from PCB poisoning (Stickel *et al.*, 1984). Applying this line of argument to exposed populations, PCBs were judged the probable cause of death of glaucous gull from Bear Island in 1972 (Bourne, 1976), and puffin mortality of Lofoten Islands in the 1970s (Walker, 1990). At this time, relatively high levels of PCBs and other organochlorine residues were being reported in sea birds, and toxic effects may have occurred following the rapid mobilization of such compounds from the storage fat of the birds (Walker, 1990).

Several studies have been concerned with the effects of chlorinated hydrocarbons on reproduction in seabirds, and two major types of effects have been considered: eggshell thinning and direct toxicity to the embryos. Cases of eggshell thinning associated with DDE and PCB residues have been reported for gannets (*Sula bassana*) in Canada (Parslow *et al.*, 1973) and Norway (Nygard, 1983). Eggshell thickness of the peregrine falcon (*Falco peregrinus*) from Norway declined 85 percent between 1854 and 1976; addled eggs containing dead embryos collected in 1976 had 724 ppm of PCBs in lipids and up to 110 ppm on a fresh weight basis (Nygard, 1983). In the Canadian study on the gannet, DDE-induced eggshell thinning was suspected of being responsible for reduced reproductive success and consequent population decline in the 1960s. These effects were associated with DDE levels of 20–30 ppm in the eggs (Parslow *et al.*, 1973). For most avian species, a reduction in eggshell thickness of 15–20 percent is suggested as a critical value beyond which population numbers are expected to decline (Nygard, 1983).

Embryotoxicity of organochlorines in different species of birds is discussed by

Cooke (1973) and Peakall and Fox (1987). Elliott *et al.* (1988) reported higher levels of several organochlorines in gannet eggs that failed to hatch compared to levels in fresh eggs.

Several field and laboratory studies have shown that organochlorine compounds are associated with biological effects on marine mammals, freshwater and marine fish, as well as in fish-eating birds in temperate climates (Gilbertson, 1989). These effects, which are mainly associated with coplanar PCBs and chlorinated dioxins and furans, include elevated hepatic cytochrome P_{450}-associated enzyme activities, histopathological abnormalities, altered steroid hormone levels in individual animals, and reproductive failure in populations. Therefore, concern may be understandable that the presence of PCBs and other organochlorine contaminants in arctic biota could adversely affect the populations, particularly marine mammals and birds which accumulate the highest concentrations. The effects of toxaphene and chlordane-related compounds, that are generally present at similar levels to total PCBs in arctic animals (unlike fish and birds in the Great Lakes or marine mammals in the Baltic Sea), are unknown. Toxaphene has been shown to affect the mechanical strength and biochemistry of bone in fish, (Mayer and Mehrle, 1977) and to produce cancer in rats and mice (National Institutes of Health, 1979).

As consumers of arctic biota, humans are also a focus of concern for possible toxicity from exposure to organochlorine substances. Levels of PCBs, coplanar PCBs, chlordane, and toxaphene components have been found to be elevated in human milk from northern Quebec, as compared with levels in southern Quebec residents (Dewailly *et al.*, 1989). Dewailly *et al.* (1992) reported a negative and statistically significant correlation between size of newborn male children and organochlorine exposure (estimated using mother's milk) in northern Quebec. These results are parallel to findings that infants of Michigan women, whose calculated PCB consumption from fish exceeded 1 μg per kg per day, had reduced birth weight, smaller head circumference, and compromised neuromuscular development (Hwang *et al.*, 1984; Gladen *et al.*, 1985).

12.7 LABORATORY TOXICOLOGY

Laboratory studies can provide some of the clearest answers on whether contaminants at levels found in arctic biota are biologically significant. Unfortunately the larger animals important to arctic people are not well suited to laboratory studies, because of the logistics of maintaining them in captivity. Seals and small cetaceans have been studied occasionally (Engelhardt *et al.*, 1977; Geraci *et al.*, 1983; Reijnders, 1986); however, only very few such experiments are likely ever to be conducted in view of the high value placed on individual marine mammals. Arctic fish can be studied more frequently and at reasonable cost. Craddock (1977) and Lockhart *et al.* (1992b) have reviewed some of the laboratory toxicity tests with arctic fish and invertebrates. Where comparisons are possible, arctic species appear quite similar to temperate species. Given the growing body of data describing concentrations of various contaminants in arctic animals, these data cannot be interpreted clearly in terms of toxic responses (McCarty, 1991).

12.7.1 ACUTE TOXICITY TESTING

Many acute toxicity bioassay systems have been developed for different purposes. However, with the exception of petroleum hydrocarbons (Craddock, 1977), little acute toxicology has been reported for arctic species. The acute toxicity of two petroleum oils to several Alaskan species was comparable to that of temperate species (Rice *et al.*, 1979). For relevance to arctic winter conditions, the testing of petroleum oils under normal bioassay conditions offers a useful illustration. Volatile components of several oils were lost rapidly under normal bioassay conditions employing a continuous stream of air bubbles to maintain oxygen levels, giving the impression that the oils were non-toxic (Lockhart *et al.*, 1987). However, such losses would not occur readily in the Arctic during winter; so the bioassays were repeated under conditions that minimized the loss of volatile components, at which time the oils were clearly toxic. Given the nature of the contamination problem in the Arctic, many acute toxicity data have limited application. Since the residues are present ubiquitously, the exposures must be essentially continuous. Indeed, examination of PCB residues as a function of age in beluga whales has shown that residues are present throughout the entire life cycle. Kenaga (1982) has shown that chronic effects are mostly within a factor of twenty-five of acute effects, but with some unpredictable exceptions. In examples of chronic studies with cutthroat trout (*Salmo clarki*), Woodward *et al.* (1981, 1983) have shown that petroleum oils have chronic, sublethal effects not evident in short-term (96-hour) conventional bioassays.

Aquatic sediments serve as sinks for many contaminants, prompting the need for methods to assess the importance of sediment-associated materials. Several bioassay systems have also been developed to examine the toxicity of contaminants associated with sediments (Long *et al.*, 1990); Krantzberg and Boyd, 1992; Tay *et al.*, 1992). Similarly, Payne *et al.* (1988) and Truscott *et al.* (1992) have reported subtle effects in flounders (*Pseudopleuronectes americanus*) maintained for long periods in contact with sediments contaminated with petroleum.

12.7.2 BEHAVIOUR

Chemicals in the water have often been hypothesized to alter some aspect of the behaviour of fish (Sprague *et al.*, 1964), and a variety of laboratory tests have been developed to measure these effects. Scherer (1992) has reviewed these tests, and grouped them into three general types: unconditioned reflexes, locomotor behaviour, and intra- and interspecific responses. McNicol and Scherer (1991) used a preference–avoidance design to test the locomotor responses of lake whitefish (*Coregonus clupeaformis*) to cadmium dissolved in the water. Curiously, the fish responded to levels lower than 1 μg per litre and to levels greater than 8 μg per litre, but showed little response to levels in between. Furthermore, at a given concentration, some individuals avoided the cadmium solution and others preferred it. No explanation is apparent for this dichotomous response.

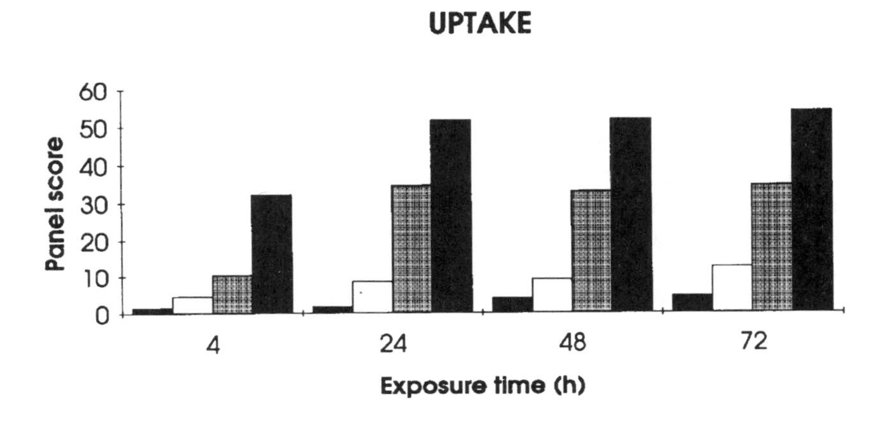

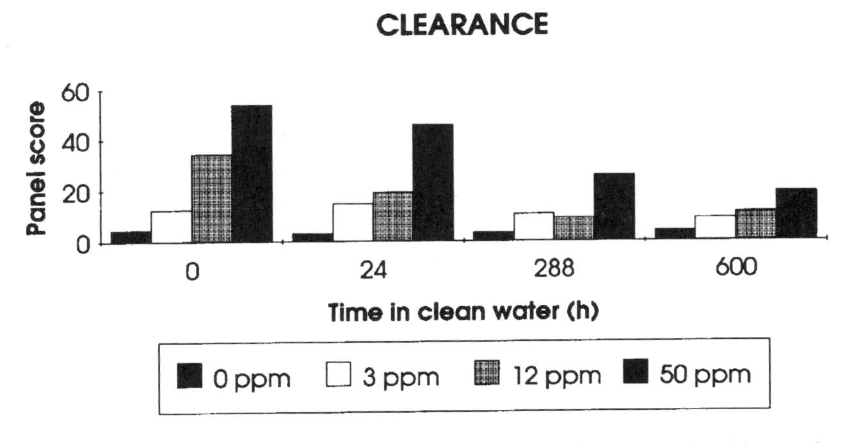

Figure 12.6. Mean taste panel scores for arctic char treated with varying amounts of Norman Wells crude oil following uptake (upper panel) and clearance (lower panel) (Lockhart and Danell, 1992)

Some behaviour testing has also been done with arctic invertebrates. Percy and Mullin (1977) reported the effects of 24-hour exposures to several crude oils on the locomotor activity of the amphipod *Onisimus affinis* and the coelenterate *Halitholus cirratus*. The activity of *O. affinis* was reduced by as much as 98 percent by the highest dosages, and was quite consistent among the three crude oils tested.

Norman Wells and Pembina crude oils inhibited the activity of *H. cirrattus*, but the other two oils had little effect. Cross and Thomson (1987) noted the rapid emergence of several benthic infaunal species following exposure to chemically dispersed oil, and their apparent recovery and return to the sediment shortly thereafter.

12.7.3 SENSORY EVALUATION

With fishery products destined for human consumption, the products must be acceptable in the market-place. For example, Krishnaswami and Kupchanko (1969) reported tainting of fish by effluent from an oil refinery. Recently tainting by several types of wastewater from a plant producing synthetic crude oil from oil sands has been detected. Exposure to petroleum products has often been associated with the production of oily tastes and odours in fish (Motohiro, 1983). Similarly, fish undergoing exposure to effluent from bleached kraft pulp mills have developed undesirable tastes and odours (Whittle and Flood, 1977). These effects are detected best by sensory evaluation panels trained to detect the presence and severity of off-flavours (York and Sereda, in press). Several different protocols for such tests have been used (Poels *et al.*, 1988); they all provide valid evaluations.

Petroleum tainting tests with arctic charr (*Salvelinus alpinus*) exposed experimentally to extracts of Norman Wells crude oil were described by Lockhart and Danell (1992). Oil and water were mixed in a mixing tank from which an outflow at the bottom drew off the water phase to be used (after appropriate dilution exposure). Charr were taken for sensory testing at intervals during the 72-hour oil uptake phase of the experiments, and the remaining fish were transferred to clean flowing water for a clearance phase lasting several weeks. Initial results of taste panel evaluations are shown in Figure 12.6, and dose-dependent tainting was produced; the low exposure threshold to produce tainting was below the lowest mixing ratio of 3 ppm. Furthermore, the tainting effect was not entirely lost by the end of the clearance phase at 600 hours. This observation clearly has practical implications for the time a fishery might remain unusable after an oil spill.

12.8 FIELD STUDIES

Some populations of fish and marine mammals are monitored by resource management agencies, but population effects become evident only after the fact. Nonetheless, adverse changes at the population level are the primary concern. Field experiments in which a known contaminant is introduced into an experimental site can give the best description of effects at ecological levels, but relatively few such experiments have been done, especially where the species of interest are large in size.

12.8.1 ECOSYSTEM EXPERIMENTS

At the Experimental Lakes Area in northwestern Ontario, several whole lake experiments have been done to evaluate ecosystem scale effects (Schindler, 1988). Lake trout responded to acidification with decreased growth rates and decreased condition factors, and, finally, with failure of recruitment of young into the population. Similarly, experimental treatments of northern streams with petroleum oils have been carried out in order to observe effects on downstream flora and fauna (Miller *et al.*, 1986). Perhaps the most complex field experiment within the tundra biome has been the Baffin Island oil spill (Sergy and Blackall, 1987), in which crude oil was intentionally introduced into a small bay near Pond Inlet, Northwest Territories, in an experiment to describe its chemical fate and biological effects, and to explore possible countermeasure options. While whole ecosystem experiments such as those mentioned above are definitive, they are increasingly difficult to do for two reasons: (1) increasingly stringent environmental protection legislation and (2) the long time over which support funding is required.

12.8.2 ENCLOSURE/MESOCOSM EXPERIMENTS

Partial ecosystems have been built to enclose small portions of freshwater or marine ecosystems, and then experimental manipulations have been performed within the enclosed portions (Schindler, 1988). These enclosures ranged in size from a few litres to partitioned embayments of lakes. They have most frequently been used for experiments with planktonic or benthic species; however, they have sometimes been used with fish. Snow and Scott (1975) described the weathering of petroleum oil spilled experimentally into enclosed areas of two lakes in the Mackenzie River Delta. They noted that adult insects were killed and that periphyton growth increased, probably as a result of oil stimulation of nitrogen-fixing organisms. Shindler *et al.* (1975) also reported increases in nitrogen-fixing bacteria following treatments of ponds with crude oil. Fish have been used less frequently in enclosure experiments, but Ramsey (1990) used a limnocorral system to show that mercury was taken up more rapidly by fish when a rich source of organic carbon was present than in control enclosures lacking supplemental carbon. These experiments are attractive for experiments designed to describe movements of some contaminants and responses of small organisms; however, they are impractical for most large, powerful animals of interest in the Arctic.

12.8.3 POPULATION EXPERIMENTS

Another approach is to manipulate the population of interest without manipulating the habitat. Lockhart *et al.* (1972) used this approach to estimate the natural rate of loss of mercury from a population of northern pike *Esox lucius*. Fish from a highly contaminated lake were tagged and transplanted to a relatively clean lake, and then recaptured at intervals to determine the rate of decline of mercury in tissues. A serious limitation of this design was the possibility of introducing parasites or diseases along with the transplanted fish, as in fact happened in this experiment; consequently, such experiments are unlikely to be done often. An extension of population manipulation approach is a partial ecosystem experiment underway at the Experimental Lakes Area. It represents an attempt to evaluate the importance of three organic contaminants, toxaphene, chlordane and 2,3,4,7,8-pentachlorodibenzofuran to fish populations without contaminating the entire lake. Lake trout and white suckers from a lake at the Experimental Lakes Area were marked and injected intraperitoneally with one of these compounds and released back to the lake. Subsequently, these fish have been recaptured on several occasions and at spawning time for use in experimental breeding crosses. In this way, information is being obtained on survival, growth, and reproduction of the treated fish as compared with untreated and sham-treated fish from the same lake. Although these experiments do not mimic "natural" exposure, they can provide information about the effect of known body burdens of these specific contaminants on the ecological performance of the fish without the experimental contamination of the whole lake.

12.8.4 EXAMINATION OF ANIMALS FROM NATURAL POPULATIONS

Perhaps the most widely applied approach has been the examination of individual animals subject to some known exposure in their natural habitat to search for indicators of exposure and pathologies (Uthe *et al.*, 1980; Dixon *et al.*, 1985). This approach, often called the "bioindicators" or "biomarkers" approach, has the advantage of describing real exposures and responses, and so avoids the uncertainties inherent in extrapolating results from a laboratory setting to real populations. It has the disadvantage of limited specificity in that multiple variables influence the animals, and so the result observed cannot be proven rigorously to be the result of the hypothesized "cause." In spite of this limitation, the biomarker approach is probably the closest an investigator can come to testing causal hypotheses in natural populations; the approach has been applied to a wide range of contamination issues including acidification (Lockhart and Lutz, 1977; Brown *et al.*, 1990), metal pollution (Munkittrick and Dixon, 1988; Klaverkamp *et al.*, 1991), organochlorine pollution (Helle *et al.*, 1976; Stegeman *et al.*, 1986; Subramanian *et al.*, 1987), hydrocarbon pollution (Dunn, 1980; Johnson *et al.*, 1988; Lockhart *et al.*, 1989), and radionuclide pollution (Swanson, 1982; Waite *et al.*, 1988, 1989).

The bioindicator approach has been applied many times in the Sub-Arctic, and studies are starting with field populations in the Arctic and in laboratory experiments with arctic species. Several correlations have been described linking biomarker responses with other toxic responses (Safe, 1990), but whether the two are linked by necessity or by coincidence is seldom clear. Nonetheless, these sublethal responses are sensitive and relatively inexpensive to measure; they can help to identify the need for more expensive ecological and chemical studies. The most obvious bioindicator is one that is visible. Physical deformities in fish have been associated with polluted habitats (Beamish *et al.*, 1975; Munkittrick *et al.*, 1992), in marine mammals (Zakharov and Yablokov, 1990), and in invertebrates (Warwick, 1985). Several populations of fish have been reported to have unusually high incidences of tumours, presumably as a result of exposure to chemicals in their habitat (Sonstegard and Leatherland, 1976; Baumann, 1984). However, the presence of tumours does not necessarily imply a chemical etiology; Yamamoto *et al.* (1985) have described skin tumours of walleye (*Stizostedion vitreum)* with a viral etiology. The most convincing approach has been the experimental demonstration that tumours are induced by treatment with the suspected causative agents. For example, polluted sediments extracted from bullheads were shown to induce cancer. Metcalfe *et al.* (1988) extracted sediment from Hamilton harbour (Lake Ontario, Canada), and injected it into rainbow trout sac fry; a year later, the fish were sacrificed and examined visually and microscopically for liver cancers. The extracts from Hamilton harbour induced cancer in 3–9 percent of the fish, while extracts from a clean site induced none. These active extracts were prepared from sediments containing high levels of polycyclic aromatic hydrocarbons (PAH), and so these probably contributed to the biological activity of the extracts; however, they may not have been the exclusive source of carcinogenic activity.

A most widely applied bioindicator has been the cytochrome P_{450} system; it has been used to indicate exposure to several ubiquitous contaminants (Stegeman and Kloepper-Sams, 1987). Laboratory experiments have shown that the system responds to PAH (James and Bend, 1980), and to some chlorinated compounds including some PCBs (Förlin and Lidman, 1981; Gooch *et al.*, 1989) and chlorinated dioxins and furans (Hahn *et al.*, 1989; van der Weiden *et al.*, 1990). All of these contaminants have been detected in northern fish. Cytochrome P_{450} activities have been used to detect subtle responses to oil spills (Payne, 1976), and petroleum hydrocarbons are found widely in parts of the Arctic, most notably the Mackenzie River drainage (Carey *et al.*, 1990). Several recent studies have shown that the system responds to effluent of bleached kraft pulp mills in Scandinavia (Andersson *et al.*, 1988) and Canada (Rogers *et al.*, 1989; McMaster *et al.*, 1991; Hodson *et al.*, 1992), although the identities of all the inducers present are still unclear. Chlorinated dioxins and furans are present in these effluents, but it seems unlikely that they are the only inducers. The induction of enzymes has also been reported in several species of birds (Ronis and Walker, 1989).

Other sublethal effects (including behavioural disturbances and immunotoxicity) on seabirds may be caused by chlorinated hydrocarbons, but sound evidence for them is still lacking (Walker, 1990). Mutagenic, carcinogenic, and teratogenic effects of PCBs are not documented for arctic seabirds.

Similar in concept to the cytochrome P_{450} system is the metallothionein response to metals (Klaverkamp *et al.*, 1984). Fish exposed to several metals increase the production of this protein, and the increased levels can serve as a bioindicator of previous exposure. However, metals are not the only vectors that can cause increased synthesis of this protein. Fish from a series of lakes with differing levels of metal contamination were analyzed and found to have a very high correlation between the metallothionein content and zinc in the water. For biomarkers generally in field settings, the preferred design has been the comparison of samples from populations exposed to higher exposures of the contaminants of interest with samples from otherwise comparable populations exposed to lower quantities. Often species, sexual, and seasonal differences in biomarkers are seen, but these are relatively simple to eliminate as sources of treatment effects through well-designed sampling programmes.

A promising approach is the examination of the same individual animals both for chemical residues and for evidence of biological responses. For example, Giesy *et al.* (1986) examined statistical associations between organochlorine residues and rearing mortality in chinook salmon from Lake Michigan. Galgani *et al.* (1991) reported a correlation between ethoxyresorufin-*O*-deethylase (EROD) activities and total PCBs (r=0.620) in postmitochondrial supernatants of plaice (*Pleuronectes platessa*) from the Bay of Seine, France. Within the Arctic, two microsomal monooxygenase activities, EROD and aryl hydrocarbon hydroxylase (AHH), were measured in burbot liver microsomes, and the results failed to show a geographic trend similar to that shown by PCB residues (Lockhart and Metner, 1992). However, further residue analysis for mono-ortho substituted congeners indicated that enzymatic activities were actually correlated with some of these residues, notably with congener 156 (Lockhart *et al.*, in press). The apparent relationship between mono-ortho PCBs and EROD seems likely to reflect the presence of non-ortho PCBs, especially PCB 126 (3,3′,4,4′,5-pentachlorobiphenyl), a potent inducer in fish (Janz and Metcalfe 1991) that was not determined in burbot liver. Much more striking correlations were found between cytochrome P_{450} catalytic activities and several non-ortho and mono-ortho PCB congeners in a group of beluga whales that became trapped in freshwater lakes in the Mackenzie Delta (Lockhart *et al.*, 1972, 1992b). These whales were, on average, about 200 kg lighter in weight than other arctic belugas of the same age. The working hypothesis is that these animals had mobilized blubber during the starvation, thereby releasing contaminants stored in the blubber.

A surprising result with biomarkers in beluga whales has been the observation that aromatic DNA adducts were no different in arctic whales than in whales from the Gulf of St. Lawrence where pollution levels are generally several times higher (Ray *et al.*, 1991).

Many other biomarkers have been used to detect biological changes due to chemical inputs. For example, brain cholinesterase activities in young walleye (*Stizostedion vitreum*) were sensitive to aerial spraying with the organophosphorous pesticide, malathion (Lockhart *et al.*, 1985). Circulating levels of several steroid hormones are influenced by exposure to effluent from some pulp mills (McMaster *et al.*, 1991; Munkittrick *et al.*, 1991; Hodson *et al.*, 1992). Energy stores have been depleted in white suckers from lakes where food supplies have been reduced by metal pollution (Munkittrick and Dixon, 1988). Lockhart *et al.* (1989) reported reduced lipid stores in burbot from the Mackenzie River where they were rejected for human consumption due to the appearance of their livers. Neff *et al.* (1987) determined energy stores (glycogen, other carbohydrates, lipids) in clams (*Mya truncata*) subjected to experimental oil spills on Baffin Island, but found that the oil had little apparent effect. The authors noted that the biochemical values were highly variable, possibly the result of sampling techniques, and so large differences were required in order to isolate them from normal variation.

In summary, chemical contaminants are ubiquitous throughout the Arctic, mainly as a result of aerial transport from lower latitudes and deposition in the Arctic. Chemical analyses have revealed the circumpolar dispersal of organochlorines and several toxic metals in animals at the top of aquatic food chains. These tissue analyses provide the best measures of intakes by people who hunt and consume the animals, and they also provide the best measures of dosages experienced by the animals. Cores of lake sediments have shown that inputs of several pollutants (mercury, polycyclic aromatic hydrocarbons) have been continuous for decades, but that levels are often relatively low compared with those at sites at lower latitudes.

The existing data on contamination levels in arctic biota generally serve as a good base for future toxicological investigations in this region. The sensitivity of arctic animals to these loadings is largely unknown. With the exception of petroleum, very little experimental toxicology has been done with arctic species, but the limited data suggest that individual sensitivities are comparable to species from further south. The application of "bioindicators" is just starting, but already some correlations with levels of some PCB congeners have been detected in fish and whales. Many unknown aspects remain concerning the effects of contaminants on Arctic seabirds; clearly studies of biological effects are needed.

12.9 REFERENCES

Addison, R.F., and Smith, T.G. (1974) Organochlorine residue levels in arctic ringed seals: variation with age and sex. *Oikos* **25**, 335–337.

Addison, R.F., Zinck, M.E., and Smith, T.G. (1986) PCBs have declined more than DDT-group residues in arctic ringed seals (*Phoca hispida*) between 1972 and 1981. *Environ. Sci. Technol.* **20**, 253–256.

Anderson, R.F., Schiff, S.L., and Hesslein, R.H. (1987) Determining sediment accumulation and mixing rates using Pb–210, Cs–137, and other tracers: problems due to postdepositional mobility or coring artifacts. *Can. J. Fish. Aquat. Sci.* **44**, 231–250.

Andersson, T., Förlin, L., Härdig, J., and Larsson, Ä. (1988) Physiological disturbances in fish living in coastal water polluted with bleached kraft pulp mill effluent. *Can. J. Fish. Aquat. Sci.* **45**, 1525–1536.

Barrett, R.T., Skaare, J.U., Norheim, G., Vader, W., and Froslie, A. (1985) Persistent organochlorines and mercury in eggs of Norwegian seabirds. *Environ. Pollut.* **A39**, 79–93.

Baumann, P.C. (1984) Cancer in wild freshwater fish populations with emphasis on the Great Lakes. *J. Great Lakes Res.* **10**, 251–253.

Beamish, R.J., Lockhart, W.L., VanLoon, J.C., and Harvey, H.H. (1975) Long-term acidification of a lake and resulting effects on fishes. *Ambio* **4**, 98–102.

Becker, P.R., Wise, S.A., and Zeisler, R. (1989) *Alaskan Marine Mammal Tissue Archival Project: Acquisition and Curation of Alaskan Marine Mammal Tissues for Determining Levels of Contaminants Associated with Offshore Oil and Gas Development*. Report for National Oceanic and Atmospheric Adminstration, Anchorage, Alaska, 98 pp.

Bergman, A., Olsson, M., and Reutergårdh, L. (1981) *Lowered Reproduction Rate in Seal Population and PCB. A Discussion of Comparability of Results and a Presentation of Some Data from Research on the Baltic Seals*. International Council for the Exploration of the Sea, Marine Environ. Quality Comm., 18 pp.

Borch-Iohnsen, B. Holm, H., Jorgensen, A., and Norheim G. (1991) Seasonal siderosis in female eider nesting in Svalbard. *J. Compar. Pathol.* **104**, 7–15.

Bourne, W.R.P. (1976) Seabirds and pollution. In: Johnston, R. (Ed.) *Marine Pollution*, pp. 318–424. Academic Press, London.

Bourne, W.R.P., and Bogan, J.A. (1976) Polychlorinated biphenyls in North Atlantic seabirds. *Mar. Pollut. Bull.* **3**, 171–175.

Braestrup, L., Clausen, J., and Berg, O. (1974) DDE, PCB and aldrin in arctic birds in Greenland. *Bull. Environ Contam. Toxicol.* **11**, 326–332.

Brouwer, A., Reijnders, P.J.H., and Koeman, J.H. (1989) Polychlorinated biphenyl (PCB) contaminated fish induces vitamin A and thyroid hormone deficiency in the common seal (*Phoca vitulina*). *Aquat. Toxicol.* **15**, 99–106.

Brown, S.B., Evans, R.E., Majewski, H.S., Sangalang, G.B., and Klaverkamp, J.F. (1990) Responses of plasma electrolytes, thyroid hormones, and gill histology in Atlantic salmon (*Salmo salar*) to acid and limed river waters. *Can. J. Fish. Aquat. Sci.* **47**, 2431–2440.

Carey, J.H., Ongley, E.D., and Nagy, E. (1990) Hydrocarbon transport in the Mackenzie River. *Can. Sci. Tot. Environ.* **97/98**, 69–88.

Campbell, P.G.C., Tessier, A., Bisson, M., and Bougie, R. (1985) Accumulation of copper and zinc in the yellow water lily *Nuphar variegatum*: Relationships to metal partitioning in the adjacent lake sediments. *Can. J. Fish. Aquat. Sci.* **42**, 23–32.

Cooke, A.S. (1973) Shell thinning in avian eggs by environmental pollutants. *Environ. Pollut.* **4**, 85–150.

Craddock, D.R. (1977) Acute toxic effects of petroleum on arctic and subarctic marine organisms. In: Malins, D.C. (Ed.) *Effects of Petroleum on Arctic and Subarctic Marine Environments and Organisms*, Vol. II. *Biological Effects*, pp. 1–93. Academic Press, New York.

Cross, W.E., and Thomson, D.H. (1987) Effects of experimental releases of oil and dispersed oil on Arctic nearshore macrobenthos. I. Infauna. *Arctic* **40** (Suppl. 1), 184–200.

Crusius, J., and Anderson, R.F. (1991) Core compression and surficial sediment loss of lake sediments of high porosity caused by gravity coring. *Limnol. Oceanogr.* **36**, 1021–1031.

Daelemans, F.F., Mehlum, F., and Schepens, J.C. (1992) Polychlorinated biphenyls in two species of arctic seabirds from Svalbard area. *Bull. Environ. Contam. Toxicol.* **48**, 828–834.

Devault, D.S., Willford, W.A., Hesselberg, R.J., Nortrupt, D.A., Rundberg, E.G.S., Alwan, A.K., and Bautista, C. (1986) Contaminant trends in lake trout (*Salvelinus namaycush*) from the upper Great Lakes. *Arch. Environ. Contam. Toxicol.* **15**, 349–356.

Dewailly, E., Nantel, A., Weber, J.P., and Meyer, F. (1989) High levels of PCBs in breast milk of Inuit women from arctic Canada. *Bull. Environ. Contam. Toxicol.* **43**, 641–646.

Dewailly, E., Bruneau, S., Laliberté, C., Bélanger, D., Gringras, S., Ayotte, P., and Nantel, A. (1992) Weight, size, head circumference and TSH of Inuit newborn prenatally exposed to high levels of organochlorines. In: *Proceedings of the 12th International Symposium on Dioxins and Related Compounds 10*, pp. 257–259. Finnish Institute of Occupational Health, Helsinki.

Dixon, D.G., Hodson, P.V., Klaverkamp, J.F., Lloyd, K.M., and Roberts, J.R. (1985) *The Role of Biochemical Indicators in the Assessment of Aquatic Ecosystem Health—Their Development and Validation.* Report No. NRCC 24371. National Research Council of Canada, Ottawa.

Dunn, B.P. (1980) Benzo(a)pyrene in the marine environment: Analytical techniques and results. In: Afghan, B.K., and Mackay, D. (Eds.) *Hydrocarbons and Halogenated Hydrocarbons in the Aquatic Environment*, pp. 109–119. Plenum Press, New York.

Eisenreich, S.J., Capel, P.D., Robbins, J.A., and Bourbonniere, R. (1989) Accumulation and diagenesis of chlorinated hydrocarbons in lacustrine sediments. *Environ. Sci. Technol.* **23**, 1116–1126.

Eisler, R. (1986) *Polychlorinated Biphenyl Hazards to Fish, Wildlife and Invertebrates: A Synoptic Review.* Biological Report No. 85(1.7). US Department of the Interior, Fish and Wildlife Service, Laurel, Maryland, 72 pp.

Elliott, J.E., Norstrom, R.J., and Keith, J.H.A. (1988) Organochlorines and eggshell thinning in northern gannets from Eastern Canada 1968–1984. *Environ. Pollut.* **52**, 81–102.

Elliott, J.E., Scheuhammer, A.M., Leighton, F.A., and Pearce, P.A. (1992) Heavy metal and metallothionein concentrations in Atlantic Canadian seabirds. *Arch. Environ. Contam. Toxicol.* **22**, 63–73.

Engelhardt, F.R., Geraci, J.R., and Smith, T.G. (1977) Uptake and clearance of petroleum hydrocarbons in the ringed seal, *Phoca hispida. J. Fish. Res. Board Canada* **34**, 1143–1147.

Fimreite, N., and Bjerk, J.E. (1979) Residues of DDE and PCBs in Norwegian seabirds' fledglings, compared to those in their eggs. *Astarte* **12**, 49–51.

Förlin, L., and Lidman, U. (1981) Effects of Clophen A50 and 3-methylcholanthrene on the hepatic mixed function oxidase system in female rainbow trout *Salmo gairdneri. Compar. Biochem. Physiol.* **70C**, 297–300.

Gabrielsen, G.W., Mehlum, F., and Nagy, K.A. (1987) Daily energy expenditure and energy utilization of free-ranging black-legged kittiwakes. *Condor* **89**, 126–132.

Galgani, F., Bocquenne, G., Lucon, M., Grzebyk, D., Letrouit, F., and Claisse, D. (1991) EROD Measurements in fish from the northwest part of France. *Mar. Pollut. Bull.* **22**, 494–500.

Geraci, J.R., St Aubin, D.J., and Reisman, R.J. (1983) Bottlenose dolphins, *Tursiops truncatus*, can detect oil. *Can. J. Fish. Aquat. Sci.* 40, 1516–1522.

Giesy, J.P., Newsted, J., and Garling, D.L. (1986) Relationships between chlorinated hydrocarbon concentrations and rearing mortality of chinook salmon (*Oncorhynchus tshawytscha*) eggs from Lake Michigan. *J. Great Lakes Res.* **12**, 82–98.

Gilbertson, M. (1989) Effects on fish and wildlife populations. In: Kimbrough, R.D., and Jensen, A.A. (Eds.) *Halogenated Biphenyls, Terphenyls, Naphthalenes, Dibenzodioxins and Related Products* 2nd edn., pp. 103–127. Elsevier Science, New York.

Gladen, B.C., Rogan, W.J., and Wilcox, A.J. (1985) Potential reproductive and postnatal morbidity from exposure to polychlorinated biphenyls: Epidemiologic considerations. *Environ. Health Perspect.* **60**, 233–239.

Goldblatt, C.J., and Anthony, R.G. (1983) Heavy metals in northern fur seal (*Callorhinus ursinus*) from the Pribilof Islands, Alaska. *J. Environ. Qual.* **12**, 478–482.

Gooch, J.W., Elskus, A.A., Kloepper-Sams, P.J., Hahn, M.E., and Stegeman, J.J. (1989) Effects of ortho- and non-ortho-substituted polychlorinated biphenyl congeners on the hepatic monooxygenase system in scup *Stenotomus chrysops*. *Toxicol. Appl. Pharmacol.* **98**, 422–433.

Gschwend, P.M., and Hites, R.A. (1981) Fluxes of polycyclic aromatic hydrocarbons to marine and lacustrine sediments in the northeastern United States. *Geochim. Cosmochim. Acta* **45**, 2359–2367.

Hahn, M.E., Woodin, B.R., and Stegeman, J.J. (1989) Induction of cytochrome P-450E (P-450IA1) by 2,3,7,8-tetrachlorodibenzofuran (2,3,7,8-TCDF) in the marine fish scup *Stenotomus chrysops*. *Mar. Environ. Res.* **28**, 61–65.

Hart, L.E., Cheng, K.M., Whitehead, P.E., Shah, R.M., Lewis, R.J., Ruschkowski, S.R., Blair, R.W., Bennett, D.C., Bandiera, S.M., Norstrom, R.J., and Bellward, G.D. (1991) Dioxin contamination and growth and development in Great Blue Heron embryos. *J. Toxicol. Environ. Health* **32**, 331–344.

Helle, E., Olsson, M., and Jensen, S. (1976) PCB levels correlated with pathological changes in seal uteri. *Ambio* **5**, 261–263.

Helle, E., Hyvärinen, H., Pyysalo, H., and Wickstrom, K. (1983) Levels of organochlorine compounds in an inland seal population in eastern Finland. *Mar. Pollut. Bull.* **14**, 256–260.

Hodson, P.V., McWhirter, M., Ralph, K., Gray, B., Thivierge, D., Carey, J.H., Van Der Kraak, G., Whittle, D.M., and Levesque, M.C. (1992) Effects of bleached kraft mill effluent on fish in the St Maurice River, Quebec. *Environ. Toxicol. Chem.* **11**, 1635–1651.

Holt, G., Froslie, A., and Norheim, G. (1979) Mercury, DDE and PCB in the avian fauna in Norway. *Acta Vet. Scand.* **70** (Suppl.), 1–28.

Hwang, H.L., Lawrence, C., Paulson, A., and Taylor, P. (1984) Polychlorinated biphenyls: influence on birth weight and gestation. *Am. J. Public Health* **74**, 1153–1154.

Ingebrigtsen, K., Skaare, J.U., and Teigen, S.W. (1984) Organochlorine residues in two Norwegian puffin (*Fratercula arctica*) colonies. *J. Toxicol. Environ. Health* **14**, 813–828.

James, M.O., and Bend, J.R. (1980) Polycyclic aromatic hydrocarbon induction of cytochrome P_{450} mixed-function oxidases in marine fish. *Toxicol. Appl. Pharmacol.* **54**, 117–133.

Janz, D.M., and Metcalfe, C.D. (1991) Relative induction of aryl hydrocarbon hydroxylase by 2,3,7,8-TCDD and two coplanar PCBs in rainbow trout (*Orcorhynchus mykiss*). *Environ. Toxicol. Chem.* **10**, 917–923.

Johnson, L.L., Casillas, E., Collier, T.K., McCain, B.B., and Varanasi, U. (1988) Contaminant effects on ovarian development in English sole, *Parophrys vetulus*, from Puget Sound, Washington. *Can. J. Fish. Aquat. Sci.* **45**, 2133–2146.

Johnson, M.G. (1987) Trace element loadings to sediments of fourteen Ontario lakes and correlations with concentrations in fish. *Can. J. Fish. Aquat. Sci.* **44**, 3–13.

Kelso, J.R.M. (1977) Density, distribution, and movement of Nipigon Bay fishes in relation to a pulp and paper mill effluent. *J. Fish. Res. Board Can.* **34**, 879–885.

Kenaga, E.E. (1982) Predictability of chronic toxicity from acute toxicity of chemicals in fish and aquatic invertebrates. *Environ. Toxicol. Chem.* **1**, 347–358.

Kinloch, D., Kuhnlein, H., and Muir, D.C.G. (1992) Inuit foods and diet: a preliminary assessment of benefits and risks. *Sci. Total Environ.* **122**, 247–278.

Klaverkamp, J.F., MacDonald, W.A., Duncan, D.A., and Wagemann, R. (1984) Metallothionein and acclimation to heavy metals in fish: a review. In: Cairns, V.W.,

Hodson, P.V., and Nriagu, J.O. (Eds.) *Contaminant Effects on Fisheries*, pp. 99–113. John Wiley & Sons, New York.

Klaverkamp, J.F., Dutton, M.D., Majewski, H.S., Hunt, R.V., and Wesson, L.J. (1991) Evaluating the effectiveness of metal pollution controls in a smelter by using metallothionein and other biochemical responses in fish. In: Newman, M.C., and McIntosh, A.W. (Eds.) *Metal Ecotoxicology Concerns and Applications*, pp. 33–64. Lewis, Chelsea, Michigan.

Krantzberg, G., and Boyd, D. (1992) The biological significance of contaminants in sediment from Hamilton Harbour, Lake Ontario. *Environ. Sci. Technol.* **11**, 1527–1540.

Krishnaswami, S.K., and Kupchanko, E.E. (1969) Relationship between odor of petroleum refinery wastewater and occurrence of "oily" taste-flavor in rainbow trout, *Salmo gairdneri*. *J. Water Pollut. Control Fed.* **41**, 189–196.

Landrum, P.F., Lee, H., and Lydy, M.J. (1992) Toxicokinetics in aquatic systems: model comparisons and use in hazard assessment. *Environ. Toxicol. Chem.* **11**, 1709–1725.

Lockhart, W.L., and Danell, R.W. (1992) Field and experimental tainting of arctic freshwater fish by crude and refined petroleum products. *Proceedings of the 15th Arctic and Marine Oil Spill Program Technical Seminar, Edmonton, Alberta*, pp. 763–771.

Lockhart, W.L., and Lutz, A. (1977) *Preliminary biochemical observations of fishes inhabiting an acidified lake in Ontario. Can. Water Air Soil Pollut.* **7**, 312–332.

Lockhart, W.L., and Metner, D.A. (1992) Applications of hepatic mixed function oxidase enzyme activities to northern freshwater fish. I. Burbot, *Lota lota*. *Mar. Environ. Res.* **34**, 175–180.

Lockhart, W.L., Uthe, J.F., Kenny, A.R., and Mehrle, P.M. (1972) Methyl-mercury in northern pike (*Esox lucius* L.): distribution, elimination, and some biochemical characteristics of contaminated fish. *J. Fish. Res. Board Canada* **29**, 1519–1523.

Lockhart, W.L., Metner, D.A., Ward, F.J., and Swanson, G.M. (1985) Population and cholinesterase responses in fish exposed to malathion sprays. *Pest. Biochem. Physiol.* **24**, 12–18.

Lockhart, W.L., Danell, R.W., and Murray, D.A.J. (1987) Acute toxicity bioassays with petroleum products: influence of exposure conditions. In: Vandermeulen, J.H., and Hrudey, S.R. (Eds.), *Oil in Fresh Water: Chemistry, Biology, Countermeasure Technology*, pp. 335–344. Pergamon Press, Oxford, Kronberg, New York.

Lockhart, W.L., Metner, D.A., Murray, D.A.J., Danell, R.W., Billeck, B.N., Baron, C.L., Muir, D.C.G., and Chang-Kue, K. (1989) Studies to Determine whether the Condition of Fish from the Lower Mackenzie River is Related to Hydrocarbon Exposure. Department of Indian Affairs and Northern Development, Environmental Studies No. 61. 84 pp.

Lockhart, W.L., Stewart, R.E.A., and Muir, D.C.G. (1992a) Biochemical stress indicators in marine mammals. In: Murray, J.L., and Shearer, R.G. (Eds.) *Synopsis of Research Conducted under the 1991/92 Northern Contaminants Program*, pp. 158–164. Environmental Studies 68. Canada Department of Indian Affairs and Northern Development, Northern Affairs Program, Ottawa.

Lockhart, W.L., Wagemann, R., Tracey, B., Sutherland, D., and Thomas, D.J. (1992b) Presence and implications of chemical contaminants in the freshwaters of the Canadian Arctic. *Sci. Total Environ.* **122**, 165–245.

Lockhart, W.L., Muir, D.C.G., Ford, C.A., and Metner, D.A. (in press) Mono-ortho polychlorinated biphenyls in burbot (*Lota lota*) from northwestern Canada and their possible relationship to hepatic microsomal mixed-function oxidase enzyme activities. In: Roots, F., and Shearer, R. (Eds.) *Proceedings of the Conference of the Comité Arctique*

International on the Global Significance of the Transport and Accumulation of Polychlorinated Hydrocarbons in the Arctic. Plenum, New York.

Long, E.R., Buchman, M.F., Bay, S.M., Breteler, R., Carr, R.S., Chapman, P.M., Hose, J.E., Lissner, A.L., Scott, J., and Wolfe, D.A. (1990) Comparative evaluation of five toxicity tests with sediments from San Francisco Bay and Tomales Bay, California. *Environ. Toxicol. Chem.* **9**, 1193–1214.

McCarty, L.S. (1991) Toxicant body residues: implications for aquatic bioassays with some organic chemicals. In: Mayes, M.A., and Barron, M.G. (Eds.) *Aquatic Toxicology and Hazard Assessment*, Vol. 14, pp. 183–192. Report No. ASTM STP 1124. American Society for Testing and Materials, Philadelphia.

MacDonald, R.W., Cretney, W.J., Crewe, N., and Paton, D. (1992) A history of octachlorodibenzo-*p*-dioxin, 2,3,7,8-tetachlorodibenzofuran, and 3,3′,4,4′-tetrachlorobiphenyl contamination in Howe Sound, British Columbia. *Environ. Sci. Technol.* **26**, 1544–1550.

McMaster, M.E., Van Der Kraak, G.J., Portt, C.B., Munkittrick, K.R., Sibley, P.K., Smith, I.R., and Dixon, D.G. (1991) Changes in hepatic mixed-function oxygenase (MFO) activity, plasma steroid levels and age at maturity of a white sucker (*Catostomus commersoni*) population exposed to bleached kraft pulp mill effluent. *Aquat. Toxicol.* **21**, 199–218.

McNicol, R.E., and Scherer, E. (1991) Behavioral responses of lake whitefish (*Coregonus clupeaformis*) to cadmium during preference-avoidance testing. *Environ. Toxicol. Chem.* **10**, 225–234.

Mayer, F.L., and Mehrle, P.M. (1977) Toxicological aspects of toxaphene in fish: A summary. *Trans. N. Am. Wildlfe Natural Resources Conf.* **42**, 365–373.

Metcalfe, C.D., Cairns, V.W., and Fitzsimons, J.D. (1988) Experimental induction of liver tumors in rainbow trout (*Salmo gairdneri*) by contaminated sediment from Hamilton Harbour, Ontario. *Can. J. Fish. Aquat. Sci.* **45**, 2161–2167.

Miller, M.C., Stout, J.R., and Alexander, V. (1986) Effects of a controlled under-ice oil spill on invertebrates of an arctic and sub-arctic stream. *Environ. Pollut.* **A42**, 99–132.

Motohiro, T. (1983) Tainted fish caused by petroleum compounds—a review. *Water Sci. Technol.* **15** (Finland), 75–83.

Muir, D.C.G., Norstrom, R.J., and Simon, M. (1988) Organochlorine contaminants in arctic marine food chains: accumulation of specific polychlorinated biphenyls and chlordane-related compounds. *Environ. Sci. Technol.* **22**, 1071–1079.

Muir, D.C.G., Ford, C.A., Grift, N.P., Metner, D.A., and Lockhart, W.L. (1990) Geographic variation of chlorinated hydrocarbons in burbot (*Lota lota*) from remote lakes and rivers in Canada. *Arch. Environ. Contam. Toxicol.* **19**, 530–542.

Muir, D.C.G., Wagemann, R., Hargrave, B.T., Thomas, D., Peakall, D.B., and Norstrom, R.J. (1992) Arctic marine ecosystem contamination. *Sci. Total Environ.* **122**, 75–134.

Munkittrick, K.R., and Dixon, D.G. (1988) Growth, fecundity, and energy stores of white sucker *Catostomus commersoni* from lakes containing elevated levels of copper and zinc. *Can. J. Fish. Aquat. Sci.* **45**, 1355–1365.

Munkittrick, K.R., Portt, C., Van Der Kraak, G.J., Smith, I., and Rokosh, D. (1991) Impact of bleached kraft mill effluent on liver MFO activity, serum steroid levels and population characteristics of a Lake Superior white sucker (*Catostomus commersoni*) population. *Can. J. Fish. Aquat. Sci.* **48**, 1371–1380.

Munkittrick, K.R., McMaster, M.E., Portt, D.B., Van Der Kraak, G.J., Smith, I.R., and Dixon, D.G. (1992) Changes in maturity, plasma sex steroid levels, hepatic mixed-function oxidase activity, and the presence of external lesions in lake whitefish (*Coregonus clupeaformis*) exposed to bleached kraft mill effluent. *Can. J. Fish. Aquat. Sci.* **49**, 1560–1569.

National Institutes of Health (1979) *National Institute of Cancer, Carcinogenesis Technical Report Series 37*, Publ. No. 920. Department of Health, Education and Welfare, Washington, D.C.

Neff, J.M., Hillman, R.E., Carr, R.S., Buhl, R.L., and Lahey, J.I. (1987) Histopathologic and biochemical responses in arctic marine bivalve molluscs exposed to experimentally spilled oil. *Arctic* **40** (Suppl. 1), 220–229.

Nettleship, D.N., and Peakall, D.B. (1987) Organochlorine residue levels in three high arctic species of colonially-breeding seabirds from Prince Leopold Island. *Mar. Pollut. Bull.* **18**, 434–438.

Neuhold, J.M. (1987) The relationship of life history attributes to toxicant tolerance in fishes. *Environ. Toxicol. Chem.* **6**, 709–716.

Nielsen, C.O., and Dietz, R. (1989) Heavy metals in Greenland seabirds. Meddelelser om Greenland. *Bioscience* **29**, 3–26.

Norheim, G. (1987) Levels and interactions of heavy metals in sea birds from Svalbard and the Antarctic. *Environ. Pollut.* **47**, 7–13.

Norheim, G., and Borch-Iohnsen, B. (1990) Chemical and morphological studies of liver from eider (*Somateria mollissima*) in Svalbard with special reference to the distribution of copper. *J. Comp. Pathol.* **102**, 457–466.

Norheim, G., and Kjos-Hansen, B. (1984) Persistent chlorinated hydrocarbons and mercury in birds caught off the west coast of Spitsbergen. *Environ. Pollut.* **33A**, 143–152.

Norstrom, R.J., Schweinsberg, R.E., and Collins, B.T. (1986) Heavy metals and essential elements in livers of the polar bear (*Ursus maritimus*) in the Canadian Arctic. *Sci. Total Environ.* **48**, 195–212.

Nygard, T. (1983) Pesticide residues and shell thinning in eggs of peregrines in Norway. *Ornis. Scand.* **14**, 161–166.

Oehme, M., Fürst, P., Krüger, Chr., Meemken, H.A., and Groebel, W. (1988) Presence of polychlorinated dibenzo-*p*-dioxins, dibenzofurans and pesticides in arctic seal from Spitzbergen. *Chemosphere* **18**, 1291–1300.

Ohlendort, H.M., Bartonek, J.C., Diivoky, G.J., Klaas, E.E., and Krynitsky, A.J. (1982) *Organochlorine Residues in Eggs of Alaskan Seabirds.* Spec. Sci. Report—Wildlife No. 245. US Department of the Interior, Patuxent Wildlife Research Center, Laurel, Maryland, 41 pp.

Olsson, M., Bergman, A., Jensen, S., and Kihlström, J.E. (1990) Effect of various fractions of PCB on mink reproduction: preliminary results from experimental studies within the Swedish seal project. In: *Proceedings of the Dioxin 1990 Symposium*, Vol. 1, pp. 393–396. Eco-Informa, Bayreuth, Germany.

Parslow, J.L.F., Jeffries, D.J., and Hanson, J.M. (1973) Gannet mortality incidents in 1972. *Mar. Pollut. Bull.* **4**, 41–44.

Payne, J.F. (1976) Field evaluation of benzopyrene hydroxylase induction as a monitor for marine petroleum pollution. *Science* **191**, 945–946.

Payne, J.F., and McNabb, G.D., Jr (1984) Weathering of petroleum in the marine environment. *Mar. Technol. Soc. J.* **18**, 24–42.

Payne, J.F., Kiceniuk, J., Fancey, L.L., Williams, U., Fletcher, G.L., Rahimtula, A., and Fowler, B. (1988) What is a safe level of polycyclic aromatic hydrocarbons for fish: subchronic toxicity study on winter flounder *Pseudopleuronectes americanus*. *Can. J. Fish. Aquat. Sci.* **45**, 1983–1993.

Peakall, D.B., and Fox, G.H. (1987) Toxicological investigations of pollutant-related effects in Great Lakes gull. *Environ. Health Perspect.* **71**, 187–193.

Pearce, P.A., Elliott, J.E., Peakall, D.B., and Norstrom, R.J. (1989) Organochlorine contaminants in eggs of seabirds in the northwest Atlantic, 1968–1984. *Environ. Pollut.* **56**, 217–235.

Percy, J.A.,, and Mullin, T.C. (1977) Effects of crude oil on the locomotory activity of arctic marine invertebrates. *Mar. Pollut. Bull.* **8**, 35–40.

Poels, C.L.M., Fischer, R., Fukawa, K., Howgate, P., Maddock, B.G., Persoone, G., Stephenson, R.R., and Bontinck, W.J. (1988) Establishment of a test guideline for the evaluation of fish tainting. *Chemosphere* **17**, 751–765.

Rapport, D.J., Regier, H.A., and Hutchinson, T.C. (1985) Ecosystem behaviour under stress. *Am. Naturalist* **125**, 617–640.

Ramsey, D.J. (1990) Experimental studies of mercury dynamics in the Churchill river diversion, Manitoba. In: Delisle, C.E., and Bouchard, M.A. (Eds.) *Managing the Effects of Hydroelectric Development*, pp. 147–173. Canadian Society for Environmental Biology, Montreal.

Rasmussen, J.B., Rowan, D.J., Lean, D.R.S., and Carey, J.H. (1990) Food chain structure in Ontario lakes determines PCB levels in lake trout (*Salvelinus namaycush*) and other pelagic fish. *Can. J. Fish. Aquat. Sci.* **47**, 2030–2038.

Ray, S., Dunn, B.P., Payne, J.F., Fancey, L., Helbig, R., and Béland, P. (1991) Aromatic DNA-carcinogen adducts in beluga whales from the Canadian Arctic and Gulf of St. Lawrence. *Mar. Pollut. Bull.* **22**, 392–396.

Reijnders, P.J.H. (1986) Reproductive failure in common seals feeding on fish from polluted coastal waters. *Nature* **324**, 456–457.

Rice, S.D., Moles, A., Taylor, T.L., and Karinen, J.F. (1979) Sensitivity of 39 Alaskan marine species to Cook Inlet crude oil and No. 2 fuel oil. In: *Proceedings of the 1979 Oil Spill Conference*, pp. 549–554. American Petroleum Institute, Washington, D.C.

Robbins, J.A. (1986) A model for particle-selective transport of tracers in sediments with conveyor belt deposit feeders. *J. Geophys. Res.* **91**(C7), 8542–8558.

Rogers, I.H., Levings, C.D., Lockhart, W.L., and Norstrom, R.J. (1989) Observations on overwintering juvenile chinook salmon (*Oncorhynchus tshawytscha*) exposed to bleached kraft mill effluent in the upper Fraser River, British Columbia. *Chemosphere* **19**, 1853–1868.

Ronis, M.J., and Walker, C.H. (1989) The monooxygenases of birds. *Rev. Biochem. Toxicol.* 301–384.

Rosenberg, D.M., and Wiens, A.P. (1976) Community and species responses Chironomidae (Diptera) to contamination of fresh waters by crude oil and petroleum products, with special reference to the Trail River, Northwest Territories. *J. Fish. Res. Board Can.* **33**, 1955–1963.

Safe, S. (1990) Polychlorinated biphenyls (PCBs), dibenzo-*p*-dioxins (PCDDs), dibenzofurans (PCDFs), and related compounds: environmental and mechanistic considerations which support the development of toxic equivalency factors (TEFs). *Crit. Rev. Toxicol.* **21**, 51–88.

Savinova, T.N. (1991) *Chemical Pollution of the Northern Seas*. Canadian translation of *Fisheries and Aquatic Sciences*, Report No. 5536, USSR Academy of Sciences, Apatity, USSR. Canada Institute for Scientific and Technical Information, Ottawa, Ontario. 174 pp.

Savinova, T.N. (1992) Soderzhanie zagryaznyayushchich veshchestv v morskich plicach Barentseva Morya: rezultaty i perspektivy issledovanij (The content of pollutants in seabirds from the Barents Sea: results and investigation's perspectives). In: *Teoretycheskie podhody k izucheniyu ekosystem morey Arktyki i Subarktiki.* (*Theoretical Approaches to the Study of Ecosystems at Arctic and Subarctic Seas*), pp. 113–116. Kola Scientific Centre Publishing House, Apatity, USSR.

Scherer, E. (1992) Behavioral responses as indicators of environmental alterations: approaches, results, developments. *J. Appl. Ichthyol.* **8**, 122–131.

Schindler, D.W. (1988) Experimental studies of chemical stressors on whole lake ecosystems. *Verh. Internat. Verein. Limnol.* **23**, 11–41.

Sergy, G.A., and Blackall, P.J. (1987) Design and conclusions of the Baffin Island Oil Spill Project. *Arctic* **40** (Suppl. 1), 1–9.

Shindler, D.B., Scott, B.F., and Carlisle, D.B. (1975) Effect of crude oil on populations of bacteria and algae in artificial ponds subject to winter weather conditions and ice formation. *Verh. Internat. Verein. Limnol.* **19**, 2138–2144.

Slemr, F., and Langer, E. (1992) Increase in global atmospheric concentrations of mercury inferred from measurements over the Atlantic Ocean. *Nature* **355**, 434–437.

Smith, J.N., and Levy, E.M. (1990) Geochronology for polycyclic aromatic hydrocarbon contamination in sediments of the Saguenay Fjord. *Environ. Sci. Technol.* **24**, 874–879.

Smith, T.G., and Armstrong, F.A.J. (1975) Mercury in seals, terrestrial carnivores, and principal food items of the Inuit from Holman, N.W.T. 1975. *J. Fish. Res. Board Can.* **32**, 795–801.

Snow, N.B., and Scott, B.F. (1975) The effect and fate of crude oil spilt on two arctic lakes. In: *Proceedings of the 1975 Conference on Prevention and Control of Oil Pollution*, pp. 527–534. American Petroleum Institute, Washington, D.C.

Sonstegard, R., and Leatherland, J.F. (1976) The epizootiology and pathogenesis of thyroid hyperplasia in coho salmon *Oncorhynchus kisutch* in Lake Ontario. *Cancer Res.* **36**, 4467–4475.

Sprague, J.B., Elson, P.F., and Saunders, R.L. (1964) Sublethal copper-zinc pollution in a salmon river—a field and laboratory study. In: Jaag, O. (Ed.) *Advances in Water Pollution Research*, pp. 61–82. Pergamon Press, Oxford, England.

Stegeman, J.J., and Kloepper-Sams, P.J. (1987) Cytochrome P_{450} isozymes and monooxogenase activity in aquatic animals. *Environ. Health Perspect.* **71**, 87–95.

Stegeman, J.J., Kloepper-Sams, P.J., and Farrington, J.W. (1986) Monooxygenase induction and chlorobiphenyls in the deep-sea fish *Coryphaenoides armatus*. *Science* **231**, 1287-1289.

Stickel, W.H., Stickel, L.F., Dyrland, R.A., and Hughes, D.L. (1984) Aroclor 1254 residues in birds: lethal levels and loss rates. *Arch. Environ. Contam. Toxicol.* **13**, 7–13.

Subramanian, A., Tanabe, S., Tatsukawa, R., Saito, S., and Miyazaki, N. (1987) Reduction in the testosterone levels by PCBs and DDE in Dall's porpoises of northwestern north Pacific. *Mar. Pollut. Bull.* **18**, 643-646.

Swanson, S.M. (1982) *Levels and Effects of Radionuclides in Aquatic Fauna of the Beaverlodge Lake Area (Saskatchewan).* SRC Publication C-806-5-E-82. Saskatchewan Research Council, Saskatoon.

Tay, K.L., Doe, K.G., Wade, S.J., Vaughan, D.A., Berrigan, R.E., and Moore, M.J. (1992) Sediment bioassessment in Halifax harbour. *Environ. Sci. Technol.* **11**, 1567–1581.

Thomann, R.V., Connolly, J.P., and Parkerton, T.F. (1992) An equilibrium model of organic chemical accumulation in aquatic food webs with sediment interaction. *Environ. Toxicol. Chem.* **11**, 615–629.

Truscott, B., Idler, D.R., and Fletcher, G.L. (1992) Alteration of reproductive steroids of male winter flounder (*Pleuronectes americanus*) chronically exposed to low levels of crude oil in sediments. *Can. J. Fish. Aquat. Sci.* **49**, 2190–2195.

Uthe, J.F., Freeman, H.C., Mounib, S., and Lockhart, W.L. (1980) Selection of biochemical techniques for detection of environmentally induced sublethal effects in organisms. *Rapp. P.-v. Reun. Cons. Int. Explor. Mer.* **179**, 39–47.

van der Weiden, M.E.J., van der Kolk, J., Penninks, A.H., Seinen, W., and van den Berg, M. (1990) A dose/response study with 2,3,7,8-TCDD in the rainbow trout *Onchorhynchus mykiss*. *Chemosphere* **20**, 1053–1058.

Vermeer, K., and Peakall, D.B. (1977) Toxic chemicals in Canadian fish-eating birds. *Mar. Pollut. Bull.* **8**, 205–210.

Vermeer, K., and Reynolds, L.M. (1970) Organochlorine residues in aquatic birds in the Canadian provinces. *Can. Field Nat.* **84**, 117–130.

Wagemann, R. (1989) Comparison of heavy metals in two groups of ringed seals (*Phoca hispida*) from the Canadian Arctic. *Can. J. Fish. Aquat. Sci.* **46**, 1558–1563.

Wagemann, R., Snow, N.B., Lutz, A., and Scott, D.P. (1983) Heavy metals in tissues ond organs of the narwhal (*Monodon monoceros*). *Can. J. Fish. Aquat. Sci.* **40** (Suppl.2), 206–216.

Wagemann, R., Stewart, R.E.A., Béland, P., and Desjardins, C. (1990) Heavy metals and selenium in tissues of beluga whales, *Delphinapterus leucas*, from the Canadian Arctic and the St. Lawrence estuary. In: Smith, D., St. Aubin, J., and Geraci, J.R. (Eds.) *Advances in Research on the Beluga Whale, Delphinapterus leucas*. *Can. Bull. Fish. Aquat. Sci.* **224**, 191–206.

Waite, D.T., Joshi, R., and Sommerstad, H. (1988) The effect of uranium mill tailings on radionuclide concentrations in Langley Bay, Saskatchewan. *Can. Arch. Environ. Contam. Toxicol.* **17**, 373–380.

Waite, D.T., Joshi, S.R., and Sommerstad, H. (1989) Movement of dissolved radionuclides from submerged uranium mine tailings into the surface water of Langley Bay, Saskatchewan, Canada. *Arch. Environ. Contam. Toxicol.* **18**, 881–887.

Walker, C.H. (1990) Persistent pollutants in fish-eating sea birds—bioaccumulation, metabolism, and effects. *Aquat. Toxicol.* **17**, 293–324.

Warwick, W.F. (1985) Morphological abnormalities in Chironomidae (Diptera) larvae as measures of toxic stress in freshwater ecosystems: indexing antennal deformities in *Chironomus* Meigen. *Can. J. Fish. Aquat. Sci.* **42**, 1881–1914.

Webb, M. (1979) The metallothioneins. In: Webb, M. (Ed.) *The Chemistry, Biochemistry and Biology of Cadmium*, pp. 195–283. Elsevier/North Holland Biomedical Press, New York.

Whittle, D.M., and Flood, K.W. (1977) *Nipigon Bay: An Assessment of the Acute Toxicity, Growth Impairment and Flesh Tainting Potential of a Bleached Kraft Mill Effluent (BKME) on Rainbow Trout (Salmo gairdneri)*. Ontario Ministry of the Environment, Water Resources Branch, 36 pp.

Woodward, D.F., Mehrle, P.M., Jr, and Mauk, W.L. (1981) Accumulation and sublethal effects of a Wyoming crude oil in cutthroat trout. *Trans. Am. Fish. Soc.* **110**, 437–445.

Woodward, D.F., Riley, R.G., and Smith, C.E. (1983) Accumulation, sublethal effects, and safe concentration of a refined oil as evaluated with cutthroat trout. *Arch. Environ. Contam. Toxicol.* **12**, 455–464.

Yamamoto, T., Kelly, R.K., and Nielsen, O. (1985) Morphological differentiation of virus-associated skin tumors of walleye *Stizostedion vitreum vitreum*. *Fish Pathol.* **20**, 361–372.

York, R.K., and Sereda, L.M. (in press) Seafoods—Chemistry, processing technology and quality. Sensory assessment of quality in fish and seafoods. *Proc. Int. U. Food Sci. Technol.*, Toronto.

Zakharov, V.M., and Yablokov, A.V. (1990) Skull asymmetry in the Baltic grey seal: effects of environmental pollution. *Ambio* **19**, 266–269.

13 Methods to Assess the Effects of Chemicals on Aquatic and Terrestrial Wildlife, Particularly Birds and Mammals

Joanna Burger
Rutgers University, USA

David Peakall
Monitoring and Assessment Research Centre, United Kingdom

13.1 INTRODUCTION

This chapter is devoted to birds and mammals, since many species in these classes are capable of moving long distances and thus moving in and out of study areas. Although some species are sedentary, many bird and some mammal species in temperate latitudes are migratory, often moving for hundreds or even thousands of miles from one ecosystem to another. In the tropics, movements may be triggered by dry and wet season cycles rather than by temperature and day length. In some parts of the world, animals are nomadic and their movements seem unpredictable, making an analysis of the problems they face difficult. In all cases, having a detailed knowledge of the movements of the population under study is necessary. The requisite knowledge can vary even within a species: for example, the arctic population of the peregrine falcon (*Falco peregrinus*) migrates to South America, whereas the race in eastern North America moves only short distances. For studies of pollution, mobility has the advantage that examination of a few specimens gives some information on a wide area; conversely, it provides no detailed information on point sources of pollution. Overall the mobility of birds and mammals makes it challenging to develop methods to examine the effects of chemicals on their populations.

Nonetheless, birds and mammals are important as bioindicators of the effects of chemicals, because they are often at the top of the food web and because of wide interest in, and knowledge of, these higher vertebrates. Chemicals that enter the food chain at low concentrations can become amplified as they move up the food chain. By carefully selecting the indicator, bird or mammal, they can be used to

Methods to Assess the Effects of Chemicals on Ecosystems
Edited by R. A. Linthurst, P. Bourdeau, and R. G. Tardiff

detect traces of chemicals in the environment long before they are apparent in most other organisms, and to detect potential deleterious effects before they become widespread.

In this paper, a framework for evaluating the effects of chemicals on wildlife is presented, the general methods to evaluate the effects of chemicals are reviewed, the advantages and disadvantages of using birds and mammals to evaluate the effects of chemicals are examined, and several case studies illustrate the methods of assessment. Although the methods to evaluate effects on individuals will be mentioned, the focus is on levels of organizations above the individual, at the population, community, and ecosystem levels. The case studies were chosen to illustrate the methods that have been used with some success to evaluate the effects of chemicals. Most of the case studies are with birds, because these animals are diurnal, large, and obvious members of ecosystems. Thus, any changes in their health or population characteristics are obvious immediately. People notice when birds start dying, are hatched without feathers or with other deformities, or entire colonies or populations may fail to hatch eggs because they all crack when the parents incubate them.

13.2 FRAMEWORK TO EVALUATE EFFECTS OF CHEMICALS

Methods to evaluate the effects of chemicals range from the molecular to the ecosystem level. Evaluating effects becomes more difficult as one moves toward the more complex organizational level (Figure 13.1). Detailed descriptions of the methods used at the level of the individual and lower can be found in Burgess *et al.* (in press) and at higher organization levels in Burger (in press).

One can start at either end of Figure 13.1. If one starts at the left-hand end, one has to extrapolate laboratory or limited field trials to ecosystems. Starting at the right-hand end, effects at the individual or physiological level are surmised by examining contaminated ecosystems. Ideally, both approaches to should be used to determine how chemicals are affecting wildlife and ecosystems at all levels.

Each approach, either from the molecular level upward or the ecosystem level downward, has advantages and disadvantages. The first approach has the advantage of laboratory testing where a clear causal relationship can be established. However, the first approach suffers from the increasing difficulty of extrapolating from lower to higher organizational levels, whereas the second requires a well-defined, contaminated ecosystem to be studied.

Ecosystems are not simply the sum of the component species, but encompass several ecosystem functions that are products of the interactions of the species. For example, predators and competitors affect the species around them. Particular predators, known as keystone predators, can maintain a high species diversity in a system by selectively preying on some species, which, if not controlled, would eliminate other species from the system by competition or predation (Paine, 1966).

Measuring ecosystem changes directly has the advantage of allowing us to determine the effects of chemicals on these systems; extrapolation is not needed. However, the costs in time and money might be prohibitive to measure the effects of even one chemical on all aspects of the structure and function of a single ecosystem. Thus, even for ecosystems, indicators of structures and functions must be selected, and that selection can prove problematic.

"Nothing is more dangerous than to leap a chasm in two jumps"
– Lloyd-George

Binding of pollutant to receptor	Bio-chemical response	Physiol-ogical alterations	Whole organism	Population and community
		Time scale		
Seconds to minutes	Minutes to days	Hours to weeks	Days to months	Months to years

Least >>>**Difficulty in relating observed effects to a specific chemical>>>**→ Greatest

Least >>>**Importance>>>** Greatest

Note: On the far right of the diagram, changes in structure and function of ecosystems occur, and the chasm that separates this impact from the stages on the left is too great to demonstrate graphically.

Figure 13.1. Levels of organization to evaluate the effects of chemicals

The effects of chemicals on ecosystem functions can only be determined by studying these functions, and cannot be directly measured or estimated by merely examining individual species. As with measuring effects on species, measuring ecosystem effects requires indicators for endpoints. These indicators should be sufficiently sensitive to provide early warnings, distributed over a broad geographical area, be capable of providing assessment over a wide range of stresses, independent of sample size, and be cost-effective (Sheehan, 1984a; Noss 1990). Ecosystem indicators must be ecologically significant phenomena (Sheehan, 1984b). Indicators for ecosystems might include indices of species diversity, relative species abundances, indices of species richness, or landscape parameters.

In an ideal world, laboratory studies should be predictive of ecosystem effects. Such extrapolation should be possible as one learns how to use microcosm and macrocosm studies to duplicate the structure and function of ecosystems. However, at least for higher vertebrates, this goal is as yet unattainable. In practice, ecologists evaluate the potential effects of chemicals on individual species, or species groups (such as songbirds, seals, ducks, whales). Even attempts to evaluate all resources are usually compilations of species effects (i.e., Bolze and Lee, 1989) rather than an evaluation of community and ecosystem effects.

13.3 ADVANTAGES AND DISADVANTAGES OF USING BIRDS AND MAMMALS TO EVALUATE THE EFFECTS OF CHEMICALS

Birds are ideal targets to evaluate the effects of chemicals, because they represent several trophic levels, they are visible and conspicuous, their populations and behaviour are observable, reproductive success can be measured relatively inexpensively, comparative data for many species are available, and the young feed on local resources. The main disadvantage of using birds (that they migrate from place to place) can be eliminated by examining the young that have received all of their food from local sources. Population and community characteristics of birds such as population and colony size, reproductive success, behavioural abnormalities, physiological and morphological abnormalities, and their effects on other species in the ecosystem can be monitored. Since birds are diurnal, they are easy to observe; many are large and conspicuous, and they remain in one place during their breeding season. Additional advantages include their high interest to the public, their ability to integrate exposure over time and space, and their ability to serve as early warning sentinels before other components of the ecosystem have been affected. The disadvantages of using birds include: (1) they are mobile, (2) their populations may be endangered, making it difficult to obtain specimens for toxic analysis, and (3) they usually feed over a large area making point-source determinations difficult.

Mammals are useful indicators of the effects of chemicals on ecosystems, because, being mammals, they share some physiological characteristics with humans. Additionally, some mammals, such as rodents, are relatively sedentary, and thus reflect local conditions over a number of years. Since rodents and other small mammals are relatively common and inconspicuous, they can be used to monitor without undue public outcry. This approach will continue to be an important consideration as the public awareness of ecosystem protection continues to increase. Mammals share many of the advantages of birds, including their role at the top of the food web, and their role as important competitors and predators on other organisms (influencing community and ecosystem functions such as species abundance and diversity, and productivity). Furthermore, they are excellent laboratory models, with an extensive literature of experiments on many aspects of their behaviour and physiology.

The disadvantages of using mammals are that they are often nocturnal, their population sizes sometimes vary widely, making interpretations difficult from year to year, and many are difficult to trap. Moreover, reproductive success is more difficult to measure than it is for birds; and few studies of reproductive success have been available over several years. Such baseline data are essential to evaluate the effects of chemicals.

13.4 CASE STUDIES

Four case studies have been selected to illustrate the methods available to evaluate the effects of chemicals on birds and mammals. These cases have been chosen because they evaluated effects at several different levels within ecosystems. These case studies are:

1. population declines of raptors caused by DDE-induced egg-shell thinning;
2. mortality of songbirds caused by forest spray programs;
3. effects of pollution in the North America Great Lakes on fish-eating birds and mammals; and
4. effects of oil on marine organisms and ecosystems

13.4.1 CASE 1: POPULATION DECLINES OF RAPTORS CAUSED BY DDE-INDUCED EGGSHELL THINNING

This particular investigation started at the population level when declines of populations of the peregrine falcon and several other raptorial birds were noted in the 1960s. Decreases in population were reported again throughout the Holarctic; and indeed in eastern North America the species had disappeared completely (Hickey, 1969). No studies appear to have been made on the effect of the decline or, in some areas, the disappearance of this species on community structure.

Even current preregistration tests are unlikely to have detected eggshell thinning that led to reproductive failure. The reason is the wide difference in the sensitivity of different species; several of the commonest avian test species—quail, pheasant, chicken—are almost completely insensitive, and others—the duck—only moderately so (Peakall, 1975). However, many species of fish-eating birds and raptors are extremely sensitive.

Only if experimental work had been carried out on a sensitive species such as the American kestrel, *Falco sparverius*, would reproductive effects have been seen (Lincer, 1975). Then egg breakage would have pointed to eggshell thinning and to the relationship between the degree of eggshell thinning and reproductive failure. Thus, the dose and residue levels of DDE could have been established.

The mechanism is considered to be via the inhibition of Ca-ATPase in the oviduct, reducing the transport of calcium to the site of eggshell formation (Peakall *et al.*, 1973), but details of the pharmacodynamics involved in inter-species variation still remain unclear. Again this effect would not likely have been detected at an early stage. Thus, even today, preregistration studies are unlikely to have found this particular adverse effect of DDT.

13.4.2 CASE 2: MORTALITY OF SONGBIRDS CAUSED BY FOREST SPRAY PROGRAMMES

Attempts to control spruce budworm (*Choristoneura fumiferana*) in the forests of eastern Canada by the use of pesticides is the longest and largest spraying programme undertaken. A detailed review of the amount of pesticide used and the area sprayed from the start of the program in 1952 to 1981 was made by Peakall and Bart (1983). During that time, 18 million kilograms of pesticide were used on some 55 million hectares. The advantage of the programme from the authors' viewpoint is that the areas were so large that problems of immigration of birds occurred after mortality vanished and pesticides were the only chemical added to the system.

Initially DDT was the pesticide used. The most obvious effect of DDT on non-target organisms was the mortality of fish and the aquatic invertebrates on which they depended. The loss of whole year classes of Atlantic salmon, *Salmo salar*, were reported from rivers in New Brunswick in the mid-1950s (Logie, 1975). Although fish are outside the scope of this paper, laboratory studies that showed that fish died at dilutions of 1:10000000 were published by Ginsburg (1945); and detailed field studies made by the Ontario Department of Lands and Forests (Langford, 1949), showing just the type of effect later found as a side effect of the forest spray programs in eastern Canada, were available well before the operational use of DDT. The second, much more subtle effect on raptorial birds was discussed in the previous section.

Table 13.1. Calculated and observed effects on songbirds of insecticides used in forest spray operations

Pesticide	Calculated area of spray needed to kill kinglet (cm^2)	Minimum dose causing mortality (gm/ha)	Usual dose of pesticide (gm/ha)	LD_{50} (mg/kg)
Phosphamidon	14	70	140	2
Fenitrothion	60	280	168	10
Carbaryl	50	>1120	1120	56
Malathion	3300	>180	90	<300
DDT	1500	2100	540	840

Modified from Peakall and Bart (1983).

A similar, although much smaller, programme, was carried on in the United States against gypsy moth, *Porthetria dispar*. DDT was last used on an operational scale for forest spraying in the US in 1967 and in Canada in 1968. The choice of pesticides in the two countries in the post-DDT era was quite different. In temporal terms it was carbaryl (1959) and malathion (1964) in the US and phosphamidon (1963) and fenitrothion (1967) in Canada. The reasons for the difference in the pesticide usage pattern between the two countries is unknown, but the choices can be examined in toxicological terms. The mechanism of action of each of these pesticides is the inhibition of the enzyme acetylcholine esterase which disrupts nerve function. The main concern is acute mortality, and this is used as the criterion to rank the pesticides. The risk is dependent on the actual toxicity (the LD_{50}) and the dosage used. The area of spray necessary to kill a small (10 g) songbird such as a kinglet, *Regulus* spp. can be calculated (as noted in Table 13.1). On this basis, phosphamidon is clearly the most hazardous material, as borne out by several events.

Two methods, one intensive and the other more extensive, were used to assess the impact of pesticides on canopy songbirds in New Brunswick in 1975. The intensive studies were based on counts of singing males along walked transects. The extensive studies were based on motored transects. This technique has been widely used to assess bird populations in North America (Robbins and Van Velzen, 1967). Even extensive techniques only cover a small proportion of the area, as can be appreciated, since the total area sprayed that year in New Brunswick was 2740000 ha. The total avian casualties were estimated at 2 million canopy songbirds and another 0.9 million of wide-ranging species (Pearce *et al.*, 1976). The initial reduction of population in some areas was 80 percent. No long-term effects on population or community structure could be demonstrated. The relationship of field studies to calculate for the forest spraying programme is shown in the last column in Table 13.1. Although the data are incomplete, the general sequence follows the calculated effect.

13.4.3 CASE 3: EFFECTS OF POLLUTANTS IN THE NORTH AMERICAN GREAT LAKES ON FISH-EATING BIRDS AND MAMMALS

The third case study is the reproductive failure of fish-eating birds and mammals in the North American Great Lakes, as an example of two entirely different lines of research—molecular biology and field investigations—that eventually blended to provide a comprehensive answer to the problem.

The initial observations in the early 1970s showed severe reproductive impairment of the herring gull, *Larus argentatus*, and the disappearance of the double-crested cormorant, *Phalacrocorax auritus*, in the lower Great Lakes (Gilman *et al.*, 1977). Trapping records indicate that population declines of two fish-eating mammals, the mink *Mustela vison* and the otter *Lutra canadensis*, also occurred (Environment Canada, 1991). The population increase of the cormorant, from ten to twenty times the pre-pollutant level, after restriction on pollutants had been put in place indicates that effects were exerted on the community structure (Price and Weseloh, 1986).

A correlation was found between polyhalogenated aromatic hydrocarbon (PHAH) levels and reproductive effects. Relating these effects to a specific chemical(s) was much more difficult. Apart from the difficulties inherent in tackling the effects of complex mixtures, two additional problems arose. First, the levels of most PHAHs cross-correlated strongly to each other, and second the effects seen—embryotoxicity, edema, structural abnormalities, behavioural changes—were known from laboratory studies to be caused by a wide range of PHAHs. Only in the case of eggshell thinning in cormorants was it possible to assign a specific cause with a high degree of certainty. Detailed field studies, which included egg injection, egg exchange experiments, and nest attentiveness studies, were undertaken (Mineau *et al.*, 1984). These investigations were all the more difficult to conduct because the levels of PHAHs rapidly decreased following bans and restrictions on their usage in the late 1970s.

The identification of the Ah receptor (Poland *et al.*, 1976) by highly specific binding to 2,3,7,8-tetrachlorodibenzo-*p*-dioxin (2,3,7,8-TCDD) was a key finding in bringing molecular biology into the realm of toxicology. This receptor controls the induction of the mixed function oxidase (MFO) systems that are of considerable importance in detoxifying a wide variety of pollutants. Studies were broadened to cover several polychlorinated biphenyls (PCBs), polychlorinated dibenzofurans (PCDFs), and polychlorinated dibenzodioxins (PCDDs). A strong correlation between potency to induce MFO activity and their toxicity was found.

The application of this complex biochemistry into field investigations by means of expressing the complex mixtures of PHAHs as "dioxin equivalents" is based on this correlation. The most toxic compound and most potent inducer is 2,3,7,8-TCDD, whose assigned value is 1. All other compounds have lower values; however, when they are multiplied by their concentration, they can give larger values than that obtained for 2,3,7,8-TCDD. This value, potency times concentration, is the "dioxin equivalent." Since the potency is correlated with the ability to induce mixed function oxidases enzymes, the induction of these enzymes can be used to calculate dioxin equivalents (TCDD-EQs).

Significant differences in the ability of egg samples from fish-eating birds from various regions around the Great Lakes to cause induction was found by Tillitt *et al.* (1991). The relative ranking of colonies correlates well with known areas of contamination. When the overall reproductive success of double-crested cormorant and Caspian tern (*Hydroprogne caspia*) colonies was plotted against TCDD-EQs of eggs from each colony, a high degree of correlation was found, suggesting that these pollutants, expressed as dioxin equivalents, are the cause of reproductive failure.

The data on mammals are not as clear. Trapping data, which have severe limitations, indicate decreases of both species, especially along the lake shore where contamination would be expected to be highest (Environment Canada, 1991). Feeding Great Lakes fish to ranch mink can cause reproductive failure, as was demonstrated many years ago. Detailed studies have shown that PCBs caused adverse reproductive effects that are at the environmental levels (Aulerich and

Ringer, 1977); however, no such examination has been undertaken in terms of dioxin equivalents.

13.4.4 CASE 4: EFFECTS OF OIL ON AQUATIC AND ADJACENT TERRESTRIAL ECOSYSTEMS

The fourth case study deals with the effects of oil on aquatic ecosystems, including birds and mammals within these systems. This is an example of three levels of research, (1) field observations, (2) field experimentation, and (3) laboratory investigations; these blend to provide a comprehensive understanding of the effects of chemicals. It is also an example where ecosystem functions have been examined extensively both in the field and in the laboratory. In this case, the presence of the oil has stimulated research on lethal and sublethal effects and on immediate and long-term effects. Oil and its breakdown petroleum hydrocarbons are of global concern, because the potential for exposure is cosmopolitan.

Another aspect of oil contamination relevant to methods to assess the effects of chemicals is the potential for acute and chronic exposure. Media and scientific attention is often directed at the massive pollution events that provide acute exposures; yet on a global scale, chronic, low-level exposure may prove to be more challenging to evaluate and regulate the effects of chemicals.

As early as the 1920s the effects of oil were noted on birds. The most obvious effect was mortality due to severe oiling. The increase in oil transport following the Korean War and the use of larger oil tankers led to major oil spills, and these spills still continue today. Although the image of oiled fish, birds, and mammals was very potent, other delayed and sublethal responses were soon suspected. Oil affects all components of the ecosystem from beach grass and Spartina to marine mammals, birds, and terrestrial animals that come to these ocean, estuary, or river waters to feed or drink.

In 1967, the *Torrey Canyon* went aground in the United Kingdom, and spilled over 80,000 tons of oil into the sea. Since then, over twenty accidents of this magnitude have occurred, including the *Amoco Cadiz*, the *Ixtoc*, and finally the *Exxon Valdez*. Although many oil spills have occurred, the variation in environmental conditions (geography, latitude, time of year) make comparisons challenging. The large number of spills facilitates hypothesis formulation to not only assess adverse effects in the field but also test in the laboratory (NRC, 1981).

Physical characteristics of the oil that determine the effects on marine resources include the volatile fraction, saturated hydrocarbon content, specific gravity, and viscosity. Initially, laboratory studies were designed to examine how variations in these characteristics affect particular organisms. Now, however, laboratory and field experiments and observations are used to understand and evaluate the effect of oil and its constituents. Species and ecosystem characteristics also determine the effects and methods of assessment including: life stage (eggs, young, adult), differential vulnerability by season, mobility, natal habitat (i.e., bottom-dwelling, shoreline), and the presence of refugia (Cairns and Elliot, 1987). The methods to evaluate the effects of oil, from the molecular to ecosystem levels, are numerous and varied (Tables 13.2, 13.3, and 13.3). These tables provide only a partial list of methods at each level of organization, to provide a sense of the type of assessments that have been conducted. Birds and mammals are the focus herein; other examples are only when they indicate the potential of the methodology. For ecosystem effects, examination of both structure and function is essential.

The methods used to assess the effects of oil on ecosystems are further developed than for other contaminants because: (1) many large oil spills have occurred in the past 20 years; (2) many places have produced chronic exposure; (3) the area of exposure is often a well-defined ecosystem such as a river, estuary, or bay; (4) the area of exposure is easy to define visually and with sophisticated oil fingerprinting techniques; and (5) since the cause of the exposure is readily apparent, blame can be established, and funds levied for scientific study, bioremediation, and rehabilitation.

Table 13.2. Illustrations of methods to assess the toxicity of oil on the molecular and physiological levels of organization for wildlife

Molecular and physiological effects	Assessment method	Example (reference)
Disruption of thermoregulation	Body temperature; microscopic exam	Sea otters (Costa and Kooyman, 1981)
Altered cell membrane structure and function	Transport across membranes	Fish (Englehardt *et al.*, 1981)
Changes in liver enzymes	Enzyme activity	Seals (Addison *et al.*, 1986); fish (Burns, 1986)
Changes in hormones	Hormone levels	(Englehardt, 1982)
Feather structure	Microscopic exam	Birds (Ridjke, 1970)
Kidney and liver damage	Microscopic exam	Birds (Leighton, 1991)

13.5 PROACTIVE VERSUS REACTIVE APPROACHES

Obviously a need exists to be proactive rather than reactive. The first case study (DDE-induced eggshell thinning) is completely reactive, and is unlikely to have been detected proactively, even with today's methods. The interspecies variation means that, in this case, experimental studies would not have detected eggshell thinning. Historically, although studies have been carried out on the interaction of DDT and its metabolites with ATPases, the extension of the findings to avian eggshell thinning is unlikely even if the chain was not blocked by interspecies variation.

Table 13.3. Illustrations of methods to assess the toxicity of oil on the individual and population levels of organization for wildlife

Effect	Assessment method	Example (reference)
Individual:		
Mortality	Census numbers	Birds (Frink, 1993); Mammals (Hutchinson and Simmons, 1991)
Changes in reproduction rates	Reproductive rates	Coral (Loya, 1975); birds (Leighton, 1991)
Changes in growth rates	Growth rates	Fish (Schwartz, 1985); birds (Eastin and Hoffman, 1979)
Changes in hatchability	Hatchability tests	Fish (Teal and Howarth, 1984); sea turtles (Fritts and McGehee, 1981); birds (Eastin and Hoffman, 1979)
Population:		
Changes in biomass	Measure biomass	Algae communities (Perez, 1978)
Inefficient feeding	Rates of predation	Fish (Weis and Khan, 1991)

The second case suggests that simple toxicological calculations on chemicals based on toxicity and dose could have prevented serious environmental effects. This case is comparatively straightforward as single chemicals of known mode of action were involved.

The third case study, despite its complexity, gives a hope that eventually a proactive approach based on biochemical mechanisms is possible. In the actual event, an approach combining field observations and molecular biology studies which was successful. However, in theory, the progression from the specific high-affinity binding of TCDD, a known environmental pollutant, to the Ah receptor to toxic effects in the field could have been anticipated. Work on receptors is moving forward rapidly. The receptor approach is being used in the case of the rodenticide flucoumafen. When the hepatic binding sites become saturated the anticoagulant effect becomes lethal (Huckle *et al.*, 1989). This basis is now being used to investigate the possible impact of rodenticides on the barn owl, *Tyto alba*. A fruitful line of investigation might be to examine the extent that known receptors can be blocked by environmentally important compounds and to examine the implications of the positive findings.

Table 13.4. Illustrations of methods to assess the toxicity of oil on the ecosystem level of organization for wildlife

Ecosystem effects	Assessment method	Example (reference)
Changes in species abundance	Comparisons of numbers in populations of several species	Arthur Kill ecosystems (Brzorad and Burger, 1993)
Changes in rates of decomposition	Decomposition rates	Baltic Sea (Lindenof *et al.*, 1979)
Erosion of marshes	Erosion rates	Marsh grass ecosystems (Sheehan, 1984a, 1984b)
Reduced primary productivity	Field measures of productivity	Grass-herb and marsh ecosystems (Kinako, 1981)

The fourth case study, although complex, spans the range from molecular to ecosystem effects, and is both reactive and proactive. Studies continue at all levels, from molecular to ecosystem, to evaluate the effects of oil. Since the potential for oil spills continues, oil represents a class of chemicals whose assessment will continue to pose problems for a long time. In theory, the development of methods of assessment, providing data on effects, will ultimately help develop methods of prevention or clean-up.

Overall, these case studies indicate that birds and mammals are important indicator species to evaluate the effects of chemicals. They have served as early warnings in the past, and are expected to continue to do so.

13.6 CONCLUSIONS

Evaluating the effects of chemicals individually or as mixtures on ecosystems is a difficult task, because of the levels of organization and complexity of ecosystems, and the matrix of interactions between all components of the ecosystem. Measurement of effects is not possible on all components or levels, because of constraints on manpower, time, and money. The task of evaluating chemical effects on ecosystems thus converges on selecting a series of indicators, regardless of whether they are at the molecular, individual, population, or ecosystem level.

Birds and mammals are ideal as indicators for several reasons. They are usually high on the food chain, and thus they can bioaccumulate chemicals, making chemicals easier to detect before they can be detected in other organisms, and making the organisms vulnerable to morphological, behavioural, and physiological effects that can be observed easily. These sublethal effects, as well as direct mortality, can result in catastrophic population declines, that is an immediate and clear sign of intolerable levels of contaminants in the environment. Because birds are so visible to the general public, as well as to conservationists and scientists, any changes in their health or population levels are likely to be noted. The public cares about birds and mammals, not only for themselves, but as indicators of environmental health, and ultimately of their own health. The disadvantages of using birds (e.g., many migrate) or mammals (e.g., some migrate, and some are nocturnal) can be overcome by examining the levels of chemicals and their effects in young, using laboratory experiments to ascertain causation and dose-response relationships, and carefully monitoring reproductive success in wild populations.

Overall, the advantages of using birds and mammals, as indicated by these case studies, indicate that each is an important indicator species to evaluate the effects of chemicals. They have provided early warnings of adverse effects of chemicals (e.g., DDT), even before any other organismic or larger ecosystem effects were noted. After birds indicated a problem, similar or other effects were noted in other organisms. Thus, birds and mammals are likely to continue to serve as early warnings of adverse chemical effects, and they should be incorporated into any suite of indicators used to evaluate the effects of chemicals on ecosystems and their component parts.

13.7 REFERENCES

Addison, R.F., Brodie, P.F., Edwards, A., and Sadler, M.C. (1986) Mixed function oxidase activity in the harbor seal (*Phoca vitulina*). *Comp. Biochem. Physiol.* **85**, 121–124.

Aulerich, R.J., and Ringer, R.K. (1977) Current status of PCB toxicity to mink, and effect on their reproduction. *Arch. Environ. Contam. Toxicol.* **6**, 279–292.

Bolze, D.A., and Lee, M.B. (1989) Offshore oil and gas development. *Marine Policy*, July, 1–6.

Brzorad, J., and Burger, J. (1993) Fish and shrimp populations in the Arthur Kill. In: Burger, J. (Ed.) *The Arthur Kill: Anatomy of an Oil Spill*. Rutgers University Press, New Brunswick, New Jersey.

Burgess P.S., Peakall, D.B., and Landa, V. (in press) Wildlife species as monitors of hazardous waste dumps. In: Moore, J.A., Subramanyam, B.V.R., and Tardiff, R.G. (Eds.) *Methods to Assess the Effects of Chemicals on Hazardous Waste Sites*. John Wiley & Sons, London.

Burger, J. (in press) Ecological effects of exposures to hazardous waste sites. In: Moore, J.A., Subramanyam, B.V.R., and Tardiff, R.G. (Eds.) *Methods to Assess the Effects of Chemicals on Hazardous Waste Sites*. John Wiley & Sons, London.

Burns, K.A. (1976) Microsomal mixed function oxidases in an estuarine fish, *Fundulus heteroclitis*, and their induction as a result of environmental contamination. *Comp. Biochem. Physiol.* **53B**, 443–446.

Cairns, D.K., and Elliot, R.D. (1987) Oil spill impact assessment for seabirds: the role of refuging and growth centres. *Biol. Conserv.* **40**, 1–9.

Costa, D.P., and Kooyman, G.L. (1982) Oxygen consumption, thermoregulation and the effect of fur oiling and washing on the sea otter, *Enhydra lutris*. *Can. J. Zool.* **60**, 2761–2767.

Eastin, W.C., and Hoffman, H.P.J. (1979) Biological effects of petroleum on aquatic birds. In: *Proceedings of the Conference on Assessment of Ecological Effects of Oil Spills* Vol. 8, pp. 561–582.

Engelhardt. F.R. (1982) Hydrocarbon metabolism and cortisol balance in oil exposed ringed seals, *Phoca hispida*. *Comp. Biochem Physiol.* **72C**, 133–136.

Englehardt, F.R., Wong, M.P., and Duiey, M.E. (1981) Hydromineral balance and gill morphology in rainbow trout *Salmo gairdneri*, activated to fresh and sea water as affected by petroleum exposure. *Aquat. Toxicol.* **1**, 175–186.

Environment Canada. (1991) *Toxic Chemicals in Great Lakes and Associated Effects. Synopsis*. Vol. II. *Effects*, pp. 495–755. Environment Canada, Ottawa.

Frink, L. (1993) Rehabilitation of oiled birds. In: Burger, J. (Ed.) *Anatomy of an Oil Spill: The Arthur Kill*. Rutgers University Press, New Brunswick, New Jersey.

Fritts, T.H., and McGehee, M.A. (1981) *Effects of Petroleum on the Development and Survival of Marine Turtle Embryos*. Report to US Department of the Interior, Fish and Wildlife Service, Contract No. 121-16-0, Washington, D.C., 41 pp.

Gilman A., Peakall, D.B., Hallett, D., Fox, G.A., and Norstrom, R.J. (1977) Herring Gulls (*Larus argentatus*) as monitors of contamination in the Great Lakes. In: National Research Council, *Animals as Monitors of Environmental Pollutants*, pp. 280–289. National Academy of Sciences, Washington, D.C.

Ginsburg, J.M. (1945) Toxicity of DDT to fish. *J. Econ. Entomol.* **39**, 274–275.

Hartsough G.R. (1965) Great Lakes fish now suspect as mink food. *Am. Fur Breeder* **38**(10), 25–27.

Hickey, J.J. (1969) *The Peregrine Falcon Populations: Their Biology and Decline.* University of Wisconsin Press, Madison, 596 pp.

Huckle, K.R., Hutson, D.H., Logan, C.J., Morrison, B.J., and Warburton, P.A. (1989) The fate of the rodenticide flocoumafen in the rat: retention and elimination of a single oral dose. *Pest. Sci.* **25**, 297–312.

Hutchinson, J., and Simmonds, M. (1991) *A Review of the Effects of Pollution on Marine Turtles.* Greenpeace International, Washington, D.C., 20 pp.

Kinako, P.D.S. (1981) Short-term effects of oil pollution on species numbers and productivity of a simple terrestrial system. *Environ. Pollut. Ser. A* **26**, 87–91.

Langford R.R. (1949) The effect of DDT on freshwater fish. In: Ontario Department of Lands and Forests. *Forest Spraying and the Effects of DDT*, pp. 19–27. Biological Bulletin 2. Ottawa, Ontario.

Leighton F.A. (1991) The toxicity of petroleum oils to birds: an over-view. In: White, J., and Frink, L. (Eds.) *The Effects of Oil on Wildlife*, pp. 78–94. IWRC, California.

Lincer, J.L. (1975) DDE-induced eggshell thinning in the American kestrel: A comparison of the field situation and laboratory results. *J. Appl. Ecol.* **12**, 781–793.

Lindenof, O., Elmgren, R., and Boehm, P. (1979) The Tsesis oil spill: its impact on the coastal ecosystem of the Baltic Sea. *Ambio* **8**, 214–253.

Logie, R.R. (1975) Effects of aerial spraying of DDT on salmon populations of the Miramichi River. In: Prebble, M.L. (Ed.) *Aerial Control of Forest Insects in Canada*, pp. 293–300. Environment Canada, Ottawa.

Loya, Y. (1975) Possible effects of water pollution on the community structure of Red Sea corals. *Mar. Biol.* **29**, 177–185.

Mineau, P., Fox, G.A., Norstrom, R.J., Weseloh, D.V., Hallett, D.J., and Ellenton, J.A. (1984) Using the herring gull to monitor levels and effects of organochlorine contamination in the Canadian Great Lakes. *Adv. Environ. Sci. Technol.* **14**, 425–452.

Noss, R. (1990) Indicator for monitoring biodiversity: a hierarchial approach. *Con. Biol.* **4**, 355–364.

NRC (National Research Council). (1981) *Testing for Effects of Chemicals on Ecosystems.* National Academy Press, Washington, D.C., 128 pp.

Paine, R.T. (1966) Food web complexity and species diversity. *Am. Natur.* **100**, 65–75.

Peakall D.B. (1975) Physiological effects of chlorinated hydrocarbons on avian species. In: Haque, R., and Freed, V.H. (Eds.) *Environmental Dynamics of Pesticides*, pp. 343–360. Plenum Press, New York.

Peakall, D.B., and Bart, J.R. (1983) Impacts of aerial applications of insecticides on forest birds. *CRC Crit. Rev. Environ. Control.* **13**, 117–165.

Peakall, D.B., Lincer, J.L., Risebrough, R.W., Pritchard, J.B., and Kinter, W.B. (1973) DDE-induced egg-shell thinning: Structural and physiological effects in three species. *Comp. Gen. Pharm.* **4**, 305–313.

Pearce P.A., Peakall, D.B., and Erskine, A.J. (1976) *Impact on Forest Birds of the 1975 Spruce Budworm Spray Operations in New Brunswick.* CWS Progress Notes No. 62, 7 pp.

Perez, P. (1979) State of exploitable algae populations in January 1979. In: *Rapport Inst. Peches Marit. Roneo.*, pp. 1–3.

Poland, A., Glover, E., and Kende, A.S. (1976) Stereospecific, high affinity binding of 2,3,7,8-tetrachlorodibenzo-*p*-dioxin by hepatic cytosol. Evidence that the binding species is receptor for induction of aryl hydroxylase. *J. Biol. Chem.* **251**, 4936–4946.

Price, I.M., and Weseloh, D.V. (1986) Increased numbers and productivity of double-crested cormorants, *Phalacrocorax auritus*, on Lake Ontario. *Can. Field-Natural.* **100**, 474–482.

Ridjke, A.M. (1970) Wettability and phylogenetic development of feather structure in waterbirds. *J. Exp. Biol.* **52**, 469–479.

Robbins, C.S., and Van Velzen, W.T. (1967) *The Breeding Bird Survey, 1966*. Special Sci. Report No. 102. US Department of the Interior, Fish and Wildlife Service, Washington, D.C. 43 pp.

Schwartz, J.P. (1985) Effect of oil contaminated prey on the feeding and growth rate of pink salmon prey (*Oncorhyncus gorbuscha*). In: Vernberg, F.J., Thurberg, F.P., Calabrese, A., and Vernberg, W. (Eds.) *Marine Pollution and Physiology, Recent Advances*, pp. 459–476. South Carolina Press, Columbia.

Sheehan, P.J. (1984a) Functional changes in ecosystems. In: Sheehan, P.J., Miller, D.R., Butler, G.C., and Bourdeau, P. (Eds.) *Effects of Pollutants at the Ecosystem Level*, pp. 101–146. SCOPE 22. John Wiley & Sons, Chichester.

Sheehan, P.J. (1984b) Effects on community and ecosystem structure and dynamics. In: Sheehan, P.J., Miller, D.R., Butler, G.C., and Bourdeau, P. (Eds.) *Effects of Pollutants at the Ecosystem Level*, pp. 51–99. SCOPE 22. John Wiley & Sons, Chichester.

Teal, J.M., and Hawarth, R.W. (1984) Oil spill studies: A review of ecological effects. *Environ. Manag.* **8**, 27–44.

Tillitt, D.E., Ankley, G.T., Verbrugge, D.A., Giesy, J.P., Ludwig, J.P., and Kubiak, T.J. (1991) H4IIE rat hepatoma cell bioassay-derived 2,3,7,8-tetrachlorodibenzo-*p*-dioxin equivalent in colonial fish-eating waterbird eggs from the Great Lakes. *Arch. Environ. Contam. Toxicol.* **21**, 91–101.

Weis, J.S., and Khan, A.A. (1991) Reduction in prey capture ability and condition of mummichogs from a polluted habitat. *Trans. Am. Fish. Soc.* **120**, 127–129.

14 Assessments of Ecological Impacts on a Regional Scale

Patrick Sheehan
McLaren/Hart, USA

14.1 INTRODUCTION

Most assessments of risks of injury to ecological species caused by chemicals are currently focused on small-scale, local problems. Local assessments may evaluate the ecological hazards of chemically contaminated sites, point-source effluent discharges to rivers, lakes, and bays, chemical air emissions from individual industrial facilities, or other chemical exposures generally confined to a spatial scale of a few metres to a few kilometres. Current ecological risk assessment methods have typically been developed largely to assess chemical effects at this spatial scale.

Exposures that result from the widespread aerial release and transport of chemicals occur at a much larger, regional scale (hundreds to thousands of kilometres). Examples include acid deposition, long-range transport and deposition of ozone and other air contaminants, and broad-scale aerial application of pesticides, such as those used for spruce budworm control in the forests of eastern Canada. Not only do chemical exposures occur at various scales, but ecological processes also operate at a variety of scales in space and time. Therefore, a need exists to identify approaches to estimate risks to ecological receptors and methods useful for larger-scale, regional assessments of the effects of broad-scale chemical exposures.

Suter (1993) identified six reasons for ecological risk assessments at the regional level. First, local sources of chemicals or radionuclides may have regional consequences from their release. The Chernobyl reactor accident is an example of this scenario. Second, the combined releases of multiple individual sources within a region, each within tolerable limits, may be unacceptable, because of the combined toxic effects of the mixture at the regional scale. An example is the combined airborne releases of chemicals in urban industrial areas leading to degradation in regional air quality. Third, regional scale processes may affect the transformation and transport of airborne chemicals in ways that are not observed at local scales. An obvious example is the formation of photochemical smog from numerous independent sources of hydrocarbons and nitrogen oxides. Fourth, emissions may have effects at regional scales that do not occur on local scales.

Methods to Assess the Effects of Chemicals on Ecosystems
Edited by R. A. Linthurst, P. Bourdeau, and R. G. Tardiff

The depletion of stratospheric ozone by chlorofluorocarbons provides an example of this phenomena. Fifth, regions possess characteristics that do not occur on local scales, and these characteristics warrant protection. These characteristics are largely associated with patterns in the landscape. Finally, the success of various broad-scale regulatory and resource management programmes can be adequately assessed only on a regional scale.

This chapter provides a rationale for regional assessments, discusses spatial and temporal scale considerations for regional scale risk assessments, provides a framework and methods for regional scale assessments, and describes a regional ecological risk assessment case study, that seeks to estimate risks to a region exposed to high levels of ozone.

The assessment of risks on a regional scale is not fundamentally different from that on a local scale. Both regional and site-specific assessments of risk can be performed within the framework recently provided by the US Environmental Protection Agency (USEPA, 1992). The paradigms of predictive and retrospective risk assessment recently presented by Suter (1993) in *Ecological Risk Assessment* are equally applicable to local and regional assessments. However, in contrast to local assessments, regional scale risk assessments require consideration of a separate set of issues that do not enter site-specific assessments. These include understanding the landscape and the relationship of biota to the landscape, characterizing exposures and effects over large spatial scales and sometimes long temporal scales, identifying endpoints and measurement metrics characteristic of the region, combining data collected at very different scales and extrapolating between scales, integrating the inputs of multiple stressors that operate on large spatial scales, and integrating exposures and effects in various terrestrial and aquatic ecosystems within the region to characterize risk for the region as a whole.

14.2 SCALE ISSUES

Several issues are associated with the spatial and temporal scales of chemical hazards and ecological response to chemical exposures that must be considered when undertaking an ecotoxicological assessment of risk at the regional level. These considerations include:

1. the spatial and temporal scales associated with various levels of biological organization for which effects may be assessed;
2. the relation of landscape scales to the distributions of populations in an area and the way in which they use the available resources; and
3. the spatial and temporal scales of potential chemical exposures.

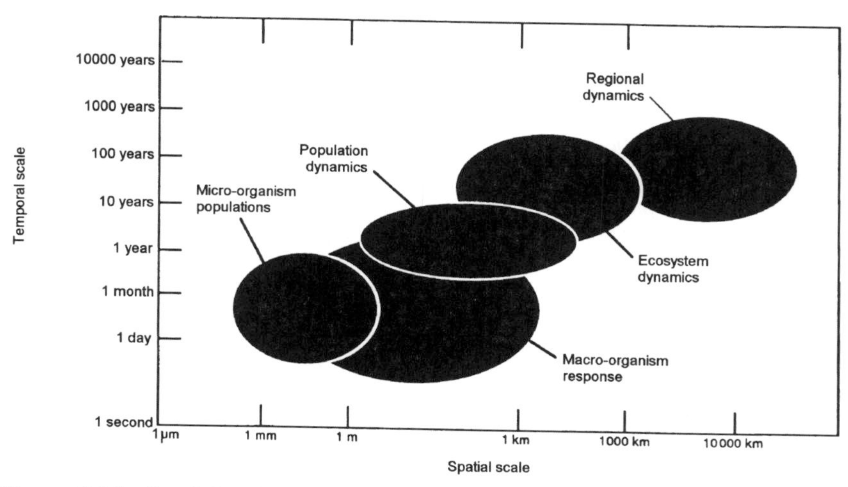

Figure 14.1. Spatial and temporal scales within which individuals, populations, ecosystems, and regions respond to environmental stressors (modified from Suter, 1993)

14.2.1 SPATIAL AND TEMPORAL SCALES RELATED TO LEVELS OF BIOLOGICAL ORGANIZATION

Effects of chemicals may be assessed at the level of the individual, population, community, ecosystem, or region, or at combinations of levels of organization. Populations are composed of individual members of a species in the same area; communities are groups of populations interacting with each other; ecosystems are communities together with their physical and chemical environment; regions, in the context of risk assessment, are spatial groupings of contiguous ecosystems.

Several reasons exist to consider the levels of biological organization and their spatiotemporal scales of operation in planning ecotoxicological risk assessments. First, although both measurement and assessment endpoints can be defined at each level of biological organization, these endpoints are not equally important at different spatial and temporal scales (Suter, 1993). This situation is shown in Figure 14.1. At short temporal scales (days to months) and small spatial scales (micrometre to metres), effects of chemicals can be assessed on micro-organism populations and micro-organism biochemistry and physiology. However, little can be estimated about the ultimate effects of chemical exposure on long-lived organisms or on the dynamics of microorganism populations or ecosystems that operate on larger spatial and longer temporal scales.

On a human time scale, the reproducing population is the smallest persistent ecological unit. The effects of chemicals on populations of organisms can be assessed on a temporal scale of months to years and a spatial scale of metres to kilometres. Populations, however, do not live in a vacuum. Direct chemical effects on one or more populations may, in turn, affect other populations in the exposed community. Such effects may occur indirectly through changes in habitat availability or in predator–prey or competitive relationships or other mechanisms. Therefore, a partial overlap exits in spatial and temporal scales for the assessment of chemical effects on population dynamics and community and ecosystem dynamics. However, because additional secondary effects may occur over longer time scales and may spread to larger areas, the range of spatiotemporal scales to evaluate chemical effects on ecosystems is broader than for the assessment of effects on populations alone (i.e., years to hundreds of years and metres to hundred of kilometres). The spatial scale of regional responses to chemical exposure overlaps with, but extends beyond, that of individual exposed ecosystems (hundreds of kilometres to thousands of kilometres). The temporal scale of regional dynamics also overlaps that of the component ecosystems, but may be longer than that for any one ecosystem in the region (tens of years to hundreds of years). Spatial and temporal scales associated with ecological effects at the various levels of biological organization are further discussed by Sheehan (1984a, 1984b) and Suter (1993).

14.2.2 TOOLS FOR REGIONAL SCALE ASSESSMENTS

A second issue of scale is the practical constraints of assessing ecotoxicological effects at the higher levels of biological organization. Implementation of small scale, short-term population studies is much easier than that for long-term, regional ecotoxicological studies. Field studies to support risk assessment at the ecosystem and regional level may require investigation on a time scale of years and a spatial scale of hundreds of kilometres. Such studies require substantial expenditures of manpower and money, and may not be practical for compliance with regulatory requirements. A recognition of costs of long-term broad-scale field studies has spurred the development of new sampling tools, such as satellite imagery, which are well suited to providing data over extended time periods and large geographical areas, but are not labour intensive. Tucker and Sellers (1986) use satellite remote sensing to assess primary productivity across broad spatial scales. More recently, Simmons *et al.* (1992) and colleagues used satellite imagery in conjunction with small-scale field studies to evaluate the dispersion of plant cover in semi-arid areas of the state of Washington and the utility of satellite imagery as tools in the assessment of landscape response to physical and chemical stressors.

Geographical information systems (GIS) are also being designed to organize and analyze data on the distribution, accumulation, and effects of chemicals on landscapes (Bartell *et al.*, 1992). In addition, the cost of large-scale field investigation has also prompted the development of models to extrapolate from smaller to larger spatial scales and to simulate regional landscape response to stressors. Solomon (1986) proposed a model to assess the response of forests in eastern North America to CO_2-induced climate change. Turner (1987) compared the utility of three different models in predicting landscape changes resulting from physical disturbance events. More recently, Graham and colleagues (1991) used a simulation model to predict the effects of ozone on Adirondack forests in the state of New York. This study is described in greater detail as a regional risk assessment case study below. Clearly, regional scale ecotoxicological risk assessments will require extensive use of these types of tools to be cost-effective.

14.2.3 THE SCALE OF LANDSCAPE STRUCTURE AND ITS INFLUENCES ON ANIMAL POPULATIONS AND RESOURCE USE

A second scale consideration for regional ecotoxicological risk assessments is the relationship of scale of landscape patterns (including vegetative structure) with the distribution of animals and the scale of their use of resources. At a regional scale, landscape responses to chemical stressors are often assessed in terms of changes in vegetation and vegetative structure, and few data are provided on the secondary effects of such changes in vegetation on animal populations. Understanding the relationship of animal population dynamics to the structure of the landscape is essential to predicting changes in animal populations resulting from changes in the vegetative structure of their habitat.

Recent studies have evaluated the proposition that a small set of plant, animal, and abiotic processes structure ecosystems across scales of time and space. Based on his studies, Holling (1992) concluded that terrestrial bird and mammal populations are dispersed according to body size by the discontinuous hierarchial structures and textures of the landscape. He found evidence for eight distinct habitat "quanta" defined by distinct textures at a specific range of scales. These eight quanta together cover tens of centimetres to hundreds of kilometres in space and months to hundred of years in time. All trophic levels of birds and mammals utilize resources in their foraging areas in the same way by measuring the spatial gain of habitat patches defined by their size (i.e., step length or some minimum unit of measurement).

Therefore, large mammals and birds have large home range areas, and these are to some degree limited by the vegetative structure and texture of the landscape. For example, a large wading bird, such as the Great Egret (*Casmerodius albus*) has a short-term foraging area that is small (a few metres), but a habitat area home range during a year that may extend from tens to hundreds of kilometres. Over several years, this area may extend over thousands of kilometres. The spatial and temporal scales within which wading birds operate are shown in Figure 14.2.

Chemically mediated changes in the vegetative structure of the landscape will influence both the distribution and abundance of these species with large home ranges. Consideration of such secondary effects is essential for regional scale risk assessments. The work of Holling (1992), Gass and Montgomerie (1981), and Orians (1980) on body size and the size of home and foraging ranges provides useful data on the mammal and bird species of interest in regional assessments and the spatial scale at which the dynamics of these populations should be assessed.

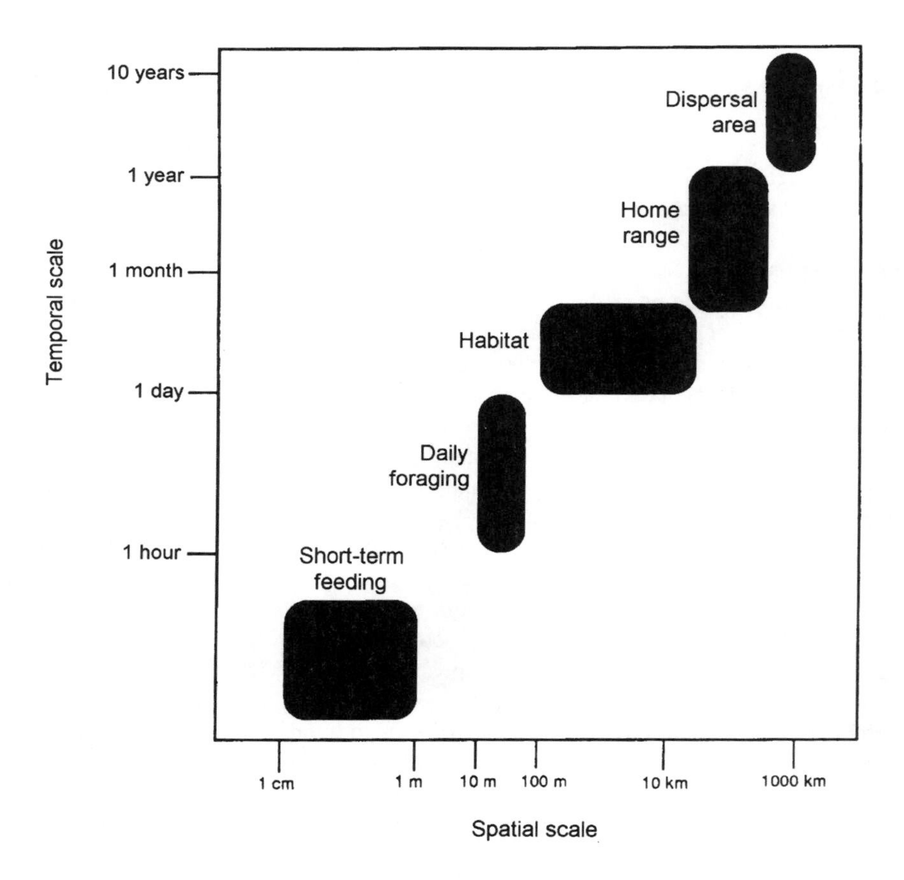

Figure 14.2. Temporal and spatial scales within which large wading birds operate daily and over their lifetimes (modified from Holling, 1992)

14.2.4 SPATIAL AND TEMPORAL SCALES OF CHEMICAL HAZARDS

Chemical exposures occur over a range of spatial and temporal scales. A representation of the spatial and temporal scales of selected chemicals hazards is shown in Figure 14.3. Single local applications of pesticides often occur on a spatial scale of metres in homes and gardens to perhaps a kilometre in agricultural settings over a temporal scale of hours to a few days. Chemical and petrochemical spill events occur on a somewhat larger spatial scale, but the residual contamination from these spills may pose exposure on a temporal scale of days to years. Chemical releases from hazardous waste sites and liquid effluent discharges occur on a spatial scale similar to spill events, but exposures from these sources occur over longer temporal scales of months to tens of years. In contrast to these small-scale chemical releases, broad-scale applications of pesticides, such as aerial spraying to control grasshoppers in the prairie regions of the United States and Canada (Sheehan *et al.*, 1987) and the spruce budworm in the forests of New Brunswick (Mitchell and Roberts, 1984), cover hundreds to thousands of kilometres in space, and may occur intermittently over years to tens of years. At the extreme, widespread aerial transport of photochemical oxidants and acids have resulted in contamination of large geographical regions (thousands to tens of thousands of kilometres). Cumulative exposures to airborne chemicals also takes place over long time periods (tens to hundreds of years).

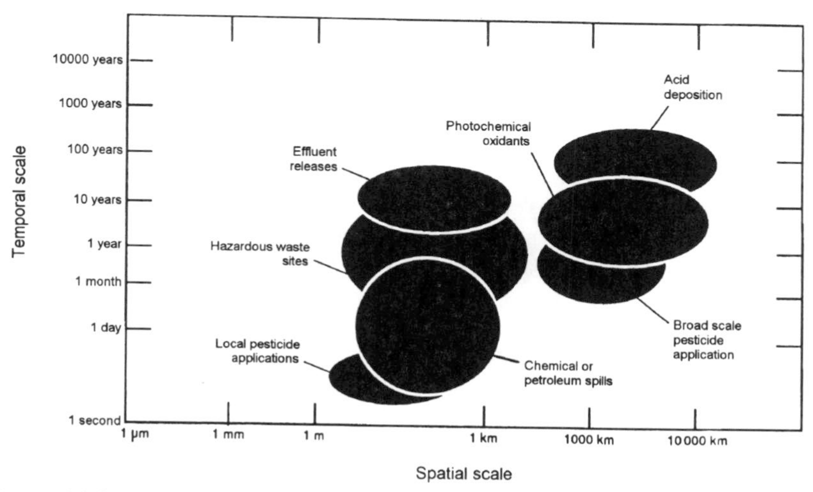

Figure 14.3. Arrangement of the spatial and temporal scales of selected chemical hazards (modified from Suter, 1993)

Clearly, regional exposures to chemicals occur as a result of widespread, long-term deposition of airborne chemicals or broad-scale pesticide applications. In both cases, exposures may cover several individual aquatic and terrestrial ecosystems, and may last for tens to hundreds of years. This scale of exposures points to the need for two types of methods: (1) measurement methods that can be used to assess effects on a regional scale and (2) extrapolation methods that can be used to scale-up local concentration-response data to predict the effects on regional dynamics. A framework and methods for regional scale ecotoxicological risk assessments are discussed in the following sections.

14.3 A FRAMEWORK AND METHODS FOR REGIONAL SCALE ECOLOGICAL RISK ASSESSMENTS

Although the need for ecotoxicological risk assessments at a regional scale is clear, little has been written about methods for such assessments, and few examples are available. As such, this discussion of framework and methods for regional scale ecotoxicological risk assessment is not so much a critique of approaches as it is a review of considerations for such assessments.

The framework for regional ecological risk assessment was first described by Hunsaker and colleagues (1990). They proposed a two-phased approach to regional assessments:

1. The definition phase, in which the regional endpoints, source terms, and reference environments are defined.
2. The solution phase, in which the exposure and effects are assessed, the exposure levels are related to levels of effects to estimate risk, and risks are extrapolated to a regional scale.

An example of the application of this two-phase approach to assess the effects of regional chemical exposures was provided by Graham *et al.* (1991). They evaluated the probabilities of significant changes in Adirondack forests as a consequence of ozone exposures and related beetle attacks on trees (discussed below). More recently, Suter (1993) described a set of considerations specific to regional scale assessment. These include scaling, landscape description, and integration of several qualitatively dissimilar stresses. In addition, while landscape ecologists have largely ignored chemical exposures, they have developed methods to assess changes in landscape patterns as the result of physical disturbances, and these methods may find applications in regional ecotoxicological assessments. Examples are provided in the work of Turner (1987), Turner and Gardner (1991), Urban *et al.* (1987), and others published in the journal *Landscape Ecology*.

Although our experience with regional scale ecotoxicological risk assessments is limited, potentially a great deal of carry-over exists from the framework and methods for local scale assessments for application to regional scale assessments.

The recently proposed generic EPA framework for ecological risk assessment is applicable to assessments at both local and regional scales (USEPA, 1992). A slightly modified version of the EPA framework that emphasizes considerations for regional scale risk assessments is presented in Figure 14.4. This diagram reinforces the need for understanding of the unique features of the region as well as the terrestrial and aquatic ecosystems within the region. A regional scale risk assessment should include problem formulation, exposure and effects assessments, and risk characterization for the component terrestrial and aquatic ecosystems and the region as a whole. The risks for the component ecosystems must be integrated over the region to characterize risks for the regional unit. The importance of airborne chemical inputs to regional-scale chemical hazards and the hydrological linkage as an important mechanism of chemical transport are emphasized in the diagram.

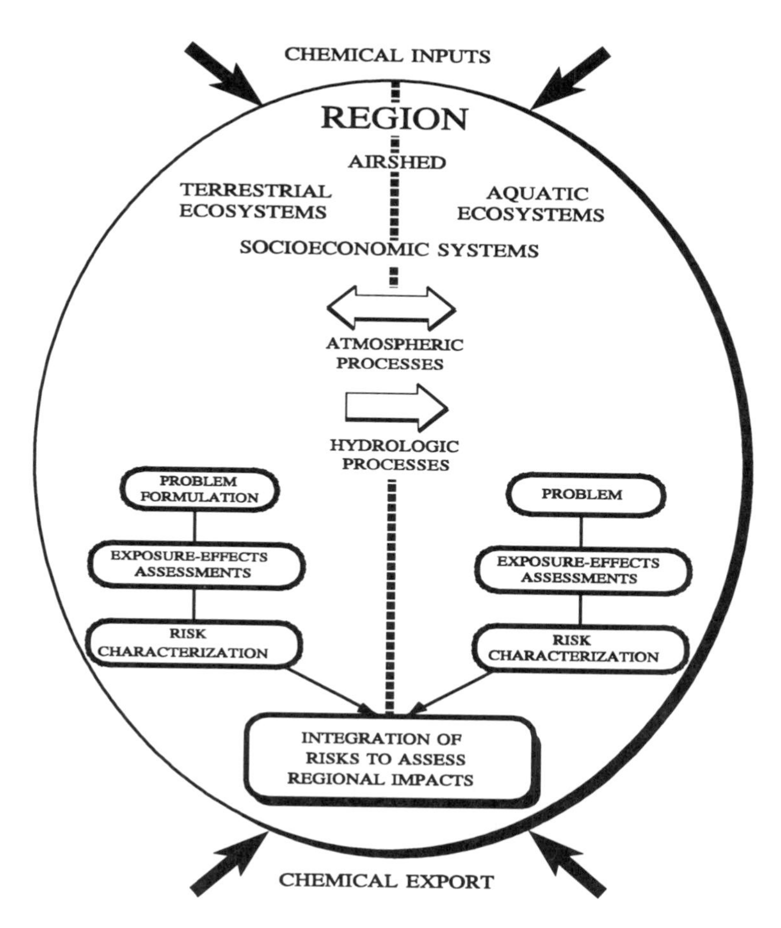

Figure 14.4. Generic framework for a regional scale ecotoxicological risk assessment

Ecological risk assessment of chemical exposures and effects at the regional scale may be either predictive or retrospective. Predictive risk assessments begin with a hypothetical chemical release scenario that could contaminate a large geographic area (such as the proposed broad-scale application of the new synthetic pyrethroid insecticides for grasshopper control in the prairie region of North America), and proceeds to the estimation of the risks of ecological effects such as direct toxicity to aquatic organisms and indirect effects on waterfowl populations that may be associated with broad-scale pyrethroid insecticide use (Sheehan *et al.*, 1987). In contrast, retrospective risk assessments at a regional scale generally begin with evidence of chemical contamination (e.g., acid deposition) and proceed to the characterization of risks associated with ecological responses to this contamination such as changes in soil chemistry and subsequent effects on soil fauna and the tolerance of trees to insect pest infestations reported for acid-stressed forests (Loucks *et al.*, in press).

The sources of information are generally quite limited for a predictive risk assessment; therefore, the evaluation of an exposure scenario is usually based on modelling and effects are characterized from toxicity data extrapolations. The tools of the predictive assessment may also be used in a retrospective assessment along with field data. Regardless of whether an ecotoxicological risk assessment is predictive or retrospective in nature, it will likely contain four components: (1) problem formulation, (2) exposure assessment, (3) effects assessment, and (4) risk characterization.

The first corresponds to the definition phase and the final three components to the solution phase described by Hunsacker *et al.* (1990). The components of a regional scale ecotoxicological risk assessment are identified in Figure 14.5. The options as to what to assess at a regional scale and how to assess it are discussed for the planning phase in the following sections. An example of exposure assessment, effects assessment, and risk characterization for a regional assessment are provided in a case study below.

14.3.1 PROBLEM FORMULATION

Problem formulation is the first phase of an ecotoxicological risk assessment. This planning phase establishes the assessment objectives and scope and provides a "blueprint" for the risk assessment process. Problem formulation should include the evaluation of existing data, identification of the region and ecosystems at risk, identification of the chemicals of interest, the establishment of risk assessment objectives and scope, selection of measurement and assessment of endpoints, and development of a conceptual risk assessment model to guide the solution phase activities.

14.3.1.1 Evaluation of existing data

Regardless of the type of risk assessment, information generally is available on actual or potential sources of chemicals and the region and types of ecosystems into which the chemicals may be or have been released. Source data may be in the form of chemical inventories and design properties for emission sources at manufacturing facilities such as that produced to meet the requirements of Hazardous Air Pollutant Provisions (Title III) of the US Clean Air Act Amendments of 1990. The modelling of these data to predict airborne chemical concentrations surrounding individual facilities was performed in California to meet

Air Quality Management District Requirements (Conner *et al.*, 1992). The combination of individual facility data could be used to provide an estimate of chemical concentrations in various California air basins. Emissions data for sulphur and nitrogen oxides (Slade, 1990) provided the focus for ecotoxicological risk assessments of the effects of acid deposition on forests in exposed regions of Ohio (Loucks *et al.*, 1993) and Pennsylvania (Nash *et al.*, 1992).

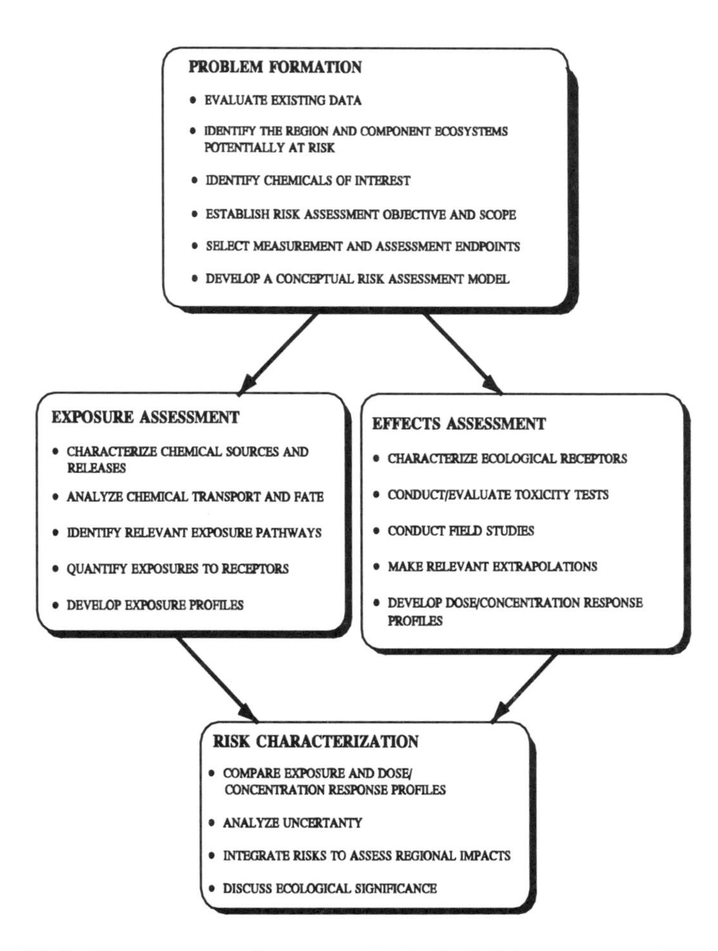

Figure 14.5. Components of an ecotoxicological risk assessment for regional scale evaluations

Data on chemical concentrations in air and other media are also available from various national and regional monitoring programmes. The USEPA monitors sulphur dioxide (SO_2), carbon monoxide (CO), nitrogen dioxide (NO_2), ozone (O_3), and lead (Pb) at over 200 fixed locations throughout the United States (USEPA,

1991). State air monitoring programmes such as that in the South Coast Air Quality Management District in Southern California provide regional data on a wider variety of chemicals commonly released to the atmosphere from mobile and point sources in that region.

Several sources may provide useful data for regional assessments of water quality in the US The US Geological Survey (USGS) and state water resource agencies have monitored the quality of surface water throughout the United States since the 1940s. Monitoring is conducted on a regular basis at fixed location within river basins. Data collected includes concentrations of nutrients, trace metals and pesticides in water. This programme has provided a 20- to 50-year record of concentration trends of selected chemical substances in more than 500 watersheds across the United States.

The USGS more recently established the National Water Quality Assessment Program (NWAQAP) which will eventually investigate about 120 study areas distributed throughout the United States (Hirsch *et al.*, 1988). These watersheds incorporate about 80 percent of the nations water use. These study units will be linked together to form a national network by using a prescribed set of study approaches and protocols for each river basin. The assessment program is perennial to provide data for trends analysis. The programme is focused on water quality conditions that are prevalent or large in scale and persistent in time. Chemical measurements focus on a set of target variables including physical measurements, inorganic constituents, and organic compounds. Biological measurements include plant and animal tissues to help determine the occurrence of trace element and organic compounds, to provide a measure of the bioavailability of these contaminants in water and sediments, and to help understand their environmental fate. Biological measurements also include aquatic toxicity tests and ecological surveys to assess toxicity of water and sediment, document the current status of the biological community, and describe and explain, to the extent possible, the relationships of the biological communities to the physical and chemical characteristics of the drainage river or streams.

A similar long-term monitoring programme providing data on chemical accumulation trends in watersheds is the National Contaminant Biomonitoring Program (formerly called the National Pesticide Monitoring Program). Since 1964, this programme has periodically analysed residues of selected organochlorine chemicals and trace metals in samples of fish and wildlife collected from a nationwide network of over 120 stations (May and McKinney, 1981; Schmitt *et al.*, 1981, 1983, 1985). The National Contaminant Biomonitoring Program data have documented the widespread distribution of DDT and PCBs in fish and the decline in the residue concentrations of these chemicals following discontinuation of their use. This programme has also provided data to identify new sources of persistent chemicals.

Recognizing its lack of an integrated approach to monitoring indicators of ecological conditions and exposures to chemicals in ecosystems, the USEPA has implemented the Environmental Monitoring and Assessment Program (EMAP)

(Messer *et al.*, 1991). This programme is focused on documenting changes at the regional level. The EMAP defines stress indicators and economic, social, and engineering data that can be used to determine the most probable sources of physical, chemical, and biological stressors. Exposure indicators are physical, chemical, and biological measurements that can be related to chemical exposure, habitat degradation, or other causes of poor ecosystem conditions. Response indicators are biological measures that quantify the condition of the region and ecosystems, and integrate the effects of various stresses. Examples include evidence of gross pathology, presence or absence of "sentinel" species, effects on keystone species that are important to maintain ecosystem structure, changes in populations in species that are of sports, commercial, or aesthetic interest, and effects on ecosystem structure (e.g., diversity) or function (i.e., primary production).

A goal of EMAP is the identification of a suite of assessment endpoints applicable to physical, chemical, and biological stressors in various types of ecosystems. The EMAP monitoring strategy is to establish a systematic grid of sampling points across the United States, including the continental shelf waters, and to make field measurements on a suite of indicators at grid points of a certain density and at a certain time period (e.g., a four-year average baseline). This programme will provide long-term regional data to assess the status and trends in condition of large-scale ecological systems.

14.3.1.2 Identification of the region and ecosystems potentially at risk

The region and ecosystems within which widespread exposures to chemicals may occur or have occurred is the geographic area of interest for an ecotoxicological risk assessment. However, several complicating factors are present in setting assessment boundaries. For a regional assessment to be effective, the spatial and temporal boundaries must be defined appropriately for both the hazard and assessment endpoints. In this sense, the assessment region may be effectively defined by the projected or actual distribution of chemicals of interest in the environment and/or the distribution of the populations or ecosystems of interest.

As an alternative, the bounds on the region may be based on the distribution or magnitude of the source. For example, a source-defined region includes the Appalachian coal mining area as a region for the assessment of acid mine drainage (Suter, 1993). Regions may also be defined in terms of natural features such a watershed or air basin boundaries, physiographic provinces, or ecoregions. Naturally defined regions have relatively uniform physical and biological properties, and are, therefore, more easily described than more heterogeneous geographic areas. These regions also include the hydrologic and atmospheric processes that, in large part, control chemical transport and exposures. The watershed is an effective unit for the study of effects of physical disturbance (e.g., clear cut logging) on terrestrial and aquatic ecosystems (Bormann *et al.*, 1968), and for the assessment of the effects of chemicals on water quality and aquatic biota. The watershed is the assessment unit for the National Water Quality Assessment Program described earlier (Hirsch *et al.*, 1988). Air basins provide a more appropriate assessment unit for chemicals transported through the atmosphere. The air basins containing coal burning plants in Ohio and Pennsylvania have been used to bound assessments of acid deposition effects on forested regions (Loucks *et al.*, 1993; Nash *et al.*, 1992).

The region may also be defined by the overlap of the chemicals and habitats of the populations of interest. For example, the assessment of potential effects of regional applications of insecticides in the Canadian prairies to control grasshopper infestations was bounded by an analysis of the area of overlap between waterfowl nesting habitat and crop areas receiving aerial applications of the specific insecticides of concern (Sheehan *et al.*, 1987).

Historically, regional assessment boundaries also have been defined by political boundaries. As the dispersal of chemicals in the environment is unconstrained by political boundaries, defining an ecotoxicological risk assessment by the political boundary is largely artificial. Where possible, assessments should be done on a region defined by appropriate ecological and anthropogenic factors with political boundaries applied secondarily as an additional overlay.

14.3.1.3 Identification of chemicals of interest

One or more chemicals may be of interest in a regional ecotoxicological risk assessment. For a pesticide application, the chemicals of concern are generally well defined and few in number. For insecticide spraying for spruce budworm control in the forests of New Brunswick, Canada, only DDT, fenitrothion, phosphamidon, trichlorfon, and aminocarb were identified as widely used. These risk assessments focused on DDT and fenitrothion (Brooks, 1974; NRCC, 1977; Mitchell and Roberts, 1984). Similarly, Sheehan *et al.* (1987) identified 13 insecticides that are widely used and aerially applied for insect pest control on prairie oil seed grain crops in Canada. Again, screening evaluations based on toxicity to aquatic invertebrates showed that only six compounds (permethrin, azinophos, methyl chlorpyrifos, deltamethrin, methoxychlor, and cypermethrin) potentially posed a high risk to aquatic organisms or waterfowl under normal application conditions.

For aqueous effluents and airborne emissions, the number of chemicals released may be tens or hundreds. Although in some cases understanding the risks posed by all of the chemicals released in effluents may be desirable, assessing the exposures and effects of hundreds of chemicals in either a human health or ecotoxicological risk assessment is generally impractical and unnecessary. The USEPA has recognized this issue in their technical risk assessment guidance (USEPA, 1989). Rather than undertake a superficial and unwieldy assessment based on quantitatively evaluating tens or hundreds of chemicals, risks should be assessed only for those chemicals that are likely to pose the greatest percentage of the risk. The USEPA suggests several screening procedures based on mobility, persistence, bioaccumulation, concentration, toxicity, and other site, or area, specific factors that can be used to identify and justify a subset of the chemicals released for risk assessment (USEPA, 1989). The procedures developed for human health assessment are also applicable to ecotoxicological assessments. The list of chemicals of primary ecological concern can be limited by screening chemicals for environmental persistence, bioaccumulation potential, and toxicity to representative species. Laskowski *et al.* (1982) describe several indices of mobility and persistence based on combinations of physical–chemical properties (leaching potential, volatility potential, on-site exposure potential) that can be used to rank chemical persistence. Bioaccumulation and toxicity data for fish are summarized in the USEPA AQUIRE Database (Environmental Research Laboratory Duluth, Minn.). Toxicity data for selected wildlife species are present in various US Fish and Wildlife Service Publications (Contaminant Hazard Reviews Reports 1 to 24). Plant toxicity data are summarized in the PHYTOTOX Database (Department of Botony, University of Oklahoma).

Examples of chemical screening for ecotoxicological assessments are available. A ranking procedure based on chemical persistence (Koc, Vp and $T_{1/2}$), bioaccumulation, potential (K_{ow} and BCF) and toxicity to fish (LC_{50}) and rodents (LD_{50}) has been used to identify 29 of 189 chemicals released from a hazardous waste incinerator as appropriate for a quantitative ecotoxicological assessment of incinerator emissions. Similar, although less well documented, procedures were used by various researchers to identify the chemicals of interest in assessing ecological effects of effluents to the Great Lakes (Evans, 1988).

For some of the more obvious regional contamination issues such as acid deposition and photochemical oxidant exposures, the chemicals of primary interest have been identified. Acid deposition exposures have been described in terms of mass loading of sulphur, nitrogen, and hydrogen ions (Loucks *et al.*, 1993; Nash *et al.*, 1992). Photochemical oxidant exposures have been further quantified using ozone as a surrogate for the oxidant mixture (Skelly, 1980; Graham *et al.*, 1991).

14.3.1.4 Establishment of risk assessment objectives and scope

To be meaningful and effective, an ecotoxicological risk assessment must be scientifically valid and relevant to regulatory needs and public concerns. Therefore, the regulatory or risk management framework within which the assessment is to be used should be considered in identifying the risk assessment objectives. In the US, for example, the Endangered Species Act requires analysis at the level of the individual of a threatened or endangered species. The Comprehensive Environmental Response, Compensation and Liabilities Act (CERCLA) requires that actions selected to remedy hazardous waste sites be protective of human health and the environment. CERCLA assessments generally require analysis of effects at several levels of biological organization and the development of risk-based remediation targets (USEPA, 1989).

Table 14.1. Examples of regulatory and risk management objectives and the scope of regional scale ecotoxicological risk assessments

Regulatory/risk management objectives:	Example risk assessment scope:
Federal Insecticide, Fungicide, and Rodenticide Act; registration of a new pesticide for broad-scale aerial application	Assess potential ecotoxicity expected from proposed aerial application of synthetic pyrethroid insecticides for grasshopper control in prairie grain regions of the US: 1. direct on non-target aquatic invertebrate populations in prairie ponds; 2. indirect on waterfowl populations directly dependent on aquatic invertebrates; 3. extrapolate estimated levels of effects for various spray scenarios to predict regional changes in waterfowl abundance
Clean Air Act; evaluate the potential effectiveness of chemical emission control strategies for regional air contaminants	Assess potential increase in forest tree production from regional ozone reductions of 20% or 50%: 1. direct effects on tolerance of trees to insect and fungal infestations; 2. relationship of insect infestation to forest production; 3. analysis of other environmental factors influencing production; 4. probabilities of concentration reduction improving forest production
Great Lakes Water Quality Initiative; evaluate the relative importance of reducing chemical and non-chemical stressors in improving the "ecological health" of the Great Lakes	Assess relative contributions of habitat loss/degradation, exotic species introduction, and persistent chemical concentrations in water and sediments on the abundance of game fish populations and fish-eating bird populations in the U.S. Great Lakes: 1. identify fish and bird species of interest; 2. analyses of relative contribution of stressors to species abundance

On the regional scale, ecotoxicological risk assessments may be conducted to meet various regulatory, risk management, and resource management requirements. Selected examples of risk management objectives and the risk assessment scope to address these needs are presented in Table 14.1. Predictive assessments can be used to evaluate the potential ecological effects of the proposed broad-scale aerial application of a new pesticide, the likely positive effects on ecological systems that might result from a reduction in regional concentrations of air contaminants due to regulatory control measures or the long-term effects of projected increases in chemical releases to air or water based on various scenarios of population growth and consumption patterns and industrial production. By contrast, retrospective approaches can be used to evaluate the relative effects of chemical, physical, and biological stressors on resources such as fisheries in the Great Lakes or to assess the effects of pesticides or air contaminants on regional ecological systems.

Although chemical deposition from airborne releases constitutes an obvious problem requiring regional-scale assessments, multiple releases of chemicals in aqueous effluents to aquatic systems may also pose regional scale exposure issues. The investigation of ecological effects in the Great Lakes associated with widespread exposures to persistent organic chemicals and metals is an example of a regional water quality issue which should be addressed with a regional scale risk assessment approach (Evans, 1988).

The scope of a regional ecotoxicological risk assessment is likely to include analyses of both population-level and ecosystem-level exposures and effects as well as larger scale effects on landscape structure of productive capacity of the region. Regional assessments will most certainly require an analysis of multiple stressors due to the wide variety of anthropogenic activities and associated chemical releases, habitat destruction, and biological resource use that occur within large geographic areas. Multiple stressors may include chemical mixtures or combinations of chemical with physical stressors such as habitat alteration or biological stressors such as hunting or fishing pressures. The key to the integration of multiple stressors into an ecotoxicological assessment is the identification of common assessment endpoints. For example, an endpoint such as recruitment abundance (i.e., the number of young added to the population) can be used to integrate the individual effects of chemicals, habitat alteration, hunting, drought, or other stressors on large mammal and bird populations (Sheehan *et al.*, 1987; Barnthouse *et al.*, 1990). The effects of both chemical and physical stressor on plants can be expressed in terms of reductions in primary production (Adams *et al.*, 1985). By contrast, individual toxicological endpoints such as LC_{50} and no observed effect levels for chemicals have no direct equivalent in the terms generally used to describe system-wide impacts. Suter (1993) suggests that ecotoxicologists look to resource managers who have developed appropriate endpoints to integrate the effects of various stressors on plant and animal populations (e.g., USFWS, 1980; Bovee and Zuboy, 1988).

In large-scale ecotoxicological risk assessments, both the direct toxic effects of chemicals on individuals and populations and indirect effects of chemicals on the environment or biological resources that may subsequently affect other populations, communities, and ecosystems should be considered in quantifying risks. For example, in studies of the effects of acid deposition on forests, researchers have shown that the oxides of sulphur and nitrogen increase soil acidity and enhance the leaching of Ca^{2+} and Mg^{2+} cations essential for plant growth (Loucks *et al.*, 1993).

The acidification of soils can also affect soil invertebrate populations, such as earthworms which are important to the decomposer food web and nutrient recycling. Loucks *et al.* (1993) reported that the accumulation of undecomposed organic matter in surface soils was positively correlated with high levels of acid deposition. The reduction in essential cations and plant nutrients in acidified soils places a physiologic stress on trees in these areas and leaves them more susceptible to insect attacks (Haack and Blank, 1991). Thus, the effects of acid deposition on forest production are largely indirect, and are the result of environmental changes that place trees under stress.

Table 14.2. Examples of possible assessment and measurement endpoints for evaluation of the toxicity of insecticides sprayed for control of spruce budworm

Problem:	Assessment endpoint:	Measurement endpoint:
Possible non-target effects of long-term application of insecticides to regional forests to control spruce budworm	Probability of >10% reduction in salmon populations in streams in the sprayed area	LC_{50} or NOAEL for salmon or related fish species
	Significant decrease in tree canopy bird populations	Dietary LD_{50} for Japanese quail egg hatch and fledgling success in treated and reference areas; population numbers for selected bird species in treated and reference areas
	20% decrease in fruit production from bee-pollinated plants	LC_{50} for bees; abundance and diversity of natural bees; populations of selected bee species in treated and reference areas; fruit production (e.g., blue berries) in treated and reference areas
	Significant decrease in forest litter decomposition	Microbial respiration in soils from treated and reference areas; soil arthropod abundance in leaf litter in treated and reference areas

Another example of the importance of considering indirect effects is provided in the work of Sheehan and colleagues (1987) on the potential impacts of pesticides on prairie-nesting duck populations. Their evaluation showed that although the new synthetic pyrethroid insecticides pose a low toxicity hazard to ducks at recommended application rates, the widespread aerial application of these insecticides could effect the recruitment of young in sprayed regions by substantially reducing aquatic invertebrate populations, an essential food resource for ducklings, during the critical growth period following hatching. In years when there was widespread aerial application of these insecticides for grasshopper control during periods of duckling hatching and rearing, insecticide effects on recruitment and abundance of regional duck populations may be equivalent in magnitude to population losses from hunting (Sheehan *et al.*, 1987, 1993).

Table 14.3. Types of potential assessment endpoints for regional ecotoxicological risk assessments (modified from Suter, 1993)

Traditional ecological endpoints	Endpoints characteristic of regions
Population	Population/species
Extinction	Range
Abundance	
Production	Productive capability
Massive mortality	Soil loss
	Nutrient loss
Community/ecosystem	Regional production
Change in type	
Production	Pollution of other regions
	Pollution of outgoing water
Anthropocentric Endpoints	Pollution of outgoing air
Population	Susceptibility
Frequent gross morbidity	Pest outbreaks
	Fire
Community/ecosystem	
Market/sport value	Landscape indices
Recreational quality	Dominance
	Contagion
Air and water quality standards	

Clearly the accurate characterization of atmospheric and hydrologic transport of chemicals will be important in quantifying chemical exposures at a regional scale. Atmospheric transport plays a key role in the acidification phenomenon noted in North America and Europe. The distance a chemical is transported from its source depends on meteorologic factors, geographic factors, and characteristics of the chemical itself. A useful feature which describes, in general terms, a chemical's atmospheric behaviour is its average residence time in the atmosphere. Pollutants such as sulphur and nitrogen oxides, that have residence times of approximately two days (Rodhe, 1978), are particularly important on a regional scale, because this time period is comparable to the time typically required for atmospheric transport across eastern North America or Western Europe. Hidy *et al.* (1979) classified meteorologic situations that lead to long-range transport and regional pollution in the Eastern United States. Several atmospheric fate and transport models for chemicals are available, and were recently reviewed by Zannetti (1990).

Precipitation is one of the primary mechanisms removing chemicals for regional air and depositing them into terrestrial and aquatic ecosystems. Precipitation runoff and erosion are also key mechanisms of transport to surface waters of chemicals initially deposited in surface soils. The importance of properly characterizing these hydrologic linkages is again key to predicting regional changes in water quality (Hunsaker *et al.*, 1990). Several models have been developed to predict the hydrologic transport of chemical contaminants. Among these are the Midwest Research Institute Nonpoint Source Loading Function Model, the Environmental Pollution Assessment Erosion Sedimentation and Rural Runoff Model, the US Department of Agriculture Hydrograph Laboratory Model, and the Chemical Runoff and Erosion from Agricultural Management Systems Model.

14.3.1.5 Selection of measurement and assessment endpoints

An endpoint is a characteristic of an ecological component that may be affected by exposure to a chemical. Two types of endpoints should be identified: assessment endpoints and measurement endpoints.

An assessment endpoint is a formal expression of the environmental value to be protected (Suter, 1989). Given the diversity of ecosystems and the various values placed on them by society, no universal list of assessment endpoints exists. However, Suter (1993) has identified five criteria that any assessment endpoint should satisfy: societal and biological relevance, unambiguous operational definition, accessibility to prediction and measurement, and susceptibility to the hazardous agent. Societal relevance implies that the endpoint should be understood and valued by the public. Biologically relevant endpoints reflect important characteristics of the system, and are related to other endpoint up the ecological hierarchy. Unambiguous operational definitions of endpoints are essential to provide direction to testing and modelling within the assessment. Goals such as "ecosystem health" are inadequate for assessments.

Measurement endpoints are the quantifiable responses to a chemical stressor that can be related to the valued characteristic chosen as the assessment endpoint (Suter, 1989). Measurement endpoints are the analyst's input to the risk assessment, and include LC_{50} and LD_{50} from toxicity tests and population estimates from field sampling studies. Although assessment and measurement endpoints can be identical, they are frequently defined differently. Several examples of assessment and measurement endpoints that might be used to evaluate the ecological effects of pesticide spraying for spruce budworm control are provided in Table 14.2.

The number of possible assessment endpoints for a region assessment is greater than for a single ecosystem risk assessment. Endpoints are unique to the component populations and ecosystems within the region as well as for the region as a whole.

A list of possible assessment endpoints is presented in Table 14.3. These include traditional ecological endpoints such as population extinction and abundance as well as endpoints characteristic of regions, and anthropocentric endpoints such as sport fisheries values. Traditional ecological endpoints are assessed in the same manner for local and regional ecotoxicological risk assessments, but results must be scaled up for the region.

Representative regional assessment endpoints, however, are less obvious. As Suter (1993) points out, biological productivity is clearly an important regional endpoint, but it is not readily defined. Crop and timber production are largely controlled by economic factors, and are heavily influenced by technology. Therefore, realized production is not a suitable measure of productive capacity or the influence of chemicals on productivity. An approach to overcoming the problem of controlling for factors other than chemicals is to focus on production of species crops or forest types and normalize results for weather, fertilizer, energy, and other inputs. An alternative approach is to assess the proportion of a region devoted to biological production that is lost, or lost from production due to chemical exposures, again controlling for other factors.

Regional assessment endpoints should be defined in terms of observations that can be made over large geographic areas and long time periods. For terrestrial systems, endpoints might include percent cover of different vegetation types. For aquatic systems, a representative endpoint might be the frequency of lakes in which a valued fish species becomes extinct.

Integrated properties of landscapes described by landscape ecologists may also be representative regional assessment endpoints. Examples of landscape properties include dominance (i.e., the degree to which the landscape is dominated by a particular feature), contagion (the degree to which the landscape is dissected into small patches or aggregated into large patches), fractal dimension (index of complexity of shapes on the landscape), and amount of edge (Krummel *et al.*, 1986; O'Neill *et al.*, 1988; Hunsaker *et al.*, 1990; Graham *et al.*, 1991). These indices can be calculated from remote sensing data, and, therefore, may be particularly useful in large-scale regional assessments.

Table 14.4. Ecological endpoint measures used in the Adirondack demonstration (modified from Graham *et al.*, 1991)

Measure	Definition
	Land cover
Forest	% of region classified as forest
Deciduous	% of forest classified as deciduous
Conifer	% of forest classified as coniferous
Mixed	% of forest classified as a mixture of coniferous and deciduous trees
	Edge Habitat
Deciduous–open	km deciduous bordering open areas
Coniferous–open	km coniferous bordering open areas
Mixed–open	km mixed forest bordering agriculture
Deciduous–agriculture	km deciduous forest near agriculture
Coniferous–agriculture	km coniferous forest near agriculture
Mixed–agriculture	km mixed forest bordering agriculture
Deciduous–wetland	km deciduous forest bordering wetlands
Coniferous–wetland	km coniferous forest near wetlands
Mixed–wetland	km mixed forest bordering wetland
Forest interior	Total amount of forest land (hectares) ≥200 m from any non-forest land
	Landscape Indices
Dominance (D)	Degree to which total region is dominated by one or two land-cover types (high values = landscape dominated by one or two land-cover types)
Contagion (C)	Degree to which land-cover types are grouped within the region (high values = landscape composed of a few large patches)

$$D = \ln(n) + \sum_{i-1}^{n} P_i \ln(P_i) \qquad (1)$$

where D is the index of dominance, n is the total number of land-use categories, and P_i is the proportion of the region in land use i;

$$C = 2n \ln(n) + \sum_{i-1}^{n} \sum_{j-1}^{n} P_u \ln(P_u) \tag{2}$$

where C is the index of contagion, n is the total number of land-use categories, and P_u is the probability that a unit (grid cell) of land use i will be found adjacent to a unit of land use j.

14.3.1.6 Development of a conceptual risk assessment model

The conceptual risk assessment model is developed from a series of working hypotheses regarding how the chemical(s) might affect the ecological components of the region and the region as a whole. Although many hypotheses may be developed for regional exposure scenarios, only those that are considered most likely to contribute to risk should be selected for quantitative evaluation. For these hypotheses, the conceptual model describes the approach and methods that are to be used in the analytical phase and the types of data and analytical tools that are needed. An example of the development of a conceptual risk assessment model and the implementation of this model to assess the potential effects of ozone on regional forests is provided in the case study below.

14.4 REGIONAL RISK ASSESSMENT CASE STUDY

Graham and colleagues (1991) provide an example of a predictive regional ecotoxicological risk assessment of the effects of ozone exposures on forests in the Adirondack region of New York (US).

14.4.1 PROBLEM FORMULATION

In forested areas, the initial effect of elevated ozone concentrations is manifested as physiological stress in trees. Tree response to ozone is a function of cumulative uptake and, as such, is related to leaf or needle lifespan. Thus, chronic exposure to elevated ozone is expected to have a greater effect on conifers with long-lived needles than on deciduous trees (Reith, 1987). Stressed trees are more susceptible to bark beetle attack than unstressed trees (Payne, 1980). Once a tree is attacked, bark beetles will then attack neighbouring trees, spreading the infestation. Infested trees frequently die as a consequence of bark beetle attacks due to subsequent infestation by blue stain fungi (Coulson, 1980). Bark beetle-induced tree mortality can alter the amount and type of forest cover in regions with substantial ozone exposures. The conceptual risk assessment model is one in which chemical exposures are related indirectly to tree death and the amount of forest cover through a decrease in tree tolerance to insect and fungal disease infestations.

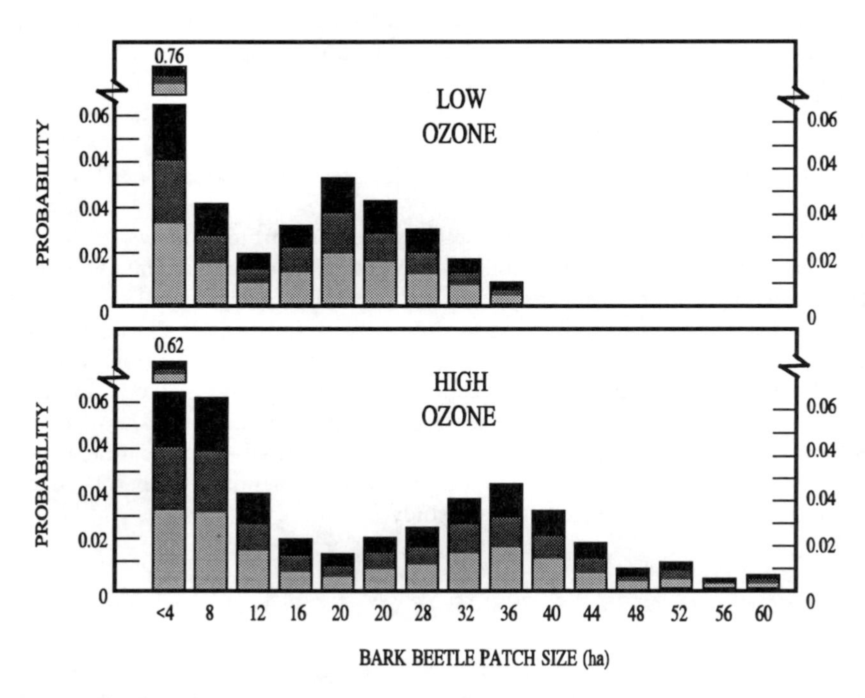

Figure 14.6. Bark beetle patch-size distribution under the high- and low-ozone scenarios (Graham *et al.*, 1991)

The assessment endpoints for the demonstration are a 10–25 percent deviation from the baseline value of the endpoints identified in Table 14.4, with the landscape indices.

A stochastic spatial simulation model of land-cover change induced by ozone-triggered bark beetle infestations was used to quantify potential ozone impacts on land cover. In the model, a uniform concentration of ozone was assumed to stress trees so that bark beetle infestations would convert coniferous forest grid cells to open-land grid cells, and mixed forest grid cells to deciduous forest grid cells.

14.4.2 EXPOSURE ASSESSMENT

Two scenarios of elevated ambient ozone concentration were used in this analysis. The high-ozone scenario assumes an average daily maximum concentration of 0.20

μL/L based on conditions slightly worse than recorded in the San Bernardino Mountains in California (Taylor, 1973). The low-ozone scenario assumes an ozone concentration of 0.04 μL/L, which is slightly higher than currently measured in the Adirondack Mountains (NYSDEC, 1986). Both scenarios assume uniform exposure across the study region.

14.4.3 EFFECTS ASSESSMENT

The probability of an initial bark beetle attack was assumed to be 0.015 under the low-ozone exposure scenario and 0.04 under the high-ozone exposure scenario. Each attack could result in a patch of dead conifer trees ranging in size from 1 grid cell (4 ha) to 15 grid cells (60 ha). The probability parameters were developed from data on ponderosa pine mortality in the San Bernardino Mountains and bark beetle patch size in the Southeastern US (Taylor, 1973; Coster and Searcy, 1981). The probability of the size distribution of patches is shown in Figure 14.6. Under the low-ozone scenario, the distribution is based on the pattern of beetle infestation patches observed in forest in North Carolina (Coster and Searcy, 1981). For thehigh-ozone scenario, the tail of the distribution was extended and the patch size increased. The modelled distribution of patch size for the high ozone scenario reflects both the survey data and field observations that, under most circumstances, initial bark beetle attacks do not spread to other trees; however, when conditions are right, the beetle infestation spreads readily (Coulson, 1980).

14.4.4 RISK CHARACTERIZATION

Risks were characterized by first calculating the baseline value for each measurement endpoint from the initial land cover pattern. For each endpoint and ozone exposure scenario, a cumulative frequency diagram of endpoint values was developed from 100 model simulation runs. Using these diagrams and baseline values for measurement endpoints, the frequency of runs in which the value of the endpoint shifted greater than 10 percent (detectable change) and greater than 25 percent (significant change) was found from the baseline value. Risk probability was calculated by dividing the frequency of runs by their total number.

The risk assessment results are summarized in Table 14.5. The analysis showed that detectable, but not significant, changes in coniferous, deciduous, and mixed-forest types were likely under the high-ozone scenario. In contrast, under the low-ozone scenario, zero risk existed for detectable change in any forest cover type. Most of the pattern-sensitive endpoints were affected by the ozone scenarios. Significant changes in forest edge habitat were highly probable under both ozone scenarios. The projected changes in forest edge habitat can be shown diagrammatically.

Table 14.5. The risk to endpoints under the low- and high-ozone scenarios, and the average values for baseline and altered landscapes under both scenarios[a] (from Graham *et al.*, 1991)

	Ozone scenarios						
	Low		High		Value of Endpoint Measure		
	Detectable Risk	Significant Risk	Detectable Risk	Significant Risk	Baseline (actual)	Low O_3 (mean)	High O_3 (mean)
Cover endpoints							
Forest	0.00	0.00	0.00	0.00	93.6	93.1	91.3
Deciduous	0.00	0.00	1.00	0.00	39.8	41.2	45.6
Coniferous	0.00	0.00	0.19	0.00	21.0	21.1	19.5
Mixed	0.00	0.00	0.36	0.00	38.6	37.8	34.9
Edge Habitat Endpoints							
Deciduous–open	1.00	0.28	1.00	1.00	67	82	132
Coniferous–open	1.00	1.00	1.00	1.00	143	238	402
Mixed–open	0.00	0.00	0.24	0.00	152	156	164
Deciduous–agriculture	0.46	0.01	1.00	0.93	15	16	21
Coniferous–agriculture	0.00	0.00	0.45	0.00	73	71	66
Mixed–agriculture	0.00	0.00	0.52	0.00	57	56	51
Deciduous–wetlands	0.01	0.00	0.52	0.00	15	15	16
Coniferous–wetlands	0.06	0.00	0.47	0.04	10	10	9
Mixed–wetlands	0.01	0.00	0.51	0.00	15	15	14
Forest Interior	0.00	0.00	0.07	0.00	260100	251400	235,600
Landscape Indices Endpoints							
Dominance	0.00	0.00	0.00	0.00	0.87	0.85	0.81
Contagion	0.00	0.00	1.00	0.00	26.43	24.13	23.14
Lake Water Quality Endpoints							
Lake pH shift	0.01	0.00	0.89	0.00	0	4%	14%
Acid improvement	0.67	0.28	0.97	0.89	0	18%	38%

[a]For this assessment, risk is defined as the probability of a detectable or significant change (a change either >10% (detectable) or >25% (significant) of the original value of the endpoint measure).

This case study demonstrates the importance of considering spatial heterogeneity in evaluating the effects of airborne chemical deposition on the original landscape. A spatial model is essential to quantifying risks to the land-cover pattern and indirectly to timber resources and wildlife habitat.

14.5 PROSPECTS FOR REGIONAL SCALE ASSESSMENT

Although few comprehensive regional scale ecotoxicological risk assessments have been conducted, the need for such assessments is obvious. The need is particular strong with respect to assessing the ecological effects of air contaminants such as ozone and oxides of sulphur and nitrogen.

To encourage the use of regional risk assessment tools to gather data over wide geographic areas, such as satellite imagery, tools to manage large data sets such as GIS, and models to accurately simulate ecological responses to broad-scale and long-duration chemical exposures, will need to be developed further, and incorporated into the process to make it more useful and cost-effective.

14.6 REFERENCES

Adams, R.M., Crocker, T.D., and Katz, R.W. (1985) Yield response in benefit cost analyses of pollution-induced vegetation damage. In: Winner, W.E., Mooney, H.A., and Goldstein, R.A. (Eds.) *Sulfur Dioxide and Vegetation*. Stanford University Press, Palo Alto, California.

Barnthouse, L.W., Suter II, G.W., and Rosen, A.E. (1990) Risks of toxic contaminants to exploited populations: influence of life history, data uncertainty, and exploitation intensity. *Environ. Toxicol. Chem.* **9**, 297–311.

Bartell, S.M., Gardner, R.H., and O'Neill, R.V. (1992) *Ecological Risk Estimation*. Lewis, Chelsea, Michigan.

Bormann, F.H., Likens, G.E., Fisher, D.W., and Pierce, R.S. (1968) Nutrient loss accelerated by clear-cutting of a forest ecosystem. *Science* **159**, 882–884.

Bovee, K.D., and Zuboy, J.R. (Eds.) (1988) *Proceedings of a Workshop on the Development and Evaluation of Habitat Suitability Criteria*. USFWS Biological Report 88(11). US Department of Interior, Fish and Wildlife Service, Fort Collins, Colorado.

Brooks, G.T. (1974) *Chlorinated Insecticides*, Vol. II. *Biological and Environmental Aspects*. CRC Press, Cleveland, Ohio.

Conner, K., Holbrow, A.M., Copeland, T.L., and Paustenbach, D.J. (1992) *Use of Quantitative Uncertainty Analysis in Air Toxic Risk Assessment*. Paper No. 92-166.04 presented at the 85th Annual Meeting of the Air and Waste Management Association, Kansas City, 21–26 June 1992. ChemRisk McLaren/Hart, Irvine, California.

Coster, J.E., and Searcy, J.L. (Eds.) (1981) *Site, Stand and Host Characteristics of Southern Pine Beetle Infestation*. USFS Technical Bulletin 1612. US Department of Agriculture, Forest Service, Washington, D.C.

Coulson, R.N. (1980) Population dynamics. In: Thatcher, R.C., Searcy, J.L., Coster, J.E., and Hertel, G.D. (Eds.) *The Southern Pine Beetle*, pp. 71–104. USFS Technical Bulletin 1631. US Department of Agriculture, Forest Service, Washington, D.C.

Evans, M.S. (Ed.) (1988) *Toxic Contaminants and Ecosystem Health: A Great Lakes Focus*. John Wiley, New York.

Gass, C.L., and Montgomeric, R.D. (1981) Hummingbird foraging behavior: Decision-making and energy regulation. In: Kamil, A.C., and Sargent, T.D. (Eds.) *Foraging Behavior: Ecological, Ethological, and Psychological Approaches*. Garland STPM Press, New York.

Graham, R.L., Hunsaker, C.T., O'Neill, R.V., and Jackson, B.L. (1991) Ecological risk assessment at the regional scale. *Ecol. Appl.* **1**(2), 196–206.

Haack, R.A., and Blank, R.W. (1991) Incidence of two-lined chestnut borer and *Hypoxylona tropunctatum* on dead oaks along an acidic deposition gradient from Arkansas to Ohio. *Proceedings, Eighth Central Hardwoods Forest Conference, University Park, Pennsylvania.* General Technical Report 148. US Department of Agriculture, Forest Service, Northeastern Forest Experiment Station, Radnor, Pennsylvania.

Hidy, G.M. Mueller, P.K., Lavery, T.F., and Warren, K K. (1979) Assessment of regional air pollution over the Eastern United States: results from the Sulfate Regional Experiment (SURE). In: *Proceedings of the WMO Symposium on the Long-Range Transport of Pollutants and its Relation to General Circulation Including Stratospheric/Tropospheric Exchange Processes*. WMO Report No. 538. World Meteorological Organization, Geneva.

Hirsch, R.M., Alley, W.M., and Wiber, W.G. (1988) *Concepts for a Natural Water-Quality Assessment Program*. USGS Circular 1021. US Geological Survey, Federal Center, Denver, Colorado.

Holling, C.S. (1992) Cross-scale morphology, geometry and dynamics of ecosystems. *Ecol. Monogr.* **62**(4), 447–502.

Hunsaker, C.T., Graham, R.L., Barnthouse, L.W., Gardner, R.H., O'Neill, R.V., and Suter II, G.W. (1990) Assessing ecological risk on a regional scale. *Environ. Manag.* **14**, 325–332.

Krummel, J.R., Gardner, R.H., Sugihara, G., and O'Neill, R.V. (1986) Landscape patterns in a disturbed environment. *Oikos* **48**, 321–324.

Laskowski, D.A., Igoring, C.A., McCall, P.J., and Swann, R.L. (1982) Terrestrial environment. In: Conway, R.A. (Ed.) *Environmental Risk Analysis for Chemicals*. Van Nostrand Reinhold, New York.

Loucks, O.L., Armentano, T.V., Foster, J., Fralish, J., Haack, R., Kuperman, R., LeBlanc, D., McCune, B., Pedorsen, B., and Somers, P. (in press) Pattern of air pollutants and response of oak-hickory ecosystems in the Ohio Corridor.

May, T.W., and KcKinney, G.L. (1981) Cadmium, mercury, arsenic and selenium concentrations in freshwater fish. 1976–1977—National Pesticide Monitoring Program. *Pest. Monit. J.* **15**, 14–38.

Mitchell, M.F., and Roberts, A.(1984) A case study of the use of fenitrothium in New Brunswick: The evolution of and ordered approach to ecological monitoring. In: Sheehan, P.J., Muller, D., Butter, G.C., and Bourdeau, P. (Eds.) *Effects of Pollutants at the Ecosystem Level*. SCOPE 22. John Wiley & Sons, Chichester, UK.

Nash, B.L., Davis, D.O., and Skelly, J.M. (1992) Forest health along a wet sulfate/pH disposition gradient in north-central Pennsylvania. *Environ. Toxicol. Chem.* **11**, 1092–1104.

National Research Council of Canada (NRCC) (1977) *Fentrothion: The Long-Term Effects of Its Use in Forest Ecosystems*. NRCC No. 16073. Associate Committee on Scientific Criteria for Environmental Quality, Ottawa, Canada.

NYSDEC (1986) *New York State Air Quality Report Ambient Air Monitoring System*. DAR-87-1. New York State Department of Environmental Conservation, Albany.

O'Neill, R.V., Krummel, J.R., Gardner, R.H., Sugihara, G., Jackson, B., DeAngelis, D.L., Milne, B., Turner, M.G., Zygmutt, B., Christensen, S.W., Dale, V.H., and Graham, R.L. (1988) Indices of landscape pattern. *Landscape Ecol.* **1**, 153–162.

Orians, G.H. (1980) General theory and applications to human behaviors. In: Lockhard, J.S. (Ed.) *Evolution of Human Social Behavior*. Elsevier, New York.

Otani, J., Dombrowski, F., Sheehan, P., Harman, C., and Brooke, K. (in preparation) A predictive ecological assessment of 30 chemicals emitted from a proposed hazardous waste treatment facility in east-central Mississippi.

Payne, T.L. (1980) Life history and habits. In: Thatcher, R.C., Searcy, J.L., Coster, J.E., and Hertel, G.D. (Eds.) *The Southern Pine Beetle*. USFS Technical Bulletin 1631. US Department of Agriculture, Forest Service, Washington, D.C.

Reith, P.B. (1987) Quantifying plant response to ozone: A unifying theory. *Tree Physiol.* **3**, 63–91.

Rodhe, H. (1978) Budgets and turnover times of atmospheric sulfur compounds. *Atmos. Environ.* **12**, 671–680.

Schmitt, C.J., Ludke, J.L., and Walsh, D. (1981) Oganochlorine residues in fish, 1970–1974: National Pesticide Monitoring Program. *Pest. Monit. J.* **14**, 136–206.

Schmitt C.J., Ribick, M.A., Ludki, J.L., and May, T.W. (1983) *Organochlorine Residues in Freshwater Fish. 1976–1979: National Pesticide Monitoring Program.* USFWS Resource Publ. No. 152.62. US Department of Interior, Fish and Wildlife Service, Washington, D.C.

Schmitt, C.J., Zajicek, J.L., and Ribick, M.A. (1985) National Pesticide Monitoring Program: residues of organochlorine chemicals in freshwater fish 1980–81. *Arch. Environ. Contam. Toxicol.* **14**, 225–260.

Sheehan, P.J. (1984a) Effects on individuals and populations. In: Sheehan, P.J., Miller, D.R., Butler, G.C., and Bourdeau, P. (Eds.) *Effects of Pollutants at the Ecosystem Level*. John Wiley & Sons, Chichester, UK.

Sheehan, P.J. (1984b) Effects on community and ecosystem structure and dynamics. In: Sheehan, P.J., Miller, D.R., Butler, G.C., and Bourdeau, P. (Eds.) *Effects of Pollutants at the Ecosystem Level*. John Wiley & Sons, Chichester, UK.

Sheehan, P.J., Baril, A., Mineau, P., Smith, K.K., Hartenist, A., and Marshall, W.K. (1987) *The Impact of Pesticides on the Ecology of Prairie Nesting Ducks*. Technical Report Series No. 19. Canadian Wildlife Service, Environment Canada, Ottawa, Canada.

Sheehan P.J., Baril, A., Mineau, P., and Paustenbach, D. (1993) Predicting the effects of insecticides on aquatic systems and the water fowl that use them: an ecotoxicological risk assessment. In: Rand, G.M., and Petrocelli, S.R. (Eds.) *Fundamentals of Aquatic Toxicology: Methods & Applications*. Hemisphere, Baltimore, Maryland.

Simmons, M.A., Cullinan, V.I., and Thomas, J.M. (1992) Salellite imagery as a tool to evaluate ecological scale. *Landscape Ecology* **7**(2), 77–85.

Skelly, J.M. (1980) Photochemical oxidant impact on mediterranean and temperate forest ecosystems: Real and potential effects. In: Miller, P.R. (Ed.) *Effects of Air Pollutants on Mediterranean and Temperate Forest Ecosystems*. General Technical Report PSW-43. Pacific Southwest Forest and Range Experimental Station, Berkeley, California.

Slade, J. (1990) Overview of sulfur and nitrogen oxide emissions in Pennsylvania. *Proceedings. Atmospheric Deposition in Pennsylvania: A Critical Assessment*. Pennsylvania State University, University Park, Pennsylvania.

Solomon, A.M. (1986) Transient repose of forests to CO_2 induced climate change: simulation modeling experiments in eastern North America. *Ecologia* **68**, 567–579.

Suter II, G.W. (1989) Ecological Endpoints. In: Warren-Hicks, W., Parkhurst, B.R., and Bakers, S.S. (Eds.) *Ecological Assessment of Hazardous Waste Sites.* A Field and Laboratory Reference Document. EPA 600/3-89/013. Environmental Research Laboratory Corvallis, Oregon.

Suter II, G.W. (1993) *Ecological Risk Assessment.* Lewis Publishers, Boca Raton, Fla.

Swank, W.T., and Grossley, D.A. (Eds.) (1987) *Forest Hydrology at Coweeta.* Springer-Verlag, New York.

Taylor, O.C. (1973) *Oxidant Air Pollution Effects on a Western Coniferous Forest Ecosystem.* Statewide Air Pollution Research Center, University of California, Riverside.

Tucker, C.J., and Sellers, P.J. (1986) Satellite remote sensing of primary production. *Int. J. Remote Sensing* **4**, 1395–1416.

Turner, M.G. (1987) Simulation of landscape changes in Georgia: A comparison of 3 transition models. *Landscape Ecology* **1**, 29–36.

Turner, M.G., and Gardner, R.H. (Eds.) (1991) *Quantitative Methods in Landscape Ecology.* Springer-Verlag, New York.

Urban, D.L., O'Neill, R.V., and Shugart, H.H., Jr (1987) Landscape ecology. *BioScience* **37**, 119–127.

USEPA (1989) *Risk Assessment Guidance for Superfund*, Vol. I, *Human Health Evaluate Manual* (Part A). Report No. EPA 510/1-89/002. US Environmental Protection Agency, Office of Emergency and Remedial Response, Washington, D.C.

USEPA (1991) *National Air Quality and Emissions. Trends Report 1989.* Report No. EPA 450/4-91/003. US Environmental Protection Agency, Office of Air Quality, Planning and Standards, Research Triangle Park, North Carolina.

USEPA (1992) *Framework for Ecological Risk Assessment.* Report No. EPA 630/12-92/001. US Environmental Protection Agency, Risk Assessment Forum, Washington, D.C.

USFWS (1980) *Habitat Evaluation Procedures (HEP).* US Department of the Interior, Fish and Wildlife Service, Division of Ecological Services, Washington, D.C.

Zannetti, P. (1990) *Air Pollution Modeling.* Van Nostrand Reinhold, New York.

15 Methods to Evaluate Whole Aquatic and Terrestrial Systems

H. E. Evans
RODA Environmental Research Limited, Canada

P. J. Dillon
Ontario Ministry of the Environment, Canada

15.1 INTRODUCTION

The traditional approach to assess the effects of chemicals on ecosystems involves the examination of either one or a few components of the system (such as an individual species), or it involves the study of only a physical portion (i.e., a micro- or mesocosm) of the system either *in situ* (e.g., enclosures) or in the laboratory. Unfortunately, the data derived from laboratory experiments apply only to the experimental conditions used; thus, they cannot be extrapolated necessarily to the natural environment. Similarly micro- and mesocosm experiments are inappropriate in situations where the effects of the chemical on organisms that are large relative to the size of the physical system are under study.

Micro- and mesocosms are unsuitable also in long-term experiments, because the natural floral and faunal assemblages may become unbalanced as a result of the experimental conditions. For example, Schindler (1987) reported that for experiments lasting more than a few weeks, even the very large aquatic enclosures (limnocorrals) used in the Experimental Lakes Area (ELA), Canada, developed abnormally high populations of periphyton.

Thus, the experimental manipulation of whole aquatic and terrestrial ecosystems is the most effective mechanism for assessing the fate of chemicals. Provided the time, the money, the manpower, and, most importantly, the appropriate "system" is available, whole system experiments can provide the opportunity to test the biological effects and responses of chemicals and to apply the scientific method *in situ*.

A long history exists for the experimental manipulation of whole ecosystems. Among limnologists, Juday *et al.* (1938) were among the earliest to manipulate a whole lake experimentally. Over a period of five years, they enriched Weber Lake, Wisconsin, with an assortment of six organic and inorganic fertilizers including phosphate, lime, potash and soybean meal. Suggesting that soybean meal was the

Methods to Assess the Effects of Chemicals on Ecosystems
Edited by R. A. Linthurst, P. Bourdeau, and R. G. Tardiff

most effective fertilizer in terms of increasing phytoplankton standing crop and stimulating fish growth, their results led them to conclude that "organic content is the limiting factor in the plankton production."

Subsequent to the experiments of Juday *et al.* (1938), Hasler *et al.* (1951) added hydrated lime to two lakes in Wisconsin, in order to test whether the transparency of the water would be improved. Based on pretreatment data and also on data from relatively similar reference lakes nearby, Hasler and his co-workers found that light penetration and alkalinity were greatly increased in Cather Lake; however, in Turk Lake, which was limed in a slightly different manner, only a small effect of the treatment was observed. These results clearly demonstrated the need for improved controls during whole-lake manipulations.

A now classic experiment involving a whole-lake manipulation was conducted in the early 1950s. Peter–Paul Lake on the Michigan–Wisconsin border was a small, brown-water lake with two basins connected by a narrow, shallow channel. In 1951, Johnson and Hasler constructed an earthen dam across the narrow part of the lake and separated it into two lakes, Peter Lake and Paul Lake. Peter Lake was treated with hydrated lime ($Ca(OH)_2$) beginning in 1951 and continuing until 1954, while Paul Lake remained as a control or reference lake. The significant feature of this experiment is that it was perhaps the first instance in which a whole ecosystem "control" was part of the experimental design.

As a result of these early experiments employing whole-lake manipulations, many other scientists have been inspired by the "ecosystem" approach for the study of the fate and behaviour of chemicals in the natural environment. The treatment of double-basin lakes as two lakes has been attempted in several other situations, e.g., in Lakes 226 and 302 of the Experimental Lakes Area of Canada (Schindler, 1974, 1975; Schindler and Fee, 1974; Schindler *et al.*, 1980a) and in Little Rock Lake, Wisconsin (Brezonik *et al.*, 1986). In addition, the experimental manipulation of systems other than lakes has become prevalent in the literature. One example is the Reducing Acidification In Norway (RAIN) project in which two pristine catchments were acidified by the addition of H_2SO_4 and H_2SO_4 plus HNO_3, respectively; while at an acidified catchment, ambient acidic precipitation was excluded by means of a "roof" and clean precipitation was added beneath.

In this chapter, some of the techniques currently in use for manipulating whole terrestrial and aquatic systems for the purposes of assessing the fate of chemicals are reviewed. These techniques will be categorized in terms of the system being manipulated; i.e., lakes, streams, and forest catchments. Since the method for manipulating a system is in many ways dependent on the chemical being tested, the discussion for each system will focus on certain categories of chemicals that have been tested historically on whole ecosystems. These include nutrients, acidifying chemicals, metals and organic contaminants. Neutralizing agents will be discussed only briefly, because generally these are added to lakes, streams, and catchments to mitigate the effects of other chemicals (i.e., strong mineral acids), and not to test the effects of the chemical itself.

15.2 METHODS FOR MANIPULATING LAKES

Since the first experiments of Juday *et al.* (1938) on Weber Lake in Wisconsin, dozens of experiments have been conducted using whole-lake manipulations. These experiments can be divided into two categories: (1) those involving only a "single" lake to which the test chemical is added and (2) those involving a "double-basin" lake which has been mechanically separated into two lakes that are then manipulated independently.

The most common approach to manipulate lakes is to add the test chemical to a single lake. The major problem with this type of lake manipulation (as indeed is the case with all whole system manipulations) is in establishing a suitable "control" or "reference." As Likens (1985) has pointed out, in whole system experiments, "reference" is perhaps the more appropriate word to use because the complexity of natural ecosystems precludes the use of an absolute experimental "control."

For single-lake manipulations, two ways exist to establish a control. The first is to use the lake itself as a reference. This approach requires that the lake be studied for one to several years prior to the experiment. Establishing the length of time needed before the treatment begins can be a problem, because determining *a priori* the natural variation in the system is impossible. A long-term (i.e., ongoing) monitoring programme is ideal, provided it is inexpensive and simple. Also, it should include measurements for abiotic and biotic parameters that are sensitive not only to changes in the chemical currently under study, but also to changes in chemicals that might be tested in the future.

Alternately, or in addition to using the lake itself as a reference, other relatively similar lakes can be used as "controls," while the study lake is being manipulated. While this approach has the advantage that potentially fewer years of background data may be required prior to the commencement of the experiment, finding two or more lakes having identical morphometries, water renewal rates, food web structures, etc., is virtually impossible. Ideally, both long-term data should be collected for the study lake prior to its manipulation, and other lakes should be monitored as references during the course of the experiment. Finally, data should continue to be collected even after the experiment has ceased.

An alternate approach to the manipulation of a single lake is to mechanically divide a lake into two or more basins, and then to manipulate each basin individually. Peter–Paul Lake was separated into two lakes by an earthen dam, while at ELA and in Little Rock Lake, Wisconsin, a vinyl curtain reinforced with nylon (ELA) or Dacron (Little Rock Lake) was installed to separate the basins. In ELA Lake 226, the north basin received annual additions of C, N, and P, while the south basin received only C and N (Schindler, 1974, 1975, 1991; Schindler and Fee, 1974; Findlay and Kasian, 1987); and in Lake 302, P, N, and C were added to the hypolimnion of the north basin of the lake, and the south basin was left as a reference (Schindler *et al.*, 1980a). In Little Rock Lake, the north basin was acidified with H_2SO_4, while the south basin remained as a reference.

The separation of a single lake into two provides a natural reference or "control." At ELA Lake 226, a problem developed with leakage of water between the two basins and movement of water over the surface collar (Findlay and Kasian, 1987). Alternately, multibasin lakes that behave as individual lakes could be used. An example is Kennedy Lake, British Columbia, Canada, which is composed of two discrete arms, separated by a very narrow, shallow sill with a depth of <10 m. Stockner and Shortreed (1985) fertilized Clayoquot Arm with P from 1978 to 1984 inclusive, while the Main Arm was fertilized only in 1979 and 1980. Similarly, at ELA, the natural sill between the two basins of Lake 302 was so effective at separating the basins, that installation of the curtain in 1974 had little effect on the chemistry or the phytoplankton of the two basins (Schindler *et al.*, 1980a).

Whatever the approach selected to manipulate the whole-lake system, the method to add the chemical to the lake is dependent on the chemical being tested. A great deal of information is available on the addition of nutrients to lakes, both in North America and Europe. For example, at ELA in Canada, Lake 227 has been fertilized with N and P since 1969; Lake 304 was treated with N, P, and C in 1971 and 1972, with N and C in 1973 and 1974, and with P and N in 1975 and 1976; and Lake 303 was treated during the summer of 1975 and 1976 with N and P, while Lake 230 was treated during the winter of 1974 and 1975 with N and P (Schindler, 1991). In other parts of Canada, P has been added to three subarctic lakes in Schefferville, Quebec (Smith *et al.*, 1984), while in British Columbia, many lakes have been fertilized with both N and P (e.g., Stockner, 1981; Stockner and Shortreed, 1985; Shortreed and Stockner, 1990). In Scandanavia, six lakes in the Telemark region of Norway were treated with P and/or N (Johannessen *et al.*, 1984) to improve fish stocks, while in Sweden N and/or P were added to Lakes Magnusjaure (1974-75), and Gunillajaure (1978–1979) (Jansson, 1978; Holmgren, 1983 as cited by Findlay and Kasian, 1987).

Phosphorus can be added to the lake in several forms. Häkanson *et al.* (1990) used the emissions from fish-cage farms to add P (and other nutrients) to two lakes in Sweden. Sodium phosphate (Na_2HPO_4) was added to Lake 227 in Canada (Schindler *et al.*, 1987) and to Lake Hymenjaure in Sweden (Jansson, 1978) in the first year of study (1969 and 1972, respectively); but in both studies, phosphoric acid (H_3PO_4) was substituted in subsequent years. Phosphoric acid would appear to be the preferred source of P for lake water additions because of its greater solubility (Schindler *et al.*, 1973). If nitrogen is being added to the lake in conjunction with the P, the two nutrients can be added as ammonium phosphate ($(NH_4)_3PO_4$) or as a commercial fertilizer (e.g., Häkanson and Andersson, 1992). Nitrogen alone is added as either ammonium chloride (NH_4Cl) (Schindler, 1975, Schindler *et al.*, 1980a) or, more commonly, as ammonium nitrate (NH_4NO_3) (Schindler, 1975; Jansson, 1978; Stockner, 1981; Johannessen *et al.*, 1984; Stockner and Shortreed, 1985; Shortreed and Stockner, 1990). Carbon was added to ELA Lakes 226, 227, 302N and 304 as sucrose (Schindler, 1975) or to ELA Lake 224 as ^{14}C-labelled $NaHCO_3$ (Hesslein *et al.*, 1980a).

During the ice-free season, the cheapest and simplest method for adding these nutrients (with the exception of the radioisotope) is by pouring the liquid or pre-dissolved fertilizer (in the case of dry chemicals) onto the surface of the lake through the prop-wash of the boat while cris-crossing the lake. To simulate point-source additions, the chemicals can be added using a semi-continuous trickle feed from a large (e.g., 200 L) barrel or drum (Findlay and Kasian, 1987; Levine and Schindler, 1989) situated on a raft in the middle of the lake or on the shore. If the funds are available, the fertilizer can be spread over the lake from an airplane (Stockner and Shortreed, 1985). During the winter, nutrients can be dispensed through a hole in the ice or through the lake inflows (Smith *et al.* 1984). However, often nutrients are not added during the winter months because productivity is low at that time of year.

While nutrients are loaded generally onto the surface or epilimnetic waters of the lake, Schindler *et al.* (1980a) injected P, N, and C into the hypolimnion of the north basin of Lake 302 to test the hypothesis that the phosphorus would be permanently transferred to the sediments before it could reach the surface waters and cause algal blooms.

In the hypolimnetic experiment of Schindler *et al.* (1980a), as in many other whole-lake fertilization experiments, the nutrient additions commonly are made weekly during the ice-free season. Loadings can be uniform throughout the season (Schindler *et al.* 1973), or they can be adjusted according to the P, N, or C concentration in the lake or the outflow discharge. If an immediate or dramatic increase in concentration is required, a large pulse of fertilizer is added at the beginning of the ice-free season. For example, after ice-out in June 1979, Smith *et al.* (1984) added 8 kg of P as H_3PO_4 to the surface of Lejeune Lake in order to double the pre-treatment concentration of total P in the lake water. Thereafter, the lake was loaded at a rate of 17.6 kg P/yr at weekly intervals.

A final point concerns the importance of pre-treatment, post-treatment, and "during" treatment sampling. The sampling regime, including the frequency of collection and the location of the sampling station(s), and the methodologies involved for both the chemical and the biological analyses must be well established prior to the commencement of the experiment, otherwise many years and much money will be wasted.

Good sampling techniques are important, not only for whole-lake manipulations involving nutrients, but also for those experiments in which other chemicals are added to the lake system. For chemicals such as sulphuric and nitric acid, some information is available on whole-lake manipulations as a result of the heightened concern over the effects of acidic precipitation. Pioneering work in this area was conducted at the ELA in Canada, where experimental acidification of Lake 223 (in the form of H_2SO_4) was conducted beginning in 1976, acidification of Lake 114 (in the form of H_2SO_4) was conducted from 1979 to 1986 (Al_2SO_4 was added in 1984), and acidification of Lake 302 (in the form of H_2SO_4 in one basin, HNO_3 in the other) had been carried out from 1982 (Schindler 1991). At Little Rock Lake, Wisconsin, the north basin was acidified with H_2SO_4 from 1985 to 1990, while the south basin remained as a reference (Sampson *et al.* 1993).

Similar to the methods employed for nutrient additions, in each lake the acid was added by slowly pouring the concentrate from a moving boat, into the prop-wash of an outboard motor. Physical mixing studies suggest that this method is sufficient to mix the acid into the epilimnion within a few hours after addition. Furthermore, despite the high density of the acid, no evidence of it sinking through the thermocline to the bottom of the lake was found (Schindler *et al.*, 1980b).

As with nutrient additions, the acidification regime for the ELA lakes was designed to reduce the pH to a predetermined value early in the ice-free season, and then to hold it at that value until the following spring. Thus, large quantities of acid were added to the lakes early in the season, followed by weekly loadings. This regime is analogous to a large pulse of acid entering the lake during spring snowmelt, followed by episodic additions from summer rains.

When acid is added to lakes, it must be relatively free of impurities such as heavy metals, otherwise, the effects of the metal contaminants might be confused with the effects of the acid. When metals have been tested on whole-lake systems, radioisotopes have been used with some success. For example, the gamma-emitting isotopes, ^{75}Se, ^{203}Hg, ^{134}Cs, ^{59}Fe, ^{65}Zn, and ^{60}Co were added to ELA Lakes 224 and 226 (northeast and southwest basin) (Hesslein *et al.*, 1980b; Hesslein, 1987) and the alpha-emitting isotope ^{226}Ra was added to Lakes 224, 226, 227, and 261 (Emerson and Hesslein, 1973; Emerson *et al.*, 1973; Hesslein and Slavicek, 1984). The isotopes were dispensed as either chloride salts or nitrate salts (^{203}Hg only) with the exception of ^{75}Se which was supplied as Na_2SeO_3. In all studies, the radionuclides were combined in a metal drum containing 10 L of water (per lake or lake basin). The drum was then mounted on a raft and towed around the lake for a period of about one hour while the contents drained out. The barrel was then refilled with water, and the operation was repeated. This procedure is the same used to dispense 14°C in the nutrient addition experiment conducted on Lake 224 (Hesslein *et al.*, 1980a).

Unlike the whole-lake manipulations involving nutrients and acids, no weekly additions were made of the metal radiotracers in the experiments discussed above. Malley *et al.* (1989) conducted an experiment in which both stable Cd (as $CdCl_2$) and the radiotracer ^{109}Cd were added to the epilimnion of Lake 382 in 33 weekly additions during the period 23 June to 29 October 1987. The large cost of radioisotopes and the hazard involved in using both radioisotopes and metals precludes their widespread use in many other lakes. Nonetheless these types of whole-lake experiments provide the opportunity to monitor the pathway of metals from the water to the biotic and the abiotic compartments.

Häkanson *et al.* (1990) and Häkanson (1991) added Se in conjunction with lime to five lakes in Sweden. The Se was added either using mixed lime, or it was encapsulated in rubber tubes placed into nets from which the selenium was successively released. However, the purpose of their work was to reduce Hg concentrations in fish by means of precipitating the Hg from the lake water as HgSe, and not to examine the behaviour of the Se. Similarly, radionuclides other than metals have been added to some ELA lakes (e.g., ^{3}H was added to Lakes 227 and 224; Hesslein, 1980; Quay *et al.*, 1980, respectively) to determine diffusion coefficients and not to evaluate the effect of the chemical itself.

15.3 METHODS TO MANIPULATE STREAMS

Experiments conducted on streams are somewhat different from those conducted on lakes, because manipulation of an entire stream is virtually impossible unless the chemical is added at the source of each tributary. Consequently, the chemical being tested is dispensed usually at a certain point in the stream, and its effects are noted downstream. Thus, the control or reference site for the experiment must be upstream of the addition. Furthermore, minimal variability (in terms of mixing characteristics, substrate type, riffle-pool sequences, etc.) must exist between the reference site and the experimental section of the study site. As with lakes, the additions of the chemical can be made over a short (i.e., less than 24 hours) or over a long period of time.

Probably the greatest number of studies on stream manipulations have involved the experimental acidification of streams. Study sites have been located in New Hampshire (Hall *et al.*, 1980, 1987; Hall and Likens, 1981, 1984), Maine (Norton *et al.*, 1992) and Colorado (McKnight and Bencala, 1989) in the United States, in Wales (Ormerod *et al.*, 1987), and in Norway (Henriksen *et al.*, 1984, 1988; Norton *et al.*, 1987; Wright *et al.*, 1988b). In many cases, the acid is added as H_2SO_4 (Hall and Likens, 1981, 1984; Norton *et al.*, 1987; McKnight and Bencala, 1989), although Hall *et al.* (1980, 1987) used HCl because the concentration of Cl^+ was low in both the biologic and geologic material at their study site and because Cl^+ is a good chemical tracer of groundwater movement (Hall *et al.*, 1987). For similar reasons, McKnight and Bencala (1989) injected LiCl (in conjunction with H_2SO_4) as a conservative tracer in their experimental stream.

Hall *et al.* (1987) added $AlCl_3$ instead of HCl in some of their experiments to compare neutralization mechanisms by the stream during acidification by a weak ($AlCl_3$) and a strong (HCl) acid, and also to produce Al levels representative of those that occur after experimental deforestation of catchments in the Hubbard Brook Watershed.

Most of the stream acidification experiments mentioned above were relatively short term (i.e., the addition of the acid lasted less than 24 hours). First-, second-, and third-order streams were manipulated, with the size of the experimental sections (i.e., the distance to the downstream sites) ranging between about 50 and 200 m (McKnight and Bencala, 1989; Norton *et al.*, 1992; Hall *et al.*, 1980, 1987; Hall and Likens, 1981, 1984). The injection rate, the concentration of acid used and the duration of the input of acid to the streams were determined for each experiment according to the flow rate of the stream and the decrease in pH required. For example, McKnight and Bencala (1989) injected a 7.25 mol/L H_2SO_4 solution into a headwater stream in Colorado at a rate of 950 mL/min for a period of 3 hours to achieve a decrease in pH from about pH 4 to about pH 3.

Samples must be collected as soon as possible and as frequently as possible at both the upstream (i.e., reference) site and at the downstream sites. To know *a priori* when steady state might be expected to occur and also to avoid unnecessary sampling, travel or mixing times and distances in these short-term experiments should be determined using a dye such as rhodamine WT, rhodamine B, or fluorescein, despite the possibility that the dyes might cause an effect independently.

For acidification experiments requiring that the pH be depressed for long period of time, frequent measurements of pH downstream of the injection site must be made so that the flow rate of the acid into the stream can be modified to account for changes in stream discharge. Hall *et al.* (1980) found that when the discharge was variable, monitoring of the pH was needed at short time intervals (i.e., every five min) to maintain a constant pH in the experimental section of Norris Brook (in the Hubbard Brook Watershed, New Hampshire, US). However, when the discharge was relatively constant, monitor pH was needed only at 6- to 8-hour intervals. They maintained approximately pH 4 in Norris Brook for a period of five months in 1977 by manually adding dilute (0.05–1 N) H_2SO_4 from a carboy, and modifying the drip rate of the acid into the stream with a Teflon stem needle valve in borosilicate.

The effects of pesticides in streams also have been studied by direct additions. For example, the effects of methoxychlor (1,1,1-trichloro-2,2-bis-(*p*-methoxyphenyl) ethane), a replacement for DDT, were investigated in the Coweeta Basin, North Carolina (Wallace *et al.*, 1987, 1989), where two first-order streams were treated with methoxychlor on nine (Wallace *et al.*, 1989) and four (Wallace *et al.*, 1987) separate occasions, and also in the province of Quebec, Canada (Wallace and Hynes, 1975). Wallace *et al.* (1989) used two hand sprayers to apply a 25 percent emulsifiable concentrate of methoxychlor for a 2- to 4-hour period at a rate of 10 mg/L (based on discharge). Wallace and Hynes (1975) compared two methods for stream manipulations. They used a Piper Pawnee 235 fixed-wing aircraft equipped with a standard boom and nozzle spray rig to spread a 15 percent methoxychlor solution from an altitude of 45 m to stream M-26, whereas stream M-11 was treated from the ground at a calculated rate of 0.075 mg/L for 15 min.

Regarding nutrient manipulations, some work has been conducted on streams located in the Walker Branch Watershed, Tennessee (Elwood *et al.*, 1981), and also in the Hubbard Brook Experimental Forest, New Hampshire (Meyer, 1979). Elwood *et al.* (1981) conducted a long-term (95 day) P-addition experiment in which they divided a 340 m reach of Walker Branch (a second-order woodland stream) into three sections: a 70 m upstream section was kept as a control (<10 µg/L PO_4–P), a 150 m section was continuously enriched to 100 µg/L PO_4–P, and a 200 m section downstream was enriched to 1000 µg/L PO_4–P. Phosphorous solutions were prepared weekly in 2 × 200 L drums containing H_3PO_4 mixed with stream water that was siphoned through surgical tubing at a rate of ~20 mL/min to the head of each stream section and introduced at points where rapid mixing occurred. As with the acidification experiments discussed above, the phosphorus concentration in the drums (~60 and 450 μg PO_4–P/L) was based on the discharge of the stream. Another similarity with the acidification experiments was the addition of a conservative tracer by Meyer (1979) who added NaCl in conjunction with the P (as KH_2PO_4) to Bear Brook, New Hampshire, over a period of 27 hours. NaCl was added as an inert tracer to monitor changes in phosphorus concentration due to dilution over the reach.

As in lake manipulations, the importance of a rigorous sampling regime cannot be over emphasized. Adequate abiotic and biotic samples must be collected before, during, and after the treatment at both the reference and the downstream sites so that the effect of the chemical can be properly evaluated.

15.4 METHODS FOR MANIPULATING CATCHMENTS AND FORESTS

Two common approaches exist to manipulate whole catchments and forests. In "addition" experiments, the test chemical is added to the system (similar to whole-lake and stream manipulations), whereas in "exclusion" experiments the chemical, in conjunction with other nutrients and contaminants, is excluded from entering the system.

Many whole-catchment and forest manipulation studies have dealt with the addition and/or exclusion of acids. Most notable among these is the RAIN project conducted in Norway (Wright, 1985; Wright and Gjessing, 1986; Wright *et al.*, 1986, 1988a). The project comprised two parallel large-scale experimental manipulations, representing both addition experiments (Sogndal) and exclusion (Risdalsheia) experiments. At Sogndal, four pristine headwater catchments were selected; two of the sites acted as controls (SOG1 and SOG3), a third site was acidified with H_2SO_4 (SOG2), and the fourth site was acidified with a 1:1 mixture of H_2SO_4 + HNO_3 (SOG4). Acid addition, which began in April 1984, consisted of application to the snowpack of 0.02 mm of water at pH 1.9 and four or five events of 11 mm at pH 3.2 during the snow-free months. Acid was mixed with lakewater from SOG1 and applied at a rate of 2 mm/hour using commercial irrigation equipment. Before and after each acid addition, 2 mm of unacidified lakewater were added to "wet" and "wash" down the vegetation, respectively (Wright *et al.*, 1988a). In conjunction with this RAIN project experiment, a similar methodology was employed by Wright *et al.* (1988b) to acidify a Sogndal catchment with sea salt, except that seawater instead of H_2SO_4 was mixed with the lakewater from SOG1.

At Risdalsheia, an acidified area in southern Norway, three natural headwater catchments were selected; a site (ROLF) acted as a control and two sites (KIM and EGIL) were covered with transparent roofs. At KIM and EGIL, precipitation was collected by means of a gutter and cistern system. At KIM, the water was pumped through a filter and ion exchange system, seawater was added to increase the concentration of salts to ambient levels, and then the water was pumped back out to a sprinkler system mounted beneath the roof. The system at the EGIL catchment was similar to that at KIM except that the water was not treated; rather, it was recycled back beneath the roof. Both catchments were watered at a rate of 2 mm/hour. During winter the systems were shut down. In 1985, artificial "acid" snow was made using commercial snow-making equipment; from 1986 onwards, ambient snow was used and added beneath the roofs with a snowblower.

Other whole-catchment experiments involving acid additions have been conducted in two subcatchments of Gärdsjön Lake, as part of the Gärdsjön Project in Sweden (Hultberg and Grennfelt, 1986), in Bear Brook Watershed, Maine (Norton *et al.*, in press) as part of the Watershed Manipulation Project (WMP) in the United States, and also in Hoglwald, Germany (Rasmussen *et al.* 1992), as part of the Experimental Manipulation of Forest Ecosystems Project (EXMAN) (Rasmussen *et al.*, 1992) in Europe. At Gärdsjön, catchment L1 received 90 kg S/ha in October 1985 and 108 kg S/ha in October 1986 (as Na_2SO_4) while catchment F5 received 112 kg S/ha (as elemental S) each year. At Bear Brook, dry $(NH_4)_2SO_4$ (1880 eq/ha yr) was applied by helicopter to the West Bear catchment while the East Bear catchment remained as a reference. At Hoglwald, the catchment is being irrigated with acid in the form of H_2SO_4.

Irrigation was also the method used by Bayley *et al.* (1987) to experimentally acidify a fen located in ELA. The fen (MIRE 239) was irrigated with a 1:1 mixture of H_2SO_4 and HNO_3 using a pipe distribution network to deliver water to 160 sprinklers, an irrigation pump to supply the water, a hydrant to regulate delivery of the water, an in-line meter to monitor pumping rates, and pressure gauges on the pump and on the distribution lines. Acid was added to the water in the experimental part of the bog, resulting in a spray with a pH of about 3.0. Irrigation lasted four to five hours, followed by 20–30 min of rinse at higher pH (5–7).

Evidently extensive, intensive, and expensive sampling networks are necessary to implement and maintain a whole-catchment manipulation study. At least one year (e.g., the RAIN project) and preferably several years (Norton *et al.*, in press) of baseline data are required before the manipulation can begin. This situation has resulted in the development of several co-operative programmes in Europe, directed towards the assessment and prediction of the impact of environmental change on whole-catchment systems. ENCORE (European Network of Catchments Organised for Research on Ecosystems) (Hornung, 1992) comprises a network of 18 sites and ~40 catchments in seven countries. As part of the baseline programme, background environmental data (e.g., vegetation, soil type) and input and output flux information are being collected at each site using standard protocols. Furthermore, a more intensive process or mechanistic study or a whole catchment manipulation experiment must be performed at each site. Past and future ENCORE whole catchment manipulations include acidification and de-acidification experiments (i.e., the RAIN project), the addition of neutral salts and fertilizers, and also liming experiments.

In addition to ENCORE, other co-operative European programmes using whole catchment manipulations include:

1. EXMAN in which two manipulations (i.e., fertilization, liming, irrigation, acidification or roof construction) must be performed at each site;
2. NITREX (NITRogen saturation EXperiment) which involves nine separate large-scale nitrogen addition or exclusion experiments (Wright *et al.*, 1992); and
3. CLIMEX (CLIMatic change EXperiment) which is a proposed project to experimentally enrich with CO_2 and raise the temperature at two entire forested headwater ecosystems (Jenkins *et al.*, 1992).

These projects are beyond the scope of this chapter; therefore, the reader is referred to Teller *et al.* (1992) and the references therein for further information.

15.5 METHODS TO ADD NEUTRALIZING AGENTS TO WHOLE AQUATIC AND TERRESTRIAL SYSTEMS

While the addition of lime or limestone ($Ca(OH)_2$ or $CaCO_3$) has long been practised as a means to increase lake productivity (Hasler *et al.*, 1951; Stross and Hassler, 1960; Stross *et al.*, 1961), the addition of lime and other neutralizing agents as a mitigation technique for restoring or protecting biota in acidified waters has become widespread only in recent years. Lake neutralization has been practised on a large scale in Sweden (Wilander and Ahl, 1972; Svendrup and Bjerle, 1983; Lindmark, 1982, 1984; Hultberg and Andersson, 1982; Hultberg and Grennfelt, 1986; Alenås *et al.*, 1991) and Norway (Wright, 1985; Hindar and Rosseland, 1988), and to a lesser extent in other European countries such as Italy (Calderoni *et al.*, 1991) and the United States (Young *et al.*, 1989; Porcella, 1989, 1991; Bukaveckas and Driscoll, 1991; McAvoy and Driscoll, 1989), and Canada (Yan and Dillon, 1984; Molot *et al.*, 1986).

While several agents have been used to neutralize lakes (Grahn and Hultberg, 1975), lime has been widely adopted, because it is readily available in different size grades, is safe to handle, and is relatively inexpensive. Although pure calcite ($CaCO_3$) may be preferred because it does not cause the pH in the lake to increase beyond the equilibrium value upon dissolution as does $Ca(OH)_2$ and CaO (Molot, *et al.*, 1986), limestone, dolomite, lime slags, and olivine all have been used with reasonable success (Bengsston *et al.*, 1980).

The disadvantage of using certain calcite-based materials is that they are not readily soluble. Some researchers (Svendrup and Bjerle, 1983) have suggested that if the $CaCO_3$ reaches the sediments after application to the lake's surface, it may form metal and humic complexes (especially in humic lakes) that decrease the dissolution rate of the calcite and render it ineffective. Consequently, they suggest that, when applying the lime, measures should be taken to maximize the dissolution of the sinking calcite and thus minimize losses to the sediments. These include using a small particle size, dispersing the lime over as large an area as possible in proportion to lake volume, and applying the lime as a slurry to separate the particles (Svendrup and Bjerle, 1983; Molot *et al.*, 1986). If the latter method is used, a small amount of surfactant (~0.15–0.20 percent by weight) such as sodium polyacrylate (Molot *et al.*, 1986; Driscoll *et al.*, 1989) is added to facilitate the suspension of the calcite and ultimately promote its wetting and dispersion in the water column.

An alternative approach is to apply a larger particle size of lime to the lake surface (McAvoy and Driscoll, 1989) in the hope that it will reach the sediments and release a slow diffusive Ca^{2+} flux across the sediment–water interface. Gubala and Driscoll (1991) compared the effectiveness of a water column application of $CaCO_3$ (i.e., mean particle size = 2 μm) to a water column–sediment application of $CaCO_3$ (i.e., a 1–2 mixture of 6–44 μm and 40–400 μm $CaCO_3$) in Woods Lake, New York. They found that while the water column–sediment application involved a 50 percent greater dose of calcite than the water treatment alone, both treatments appeared to have a similar effect in terms of the net amount of runoff and acidic inputs that were neutralized.

The method to apply the lime directly onto the lake's surface can be either from a pontoon boat equipped with high-pressure tanks (Hasselrot and Hultberg, 1984; Calderoni *et al.*, 1991) or manually from a small boat (Yan and Dillon, 1984; Alenås *et al.*, 1991), although, recently, a boat has been developed in Sweden specifically for the purpose of spreading lime (*Water/Engineering and Management*, March 1992, p. 14). In remote and relatively inaccessible areas or if a large number of lakes are to be treated, helicopters (Porcella, 1989; Bukaveckas and Driscoll, 1991; Gubala and Driscoll, 1991; Häkanson and Andersson, 1992) or other aircraft (Molot *et al.*, 1986) can be used. In addition, lime has been put onto the ice of lakes (Wilander and Ahl, 1972; Wright, 1985; Häkanson and Andersson, 1992) and onto the snow along the edge of the lake (Wright, 1985).

The successful neutralization of lake acidity by the addition of lime directly to the lake's surface requires a knowledge of the goals for the treatment (e.g., the final acid neutralizing capacity, ANC, of the water or the depth in the sediments to which neutralization should occur), the type of lime used, the method of application, the initial water column and sediment acidity, and the physical features of the lake including surface area, water depth, temperature regime, and hydrodynamics. While the amount of lime or the frequency of application required to achieve a desired ANC in Sweden cannot be stated with absolute precision, between 10 and 30 g of lime (as $CaCO_3$)/m^3 are needed to neutralize typical acid lake water (Bengsston *et al.*, 1980; Hasselrot and Hultberg, 1984).

In addition to applying lime directly onto the lake's surface, lime has been added to streams, for example in Canada (Keller and Gunn, 1982), in Sweden (Hasselrot and Hultberg, 1984), and in Norway (Abrahamsen and Matzow, 1984). An alternative approach to liming the lake directly is to apply the lime to the entire catchment of the lake or to the wetland areas within the catchment. For example, in June 1984, 1500 kg of finely ground dolomite (0–0.2 mm) was spread onto catchment F2 as part of the Gärdsjön Project in Sweden (Hultberg and Grennfelt 1986). In the "liming–mercury–cesium" project conducted in Sweden between 1986 and 1989, several types of lime or potash or Se or fertilizers were added to many lakes, wetlands, and catchments in Sweden in an attempt to reduce the Hg and ^{137}Cs concentrations in the fish (Häkanson *et al.*, 1990; Häkanson, 1991; Häkanson and Andersson, 1992). Fertilizers were added in conjunction with the lime to help mitigate the effects of the lake acidity, a procedure that has been employed by others (Yan and Lafrance, 1984).

While the magnitude of this project precludes a complete discussion here, Häkanson and his co-workers have provided a comprehensive cost-benefit analysis of these remedial measures, and they have compared the advantages and disadvantages of whole-lake versus wetland versus catchment lime additions. Häkanson and Andersson (1992) point out that wetland and catchment liming is superior to lake liming for several reasons: (1) the effect is prolonged, (2) the potential "liming shock" in the lake is avoided, (3) the biota in the streams and rivers in the catchment also benefit, and (4) the transport of metals such as Fe and Al is reduced into the lake from the catchment.

Yet, in certain types of lakes, such as humic lakes and those have a short retention time, liming has a limited effect. Therefore, Lindmark (1982, 1984) proposed an alternate approach for neutralizing lakes. The CONTRACID method, which is based on the cation exchange properties of the lake sediment, involves injecting a sodium carbonate (soda ash) solution into the sediment (by a harrow) so that the sediment becomes sodium stocked. During acidification, a reverse ion-exchange process occurs (i.e., the Na in the sediments is replaced by H^+). This process provides a long-lasting neutralizing capacity in addition to biological stimulation (from P release) that may be preferable to frequent liming. Unfortunately, no consensus is apparent in the literature as to the efficacy of this technique.

15.6 SUMMARY

The methods to manipulate whole aquatic and terrestrial ecosystems are as diffuse and abundant as the number of studies involving them. Common among all whole-lake, stream, and catchment manipulations is the necessity for good quality data before treatment, during treatment, and also post-treatment so that the effects of the chemical being tested can be accurately assessed.

While whole system manipulations provide an excellent opportunity to apply the scientific method *in situ*, co-operative programmes both among different agencies within a country and also among different countries are essential if future whole system manipulations are to be carried out to any great extent.

15.7 REFERENCES

Abrahamsen, H., and Matzow, D. (1984) Use of lime slurry for deacidification of running water. *Verh. Internat. Verein. Limnol.* **22**, 121–125.

Alenås, B.I., Andersson, H., Hultberg, A., and Rosemarin, A. (1991) Liming and reacidification reactions of a forest lake ecosystem, Lake Lysevatten, in Sweden. *Water Air Soil Pollut.* **59**, 55–77.

Bayley, S.E., Vitt, D.H., Newbury, R.W., Beaty, K.G., Behr, R., and Miller, C. (1987) Experimental acidification of a *Sphagnum*-dominated peatland: first year results. *Can. J. Fish. Aquat. Sci.* **44** (Suppl. 1), 194–205.

Bengsston, B., Dickson, W., and Nyberg, P. (1980) Liming acid lakes in Sweden. *Ambio* **9**(1), 34–36.

Brezonik, P.L., Baker, L.A., Eaton, J.R., Frost, T.M., Garrison, P., Kratz, T.K., Magnuson, J.J., Ross, W.J., Shephard, B.K., Swenson, W.A., Watras, C.J., and Webster, K.E. (1986) Experimental acidification of Little Rock Lake, Wisconsin. *Water Air Soil Pollut.* **31**, 115–121.

Bukaveckas, P.A., and Driscoll, C.T. (1991) Effects of whole-lake base addition on thermal stratification in three Adirondack lakes. *Water Air Soil Pollut.* **59**, 23–29.

Calderoni, A., Mosello, R., and Quirci, A. (1991) Chemical response of Lake Orta (Northern Italy) to liming. *Arch. Hydrobiol.* **122**, 421–439.

Driscoll, C.T., Fordham, G.F., Ayling, W.A., and Oliver, L.M. (1989) Short-term changes in the chemistry of trace metals following calcium carbonate treatment of acidic lakes. *Can. J. Fish. Aquat. Sci.* **46**, 249–257.

Elwood, J.W., Newbold, J.D., Trimble, A.F., and Stark, R.W. (1981) The limiting role of phosphorus in a woodland stream ecosystem: effects of P enrichment on leaf decomposition and primary producers. *Ecology* **62**(1), 146–158.

Emerson, S., and Hesslein, R. (1973) Distribution and uptake of artificially introduced radium-226 in a small lake. *J. Fish. Res. Board Can.* **30**, 1485–1490.

Emerson, S., Broecker, W., and Schindler, D.W. (1973) Gas-exchange rates in a small lake as determined by the radon method. *J. Fish. Res. Board Can.* **30**, 1475–1484.

Findlay, D.L., and Kasian, S.E.M. (1987) Phytoplankton community responses to nutrient addition in Lake 226, Experimental Lakes Area, northwestern Ontario. *Can. J. Fish. Aquat. Sci.* **44** (Suppl. 1), 35–46.

Grahn, O., and Hultberg, H. (1975) The neutralizing capacity of 12 different lime products used for pH adjustment of acid water. *Vatten* **2**, 120–132.

Gubula, C.P., and Driscoll, C.T. (1991) The chemical responses of acidic Woods Lake, NY to two different treatments with calcium carbonate. *Water Air Soil Pollut.* **59**, 7–22.

Häkanson, L. (1991) Measures to reduce mercury in fish. *Water Air Soil Pollut.* **55**, 193–216.

Häkanson, L., and Andersson, T. (1992) Remedial measures against radioactive caesium in Swedish lake fish after Chernobyl. *Aquat. Sci.* **54**, 141–164.

Häkanson, L., Andersson, P., Andersson, T., Bengsston, Å., Grahn, P., Johansson, J.-Å., Jönsson, C.P., Kvarnäs, H., Lindgren, G., and Nilsson, Å. (1990) *Åtgärder mot Höga Kvicksilverhalter i Insjöfisk.* Statens Naturvårdsverket Rapport 3818. 189 pp.

Hall, R.J., and Likens, G.E. (1981) Chemical flux in an acid-stressed stream. *Nature* **292**(5821), 329–331.

Hall, R.J., and Likens, G.E. (1984) Effect of discharge rate on biotic and abiotic chemical flux in an acidified stream. *Can. J. Fish. Aquat. Sci.* **41**, 1132–1138.

Hall, R.J., Likens, G.E., Fiance, S.B., and Hendrey, G.R. (1980) Experimental acidification of a stream in the Hubbard Brook Experimental Forest, New Hampshire. *Ecology* **61**, 976–989.

Hall, R.J., Driscoll, C.T., and Likens, G.E. (1987) Importance of hydrogen ions and aluminium in regulating the structure and function of stream ecosystems: an experimental test. *Freshwater Biol.* **18**, 17–43.

Hasler, A.D., Brynildson, O.M., and Helm, W.T. (1951) Improving conditions for fish in brown-water lakes by alkalization. *J. Wildl. Manag.* **15**, 347–352.

Hasselrot, B., and Hultberg, H. (1984) Liming of acidified Swedish lakes and streams and its consequences for aquatic systems. *Fisheries* **9**, 4–9.

Henriksen, A., Skagheim, O.K., and Rosseland, B.O. (1984) Episodic changes in pH and aluminum-speciation kill fish in a Norwegian salmon river. *Vatten* **40**, 255–260.

Henriksen, A., Wathne, B.M., Rogeberg, E.J.S., Norton, S.A., and Brakke, D.F. (1988) The role of stream substrates in aluminum mobility and acid neutralization. *Water Res.* **22**, 1069–1073.

Hesslein, R.H. (1980) *In situ* measurements of pore water diffusion coefficients using tritiated water. *Can. J. Fish. Aquat. Sci.* **37**, 545–551.

Hesslein, R.H. (1987) Whole-lake metal radiotracer movement in fertilized lake basins. *Can. J. Fish. Aquat. Sci.* **44**(Suppl. 1), 74–82.

Hesslein, R.H., and Slavicek, E. (1984) Geochemical pathways and biological uptake of radium in small Canadian Shield Lakes. *Can. J. Fish. Aquat. Sci.* **41**, 459–468.

Hesslein, R.H., Broecker, W.S., Quay, P.D., and Schindler, D.W. (1980a) Whole-lake radiocarbon experiment in an oligotrophic lake at the Experimental Lakes Area, northwestern Ontario. *Can. J. Fish. Aquat. Sci.* **37**, 454–463.

Hesslein, R.H., Broecker, W.S., and Schindler, D.W. (1980b) Fates of metal radiotracers added to a whole lake: sediment–water interactions. *Can. J. Fish. Aquat. Sci.* **37**, 378–386.

Hindar, A., and Rosseland, B.O. (1988) Liming acidic waters in Norway: National policy and research and development. *Water Air Soil Pollut.* **41**, 17–24.

Hornung, M. (1992) The European Network of Catchments Organised for Research on Ecosystems (ENCORE). In: Teller, A., Mathy, P., and Jeffers, J.N.R. (Eds.) *Responses of Forest Ecosystems to Environmental Changes*, pp 315–324. Elsevier Applied Science, London, 1009 pp.

Hultberg, H., and Andersson, I.B. (1982) Liming of acidified lakes, induced long-term changes. *Water Air Soil Pollut.* **18**, 311–331.

Hultberg, H., and Grennfelt, P. (1986) Gärdsjön Project: Lake acidification, chemistry in catchment runoff, lake liming and microcatchment manipulations. *Water Air Soil Pollut.* **30**, 31–46.

Jansson, M. (1978) Experimental lake fertilization in the Kuokkel area, northern Sweden: budget calculations and the fate of nutrients. *Verh. Internat. Verein. Limnol.* **20**, 857–862.

Jenkins, A., Schulze, D., van Breemen, N., Woodward, F.I., and Wright, R.F. (1992) CLIMEX—CLIMate change EXperiment. In: Teller, A., Mathy, P., and Jeffers, J.N.R. (Eds.) *Responses of Forest Ecosystems to Environmental Changes*, pp. 359–366. Elsevier Applied Science, London, 1009 pp.

Johannessen, M., Lande, A., and Rognerud, S. (1984) Fertilization of 6 small mountain lakes in Telemark, Southern Norway. *Verh. Internat. Verein. Limnol.* **22**, 673–678.

Juday, C., Schloemer, C.L., and Livingston, C. (1938) Effect of fertilization on plankton production and on fish growth in a Wisconsin lake. *Prog. Fish Culturist.* **40**, 24–27.

Keller, W., and Gunn, J.M. (1982) *Experimental Neutralization of a Small, Seasonally Acidic Stream Using Crushed Limestone*. APIOS Report No. 004/82, 24 pp.

Levine, S.N., and Schindler, D.W. (1989) Phosphorus, nitrogen, and carbon dynamics of Experimental Lake 303 during recovery from eutrophication. *Can. J. Fish. Aquat. Sci.* **46**, 2–10.

Likens, G.E. (1985) An experimental approach for the study of ecosystems. *J. Ecol.* **73**, 381–396.

Lindmark, G. (1982) Acidified lakes: sediment treatment with sodium carbonate—a remedy? *Hydrobiol.* **92**, 537–547.

Lindmark, G. (1984) Acidified lakes: ecosystem response following sediment treatment with sodium carbonate. *Verh. Internat. Verein. Limnol.* **22**, 772–779.

McAvoy, D.C., and Driscoll, C.T. (1989) The chemical response following base application to Little Simon Pond, New York State. *J. Water Pollut. Control Fed.* **61**, 1552–1563.

McKnight, D.M., and Bencala, K.E. (1989) Reactive iron transport in an acidic mountain stream in Summit County, Colorado: a hydrologic perspective. *Geochim. Cosmochim. Acta* **53**, 2225–2234.

Malley, D.F., Chang, P.S.S., and Hesslein, R.H. (1989) Whole lake addition of cadmium-109: Radiotracer accumulation in the mussel population in the first season. *Sci. Total Envir.* **87/88**, 397–417.

Molot, L.A., Hamilton, J.G., and Booth, G.M. (1986) Neutralization of acidic lakes: short-term dissolution of dry and slurried calcite. *Water Res.* **20**(6), 757–761.

Meyer, J.L. (1979) The role of sediments and bryophytes in phosphorus dynamics in a headwater stream ecosystem. *Limnol. Oceanogr.* **24**(2), 365–375.

Norton, S.A., Henriksen, A., Wathne, B.M., and Veidel, A. (1987) Aluminum dynamics in response to experimental additions of acid to a small Norwegian stream. In: *UNESCO/IHP-III Symposium on Acidification and Water Pathways, May 1987*, pp. 249–258. Norwegian National Committee for Hydrology, Bolkesjo, Norway.

Norton, S.A., Brownlee, J.C., and Kahl, J.S. (1992) Artificial acidification of a non-acidic and an acidic headwater stream in Maine, USA. *Envir. Pollut.* **77**, 123–128.

Norton, S.A., Kahl, J.S., Fernandez, I.J., Rustad, L.E., Scofield, J.P., Haines, T.A., and Lee, J.J. (in press) The Watershed Manipulation Project: two-year results at the Bear Brook watershed in Maine (BBWM). In: *Proceedings of an International Symposium on Experimental Manipulations of Biota and Biogeochemical Cycling in Ecosystems*. Commission of the European Communities, Copenhagen.

Ormerod, S.J., Boole, P., McCahon, C.P., Weatherley, N.S., Pascoe, D., and Edwards, R.W. (1987) Short-term experimental acidification of a Welsh stream: comparing the biological effects of hydrogen ions and aluminum. *Freshwater Biol.* **17**, 341–356.

Porcella, D.B. (1989) Lake Acidification Mitigation Project (LAMP): An overview of an ecosystem perturbation experiment. *Can. J. Fish. Aquat. Sci.* **46**, 246–248.

Porcella, D.B. (1991) Ecological effects of repeated treatment of lakes with limestone: An overview. *Water Air Soil Pollut.* **59**, 3–6.

Quay, P.D., Broecker, W.S., Hesslein, R.H., and Schindler, D.W.(1980) Vertical diffusion rates determined by tritium tracer experiments in the thermocline and hypolimnion of two lakes. *Limnol. Oceanogr.* **25**(2), 201–218.

Rasmussen, D.W., Beier, L.C., de Visser, P., van Breeman, N., Kreutzer, K., Schierl, R., Bredemeier, M., Raben, G., and Farrell, E.P. (1992) The "EXMAN" Project—EXperimental MANipulations of forest ecosystems. In: Teller, A., Mathy, P., and Jeffers, J.N.R. (Eds.) *Responses of Forest Ecosystems to Environmental Changes*, pp.

325–334. Elsevier Applied Science, London, 1009 pp.

Sampson, C.J., Brezonik, P.L., and Weir, E.P. (1993) Effects of acidification on chemical composition and chemical cycles in a seepage lake: Inferences from a whole-lake experiment. In: Baker, L.A. (Ed.) *Environmental Chemistry of Lakes and Reservoirs*. ACS Advances in Chemistry No. 237. American Chemical Society, Washington, D.C.

Schindler, D.W. (1974) Eutrophication and recovery in Experimental Lakes: Implications for lake management. *Science* **184**, 897–899.

Schindler, D.W. (1975) Whole-lake eutrophication experiments with phosphorus, nitrogen and carbon. *Verh. Internat. Verein. Limnol.* **19**, 3221–3231.

Schindler, D.W. (1987) Detecting ecosystem responses to anthropogenic stress. *Can. J. Fish. Aquat. Sci.* **44**, 6–25.

Schindler, D.W. (1991) Whole lake experiments at the Experimental Lakes Area. In: Mooney, H.A., Schindler, D.W., and Schutz, E.-D. (Eds.) *Ecosystem Experiments*. John Wiley & Sons, Chichester, UK, 304 pp.

Schindler, D.W., and Fee, E.J. (1974) Experimental Lakes Area: whole-lake experiments in eutrophication. *J. Fish. Res. Board Can.* **31**, 937–953.

Schindler, D.W., Kling, H., Schmidt, R.V., Prokopowich, J., Frost, V.E., Reid, R.A., and Capel, M. (1973) Eutrophication of Lake 227 by addition of phosphate and nitrate: the second, third, and fourth years of enrichment, 1970, 1971, and 1972. *J. Fish. Res. Board Can.* **30**, 1415–1440.

Schindler, D.W., Ruszczynski, T., and Fee, E.J. (1980a) Hypolimnion injection of nutrient effluents as a method for reducing eutrophication. *Can. J. Fish. Aquat. Sci.* **37**, 320–327.

Schindler, D.W., Wagemann, R., Cook, R.B., Ruszczynski, T., and Prokopowich, J. (1980b) Experimental acidification of Lake 223, Experimental Lakes Area: background data and the first three years of acidification. *Can. J. Fish. Aquat. Sci.* **37**, 342–354.

Schindler, D.W., Hesslein, R.H., and Turner, M.A. (1987) Exchange of nutrients between sediments and water after 15 years of experimental eutrophication. *Can. J. Fish. Aquat. Sci.* **44**(Suppl. 1), 26–33.

Shortreed, K.S., and Stockner, J.G. (1990) Effect of nutrient additions on lower trophic levels of an oligotrophic lake with a seasonal deep chlorophyll maximum. *Can. J. Fish. Aquat. Sci.* **47**, 262–273.

Smith, V.H., Rigler, F.H., Choulik, O., Diamond, M., Griesbach, S., and Skraba, D. (1984) Effects of phosphorus fertilization on phytoplankton biomass and phosphorus retention in subarctic Québec lakes. *Verh. Internat. Verein. Limnol.* **22**, 376–382.

Stockner, J.G. (1981) Whole-lake fertilization for the enhancement of sockeye salmon *(Oncorhychus nerka)* in British Columbia, Canada. *Verh. Internat. Verein. Limnol.* **21**, 293–299.

Stockner, J.G., and Shortreed, K.S. (1985) Whole-lake fertilization experiments in coastal British Columbia lakes: empirical relationships between nutrient inputs and phytoplankton biomass and production. *Can. J. Fish. Aquat. Sci.* **42**, 649–658.

Stross, R.G., and Hasler, A.D. (1960) Some lime-induced changes in lake metabolism. *Limnol. Oceanogr.* **5**, 265–272.

Stross, R.G., Ness, J.C., and Hasler, A.D. (1961) Turnover time and production of planktonic crustacea in limed and reference portion of a bog lake. *Ecology* **42**, 237–245.

Svendrup, H., and Bjerle, I. (1983) The calcite utilization efficiency and the long term effect on alkalinity in several Swedish liming projects. *Vatten* **39**, 41–54.

Teller, A., Mathy, P., and Jeffers, J.N.R. (1992) *Responses of Forest Ecosystems to Environmental Changes*. Elsevier Applied Science, London, 1009 pp.

Wallace, R.R., and Hynes, H.B.N. (1975) The catastrophic drift of stream insects after treatments with methoxychlor (1,1,1-trichloro-2,2-bis(*p*-methoxyphenyl) ethane). *Envir. Pollut.* **8**, 255–268.

Wallace, J.B., Cuffney, T.F., Lay, C.C., and Vogel, D. (1987) The influence of an ecosystem-level manipulation on prey consumption by a lotic dragonfly. *Can. J. Zool.* **65**, 35–40.

Wallace, J.B., Lugthart, G.J., Cuffney, T.F., and Schurr, G.A. (1989) The impact of repeated insecticidal treatments on drift and benthos of a headwater stream. *Hydrobiol.* **179**, 135–147.

Wilander, A., and Ahl, T. (1972) The effects of lime treatment to a small lake in Bergslagen, Sweden. *Vatten* **5**, 431–445.

Wright, R.F. (1985) Chemistry of Lake Hovvatn, Norway, following liming and reacidification. *Can. J. Fish. Aquat. Sci.* **42**, 1103–1113.

Wright, R.F., and Gjessing, E. (1986) *RAIN Project. Annual Report for 1985.* Acid Rain Res. Rept 9/1986, NIVA, Oslo. 33 pp.

Wright, R.F., Gjessing, E., Christophersen, N., Lotse, E., Seip, H.M., Semb, A., Sletaune, B., Storhaug, R., and Wedum, K. (1986) Project RAIN: changing acid deposition to whole catchments. The first year of treatment. *Water Air Soil Pollut.* **30**, 47–63.

Wright, R.F., Lotse, E., and Semb, A. (1988a) Reversibility of acidification shown by whole-catchment experiments. *Nature* **334**, 670–675.

Wright, R.F., Norton, S.A., Brakke, D.F., and Frogner, T. (1988b) Experimental verification of episodic acidification of freshwaters by sea salts. *Nature* **334**, 422–424.

Wright, R.F., van Breemen, N., Emmett, B., Roelofs, J.G.M., Tietema, A., Verhoef, H.A., Hauhs, M., Rasmussen, L., Hultberg, H., Persson, H., and Stuanes, A.O. (1992) NITREX—NITrogen saturation EXperiments. In: Teller, A., Mathy, P., and Jeffers, J.N.R. (Eds.), *Responses of Forest Ecosystems to Environmental Changes*, pp. 335–341. Elsevier Applied Science, London, 1009 pp.

Yan, N.D., and Dillon, P.J. (1984) Experimental neutralization of lakes near Sudbury, Ontario. In: Nriagu, J. (Ed.) *Environmental Impacts of Smelters*, pp. 417–456. Advances in Environmental Science Series. John Wiley, New York.

Yan, N.D., and Lafrance, C. (1984) Responses of acidic neutralized lakes near Sudbury, Ontario to nutrient enrichment. In: Nriagu, J. (Ed.) *Environmental Impacts of Smelters*, pp. 457–521. Advances in Environmental Science Series. John Wiley, New York.

Young, T.C., DePinto, J.V., Rhea, J.R., and Scheffe, R.D. (1989) Calcite dose selection, treatment efficiency, and residual calcite fate after whole-lake neutralization. *Can. J. Fish. Aquat. Sci.* **46**, 315–322.

16 Statistical Methods to Assess the Effects of Chemicals on Ecosystems

J. N. R. Jeffers
ESE Consultants, United Kingdom

16.1 INTRODUCTION

Determination of the effects of chemicals on ecosystems is a difficult task. Ecosystems have been defined by Tansley (1935) as "The whole system (in the sense of physics) including not only the organism-complex, but also the whole complex of physical factors forming what we call the environment of the biome—the habitat factors in the widest sense." As such, ecosystems are difficult to investigate. Established traditions of thought in science fragment reality by the application of reductionism. When comprehension of the whole is difficult, it is divided into smaller components, each of which can then be investigated separately. As a result, objects in the real world become identified as autonomous, distinct entities, inherently separate and disconnected from other objects outside, and which constitute their environment. The results of reductionism are intended to be understanding, prediction, and control.

Because of the inherent complexity of ecosystems, however, treating parts of the reality in isolation leads to unexpected, unexplained, and often unwanted consequences when the parts are considered together. An ecosystem is greater than the sum of its parts. The general inability of science to build the properties of fragments into reasonable replicas of the properties of wholes is claimed to be calling the entire reductionist–mechanistic paradigm of the physical sciences into question (Patten, 1991).

The problem of holistic science becomes one of system definition. Specifically, determination of a minimal universe sufficient to encompass the indirect effects relevant to a given scope of enquiry and no more is necessary. Having defined and bounded the system, logical methodologies capable of describing and interpreting the variability inherent in ecological systems must be used, because of the genetic mechanisms that control the inheritance and response of individuals to other organisms and to their environment. To understand ecosystems, either as complex physical entities or as a paradigm for ecological science, requires formal methodologies to represent the relationships between organisms, and between organisms and their environment. Mathematical modelling is the systematic

Methods to Assess the Effects of Chemicals on Ecosystems
Edited by R. A. Linthurst, P. Bourdeau, and R. G. Tardiff

methodology that has proven successful in discovering and understanding the underlying processes in ecology.

This chapter seeks to explore some of the important concepts in mathematical analysis and modelling as applied to the estimation of the effects of chemicals on ecosystems. First, appropriate strategies are defined to investigate the impact of chemicals on an ecosystem. The role of hierarchy theory in determining the appropriate level of search and measurement is discussed, together with the constraints imposed by statistical inference in the estimation of system parameters. Ten basic principles are presented as a guide to develop future research projects. Finally, a list of selected references is provided as a source of further information.

16.2 STRATEGIES

Studies of the effects of chemicals on ecosystems can usually be classified broadly into baseline, monitoring, or impact studies. A baseline study is one in which data are collected and analysed for the purpose of defining the present state of the ecosystem. Usually, some environmental change is anticipated, although both the nature of the change and the time of its occurrence may be unknown. However, the present state will be characterized by patterns of spatial and temporal variation. An impact study is one whose purpose is to determine whether a specified impact has caused a change in an ecosystem, and, if so, the nature of that change. The nature of the impact and the fact of its occurrence is, thus, known. A monitoring study uses data to detect change is an ecosystem as that change occurs, and commonly assumes that baseline data are available to provide a standard against which change can be measured. In the simplest cases, the nature of the change may be defined very specifically, but frequently it is not defined at all.

Appropriate strategies for environmental studies have been conveniently summarized by Green (1979) as a decision tree. The choice of strategy for the investigation of an impact on an ecosystem depends essentially on the response to three basic questions: (1) Has the impact already occurred?, (2) Are the where and when known?, and (3) Is there a control area?

If the impact has already occurred, the type of impact and its time or location are probably also known. Unless pre-existing baseline data are available, the effects of the impact and the mechanism of those effects must then be inferred from spatial differences between areas differing in their degree of impact. The methods available for such inference are not ideal, principally because the recorded differences in the degree of impact may be confounded with many factors, or, alternatively, may interact strongly with a range of environmental and ecological variables. Specific hypotheses need to be derived from an examination of spatial patterns, and then tested in additional laboratory or field experiments. The same considerations apply when the roles of chemicals in normal ecosystem processes are being investigated, but this aspect is not pursued further here. The detection of known pollution incidents or oil spills is an opposite example.

If an impact of some kind has occurred but the precise timing and location of that impact is unknown, the only available strategy is that of a survey in both time and space. The design of such a survey must take into account all available knowledge of the variability of the component ecosystems; ideally, a prepared sampling frame should be available from past research. For example, the location and timing of the radionuclide depositions in Britain that resulted from the Chernobyl explosion were made possible by the existence of a land classification that provided a stratification and sampling frame for a survey of grassland ecosystems across the whole Britain. That survey enabled the precise determination of the location of the deposition of ^{134}Ce and ^{137}Ce, and the assessment of the uptake of the caesium by vegetation and sheep.

More constructive strategies are available when the impact has not already occurred, even if the precise time and location of some future impact is unknown. Given a broad indication of the ecosystems at risk, a baseline survey can be undertaken, and then the system can be monitored to detect when and where an impact occurs. However, even this strategy is not without its difficulties, if the nature of the chemical and the changes that are likely to occur as a result of its impact are unknown in advance. Not everything can be monitored, and past history suggests that effects of chemicals have not always been anticipated correctly, or even at all. This strategy is probably, therefore, the most expensive, and perhaps the least successful.

Where the impact has not occurred, and the time and place of the future impact is known, usually because it will occur as a result of a planned intervention in the functioning of the ecosystem, the appropriate strategy will depend on the presence or absence of one or more control areas on which the impact will not occur. Without control areas, the effect of the impact will need to be inferred from changes taking place in the ecosystem over time. Such a strategy is not without its problems, especially where feedback mechanisms are present in the ecosystem processes that make it difficult to determine the point at which changes first occur, even though the precise timing of the impact is known. Although time-series analysis is a well-developed part of modern statistical methodology, considerable ambiguities persist as a result of lags in the effects of chemicals on ecosystem processes. By far the most effective strategy arises from prior knowledge of where and when an impact is likely to be of concern, again most probably because it is planned to arise from an experimental intervention, and one or more control areas are designated to provide a comparison with the treated or impacted areas. This strategy permits, at least in theory, the use of an optimum design, including replication and randomization, that will enable the parameters of an ecosystem model to be estimated. Some of the basic statistical concepts associated with such an approach are emphasized later in this chapter.

Within each strategy, several decisions remain to be made. The number and kinds of variables need to be resolved, and the means by which they are to be derived, coded, and analysed need to be defined. Special considerations that may influence choice of sample unit size and the number of replicate samples need to be explored. The procedures to be used for the preliminary screening of data for aberrant values, failures of assumption, and the estimation of ecosystem parameters require specification. These issues are considered in some detail below.

16.3 HIERARCHY THEORY

Because ecological systems are organised across a range of space and time scales, a way is needed to decide which specific mechanisms must be understood in order to predict system behaviour. The time and the resources available for research are too limited to permit the collection of data for any sizeable portion of the world's ecological systems at the level of detail implied by a reductionist philosophy of scientific investigation. Ecologists are, therefore, beginning to use hierarchy theory to help construct a link between theory and empiricism (Allen and Starr, 1982; O'Neill *et al.*, 1986).

Hierarchy theory asserts that a useful way of dealing with complex, multiscaled systems is to focus on a single phenomenon and a single time–space scale. By limiting the problem in this way, it cannot be defined clearly, or to choose the proper subsystem with which to work.

The system of interest (Level 0) will itself be a component of some higher level. The dynamics of the upper level will usually appear as constants or driving variables in a model of Level 0, so that the behaviour of Level 0 will also appear to be constrained, bounded, and controlled by this higher level.

The higher level also provides the significance of the phenomena of interest. If, for example, the effect of a chemical on an organism were being determined, behaviour difficult to explain will be observed, if one's attention is limited to the single organism. Only by reference to the higher level, the population, can the significance of that behaviour be revealed.

The next step is to divide Level 0 into components forming the next lower level (Level -1). The components of Level -1 can then be studied to explain the mechanisms operating at Level 0, and these lower level entities appear as state variables in a model of Level 0.

Defined in this way, hierarchy theory dissects a phenomenon out of its complex spatial and temporal context. Understanding the phenomenon depends on referencing the next higher and lower scales of resolution. Levels higher than +1 are too large and too slow to be seen at the 0 level, and can, therefore, usually be ignored. Levels lower than -1 are too small and too fast to appear as anything but background noise in observations of Level 0. The theory focuses attention on a defined subset of behaviour and permits systematic study of very complex systems.

O'Neill (1988) develops these concepts to provide a set of possible criteria for the study of global change. The author has adapted criteria to the study of impacts of chemicals on ecosystems:

1. Searching for the fundamental hierarchy. Although searching for one hierarchy that characterizes an ecosystem may be tempting, several different hierarchies may be necessary to address different problem areas. Requiring that a hierarchy fits *a priori* biases is an unnecessary constraint.
2. Searching for the fundamental level. Designating the one and only level to which all other phenomena must be reduced is also not fruitful. Indeed, the impact itself is likely to determine the time and space scales that will be emphasized.
3. Translating principles between levels. Generally, transposing relationships developed at one hierarchical level to higher and lower levels will be impossible. Constraints imposed at higher levels may dominate, and the overall system behaviour may have little resemblance to the behaviour of isolated components. Experimental ecologists comprehend that relationships in the field may be quite different from those measured in the laboratory.
4. Be prepared to accept innovative approaches. The approaches required to understand the impact of chemicals on ecosystems will reduce to simple repetitions of approaches used previously. For instance, measuring long-term changes at the scales currently emphasized in ecology is notoriously difficult. Limiting investigators to familiar scales of measurement may result in failure to detect significant trends until prevention of permanent changes is impossible.
5. Effects of a higher level on a lower. Higher levels of the hierarchy set constraints or boundary conditions for lower levels. Because the upper level dynamics act as forcing functions or driving variables, given sufficient difference in scale, predicting how the higher level will affect the lower is possible.
6. Predicting the higher level from the lower. While hierarchy theory predicts how higher levels affect lower ones, moving in the opposite direction is more difficult. Some higher level properties are sometimes but not always the sum or integral of lower level dynamics. This influence of the lower levels on the higher is commonly known as the "aggregation problem."
7. Interactive state variables. A useful scale for interfacing different disciplines (e.g., chemistry and ecology) can be found, provided that the state variables of a model from one discipline appear as state variables in the model of the other discipline. Once a problem area is selected, hierarchy theory provides a means of determining a specific scale at which chemical and ecological processes can be interfaced.
8. Seeking coherent levels. The previous criteria might lead to the belief that any level of resolution can be chosen arbitrarily; however, that is not the case. Scales exist at which predictive capability is improved over slightly larger or slightly smaller scales. The scale at which the predictive power is maximized is the coherent level that makes sense as an isolated object of study. Fortunately, this scale is quite likely to correspond to a traditional level of study within a discipline. While interfacing disciplines at arbitrary levels of resolution would be possible, arbitrary scales do not take advantage of the

innate organization in hierarchical systems. Only when focus is placed on coherent levels can advantage be taken of the information and insights that have developed in each discipline about their own systems.

9. Critical points in parameter space. Once a scale has been selected, the parameters to be measured must be determined. A potential solution can be offered; the normal functioning of the ecosystem is not of great concern. This normal, stable behaviour of the ecosystem is most difficult to monitor. However, concern is raised about the unusual circumstances that may perturb the system. The points of critical change are called bifurcations in the underlying mathematical theory, and a necessary and sufficient condition exists to determine when the radical change is apt to occur. The change occurs when the rapid components cease to be stable; that is, the lower level components do not return to normal behaviour following a minor perturbation.

The causes of normal behaviour by a system include the fact that the rapid portions of the system are constrained by higher levels. If the system is perturbed, the rapid components simply return to the slowly changing trajectory. The rate of recovery can be taken as an indicator of the relative stability of the system, but, as the system approaches a bifurcation, the recovery becomes slower. Thus, monitoring for a significant impact is possible by monitoring the recovery rate of lower levels in the hierarchy. If the response times are increased, the system is being moved towards a point of radical self-amplifying change. Even though the actual point of change could be precipitated by fine-scaled changes, the proximity to any radical point of change will be indicated by a change in recovery times.

16.4 STATISTICAL CONCEPTS

Any system that includes living organisms is certain to show some degree of heterogeneity, because of the genetic variation that results from sexual reproduction. Models of the reactions of organisms to applications of a chemical that do not allow for variability in the response to those applications are, therefore, mere caricatures.

The statistician regards the measurement of variation as being of greater importance than the measurement of central tendencies, whether as means or as relationships; and most of the extensive theory of mathematical statistics now deals with the measurement of variations and its characterization. Much of that theory has historically concentrated on the continuity of observations, as reflecting some idealized distribution that has desirable mathematical properties. More recently, statisticians have turned to methods that are capable of detecting discontinuities in measurements, often in multidimensional space. Increasingly statistical methods have embraced qualitative data as important extensions of purely quantitative measures.

The total set of individuals about which inferences are made is said by statisticians to constitute a population. Those individuals may be organisms, communities, societies, or whole ecosystems. While such populations will usually be finite in both time and space, they will often be so large as to make it impossible for every member of the population to be investigated, measured, or counted. Practical scientists are, therefore, usually forced to work with samples drawn from the population, and we will need to make the assumption that those samples are representative of the defined population. Only then can values calculated from the samples be regarded as unbiased estimates of the corresponding population values.

A statistician's requirement for a set of samples to be regarded as representative of a population is uncompromising. The samples must be taken by an objective and unbiased method. Selection by some form of subjective choice, guided by whatever personal consideration of the representativeness of the samples, will not satisfy the constraints of statistical inference. Two methods of objective sampling are commonly employed to meet these constraints, namely systematic or random sampling. Unless systematic sampling is repeated, severe problems occur in calculating the precision of estimates derived from the samples and in characterizing the heterogeneity of the population. The simple expedient of ensuring genuinely random choice at an appropriate part of the sampling procedure guarantees the lack of bias, and also provides a methodology to characterize the heterogeneity of the sampled population, and the precision with which population parameters are estimated from the sample (Jeffers, 1988a).

Statistically, interaction is defined as a measure of the extent to which the effect of one factor varies with changes in the strength, grade, or level of other factors in an experiment. As long ago as the mid-1920s, Fisher (1925, 1935) pointed out that interactions could be investigated experimentally only if all the factors were included in the same experiment. Together with his co-workers, he developed the concept of factorial experiments through which some or all of the combinations of factors could be used to determine the strength and character of their interactions, provided that adequate replication of the experimental treatments exists.

Genuine replicates are independent in the sense that the outcome of a given replicate has no effect on the outcome of any others. They represent the total variability affecting replicates in some specified experimental conditions. Replication improves experiments in three principal ways. First, replication is the only way in which a valid estimate can be made of experimental error. Second, experimental results become increasingly precise as the number of replicates increases. Third, replication expands the range of experimental units studied, and, therefore, the extent to which results can be generalized.

However, tension exists between the need to replicate and the need to study ecological processes at appropriately large scales. Indeed, large-scale experiments, whether planned or unplanned, are not replaceable. The serendipity of unplanned experiments usually precludes replication, and even when experiments are planned, the independent experimental units are limited in number. Candidate systems are so different ecologically that they do not constitute reasonable replicates. Funding levels and logistic limitations also often preclude replication (Carpenter, 1990). Detecting change in ecological time-series (Jassby and Powell, 1990) and Bayesian inference from non-replicated ecological studies (Reckhow, 1990) have been advocated as alternatives to replicated experiments, but the arguments are controversial.

As Walters and Holling (1990) emphasize, a major challenge to justify and design experimental management programmes is the exposure of uncertainties (in the form of alternative working hypotheses) and management decision choices in a form that will promote both reasoned choice and a search for imaginative and safe experimental options, by using tools of statistical decision analysis. Therefore, experimental designs must be identified that distinguish clearly between localized and large-scale effects, and make the best possible use of opportunities for replication and comparison. Such designs also need to permit analysts to make unambiguous assessments of transient responses by ecosystems to chemicals, in the face of uncontrolled environmental factors that may affect treated and control experimental units differently.

16.5 BACK TO BASICS

The following summarizes the important principles to be observed in any study of the effect of chemicals on ecosystems. The similarity of these principles to those suggested by Green (1979) is acknowledged:

1. Clarity is essential to the goals of investigations, and to the means of collecting data to answer the questions posed. Collecting large quantities of data in the hope that they can somehow be synthesized as a model of an ecological system has been tried many times, from the International Biological Programme onwards, and has never resulted in a useful outcome. The only effective approach is first to state the hypothesis and then to collect the data.
2. When measuring the characteristics of any ecosystem, replicate samples are important within each combination of time, location, and any other controlled variable. By comparing differences between elements with differences within elements, that statistical significance can be established.
3. To the extent possible, equal numbers of randomly allocated samples should be taken for each combination of controlled variables. Taking samples from representative or typical places does not qualify as random sampling.
4. To test whether a condition has an effect, samples must be taken from the location in which the condition is present and absent. Furthermore, interactions between factors can only be estimated from factorial combinations of those factors.
5. Pilot trials are essential to provide a basis for the evaluation of sampling designs and options for statistical analysis. Eliminating this step in project

design, for whatever reason, inevitably results in a great loss of time and resources.

6. Variation in efficiency of sampling among areas frequently biases comparisons between the areas. Therefore, sampling methods must be shown to be able to sample the population to be sampled, and with equal and adequate efficiency over the entire range of sampling conditions encountered.
7. If an area to be sampled has a large-scale environmental pattern, the area should be divided into relatively homogeneous sub-areas, and samples allocated to each sub-area in proportion to the size of the sub-area. Alternatively, if an estimate of the total abundance of some species is required, the allocation of samples should be proportional to the numbers of organisms in the sub-area.
8. Sample unit sizes must be appropriate to the size, densities, and spatial patterns of the organisms being sampled. The numbers of replicate samples required to obtain estimates with the required precision can then be determined from the information obtained from the pilot trials.
9. Appropriate methods of analysis will depend on the ways in which the data have been collected, the hypotheses the data were designed to test, and the nature of the heterogeneity displayed by the data. These "metadata"—data about data—must not be separated from the data themselves before submitting them to analysis. A fundamental lack of integration exists between database management systems and classical statistical packages.
10. Important as good computing tools are, they do not substitute for appropriate statistical knowledge about how to analyse structured data. Good tools allow bad methods to be encoded just as easily as good ones. For scientists, the key problems are in knowing what processing can be validly undertaken on the data, and, in its simplest form, what data can be legitimately combined and compared.

16.6 SELECTED TOPICAL REFERENCES

1. Systems analysis: Checkland and Scholes, 1990; Jeffers, 1978 and 1988a.
2. Hierarchy theory: Allen and Starr, 1982; O'Neill, 1988; O'Neill *et al.*, 1986.
3. Ecosystem research: Halfon,1979; Huggett *et al.*, 1992; Patten, 1991; Tansley, 1935.
4. Design: Carpenter, 1990; Fisher, 1935; Green, 1979; Reckhow, 1990; Walters and Holling, 1990.
5. Analysis: Beven and Moore, 1991; Feoli and Orlochi, 1991; Fisher, 1925; Howard, 1991; Jackson, 1991; Jassby and Powell, 1990; Lunn and McNeil, 1991; Noreen, 1989.
6. Modelling: Barnsley, 1988; Guariso and Werthner, 1989; Jeffers, 1988b; Starfield and Bleloch, 1986; Swartzman and Kaluzny, 1987; Tijms, 1986; Whittaker, 1990.

16.7 REFERENCES

Allen, T.F.H., and Starr, T.B. (1982) *Hierarchy: Perspectives for Ecological Complexity*. University of Chicago Press, Chicago.

Barnsley, M. (1988) *Fractals Everywhere*. Academic Press, London.

Beven, K.J., and Moore, I.D. (1991) *Terrain Analysis and Distributed Modelling in Hydrology*. John Wiley & Sons, Chichester, UK.

Carpenter, S.R. (1990) Large-scale perturbations: opportunities for innovation. *Ecology* **71**(6), 2038–2043.

Checkland, P., and Scholes, J. (1990) *Soft Systems Methodology in Action*. John Wiley & Sons, Chichester, UK.

Feoli, E., and Orloci, L. (1991) *Computer Assisted Vegetation Analysis*. Kluwer Academic, Dordrecht, The Netherlands.

Fisher, R.A. (1925) *Statistical Methods for Research Workers*. Oliver & Boyd, Edinburgh.

Fisher, R.A. (1935) *The Design of Experiments*. Oliver & Boyd, Edinburgh.

Green, R.H. (1979) *Sampling Design and Statistical Methods for Environmental Biologists*. John Wiley & Sons, Chichester, UK.

Guariso, G., and Werthner, H. (1989) *Environmental Decision Support Systems*. Ellis, Horwood, Ltd, Chichester, UK.

Halfon, E. (1979) *Theoretical Systems Ecology*. Academic Press, London.

Howard, P.J.A. (1991) *An Introduction to Environmental Pattern Analysis*. Parthenon, Casterton, UK.

Huggett, R.J., Kimerle, R.A., Mehrle, P.M., and Bergman, H.L. (1992) *Biomarkers: Biochemical, Physiological and Histological Markers in Anthropogenic Stress*. Lewis, Chelsea, Michigan.

Jackson, J.E. (1991) *A User's Guide to Principal Components*. John Wiley & Sons, Chichester, UK.

Jassby, A.D., and Powell, T.M. (1990) Detecting change in ecological time series. *Ecology* **71**(6), 2044–2052.

Jeffers, J.N.R. (1978) *An Introduction to Systems Analysis: With Ecological Applications*. Edward Arnold, London.

Jeffers, J.N.R. (1988a) Statistical and mathematical approaches to issues of scales in ecology. In: Rosswall, T., Woodmansee, R.G., and Risser, P.G. (Eds.) *Scales and Global Change*. John Wiley & Sons, Chichester, U.K.

Jeffers, J.N.R. (1988b) *Practitioner's Handbook on the Modelling of Dynamic Change in Ecosystems*. John Wiley & Sons, Chichester, U.K.

Lunn, A.D., and McNeil, D.R. (1991) *Computer-Interactive Data Analysis*. John Wiley & Sons, Chichester, UK.

Noreen, E.W. (1989) *Computer-Intensive Methods for Testing Hypotheses*. John Wiley & Sons, Chichester, UK.

O'Neill, R.V. (1988) Hierarchy theory and global change. In: Rosswall, T., Woodmansee, R.G., and Risser, P.G. (Eds.) *Scales and Global Change*. John Wiley & Sons, Chichester, UK.

O'Neill, R.V., DeAngelis, D.L., Waide, J.B., and Allen, T.H.F. (1986) *A Hierarchical Concept of Ecosystems*. Princeton University Press, Princeton, New Jersey.

Patten, B.C. (1991) Network ecology: indirect determination of the life-environment relationship in ecosystems. In: Higashi, M., and Burns, T.P. (Eds.) *Theoretical Studies of Ecosystems: The Network Perspective*. Cambridge University Press, Cambridge.

Reckhow, K.H. (1990) Bayesian inference in non-replicated ecological studies. *Ecology* **71**(6), 2053–2059.

Starfield, A.M., and Bleloch, A.L. (1986) *Building Models for Conservation and Wildlife*. Macmillan, London.

Swartzman, G.L., and Kaluzny, S.P. (1987) *Ecological Simulation Primer*. Macmillan, London.

Tansley, A.G. (1935) The use and abuse of vegetational concepts and terms. *Ecology* **16**, 284–307.

Tijms, H.C. (1986) *Stochastic Modelling and Analysis*. John Wiley & Sons, Chichester, U.K

Walters, C.J., and Holling, C.S. (1990) Large-scale management experiments and learning by doing. *Ecology* **71**(6), 2060–2068.

Whittaker, J. (1990) *Graphical Models in Applied Multivariate Statistics*. John Wiley & Sons, Chichester, UK.

17 A Framework for Ecological Risk Assessment

Lawrence W. Barnthouse
Oak Ridge National Laboratory, USA

17.1 INTRODUCTION

In 1983, the US National Research Council (NRC) defined a set of general principles for human health risk assessment that are now widely used as the basis for managing risks of chemicals to human health (National Research Council, 1983). The NRC's report described risk assessment as a procedure for linking scientific information about potentially hazardous substances to the decision-making process through which human exposures to these substances are regulated. The report attempted to provide a clear distinction between the roles of scientists (risk assessors) and decision makers (risk managers) in the assessment process, and to define a general procedure to identify the kinds of scientific information appropriate for quantifying human health risks of toxic chemical exposures. Although most of the report deals only with chemicals, the committee clearly intended for its major conclusions to apply to non-chemical risks as well.

Within the past several years, interest has grown in defining an analogous set of principles to assess ecological risks of toxic chemicals and other stresses. Despite the wide disparities between different kinds of ecological assessments (e.g., between the regulation of pollutant discharges and the management of fisheries), a unifying set of principles for ecological risk assessment is now emerging (USEPA, 1992; Suter, 1993a).

In this chapter, a framework for ecological risk assessment is drawn from several recent sources, and its implications for the future of chemical risk management are discussed. In addition, risk assessment will be related to the new paradigm for global environmental management implied by the term "sustainable development."

17.2 DEFINITION OF ECOLOGICAL RISK ASSESSMENT

The NRC (1983, p. 18) defined human health risk assessment as "...the characterization of the potential adverse health effects of human exposures to

Methods to Assess the Effects of Chemicals on Ecosystems
Edited by R. A. Linthurst, P. Bourdeau, and R. G. Tardiff

environmental hazards. The US Environmental Protection Agency (USEPA, 1992) recently proposed a similar definition of ecological risk assessment: "...a process that evaluates the likelihood that adverse ecological effects may occur or are occurring as a result of exposure to one or more stressors." Even more generally, ecological risk assessment can be defined by restating the 1983 NRC definition in ecological terms: the characterization of the adverse ecological effects of environmental exposures to hazards imposed by human activities.

The term "adverse ecological effects" includes all biological and non-biological environmental changes that society perceives as undesirable. The term "hazard" include both unintentional hazards such as pollution and soil erosion and deliberate management activities such as forestry and fishing which are often hazardous either to the managed resource itself or to other components of the environment.

The definitions of USEPA and the modified one of NRC emphasize that risk assessment is a decision-making process, not a computational technique. Because this perspective allows for extensive reliance on qualitative information and expert judgement, it is consistent with the current state-of-the-art in ecological science, which is much more qualitative than quantitative. More important, these definitions emphasize that the objective of risk assessment is not to provide scientific truth but to promote sound environmental decisions.

17.3 RISK ASSESSMENT FRAMEWORKS

In addition to a definition, the NRC (1983) described a framework for human health risk assessment consisting of four components: hazard identification, dose-response assessment, exposure assessment, and risk characterization. The purposes of the framework were to (1) provide for more detailed definitions of the scientific components of risk assessments, (2) define the relationship of risk assessment to risk management, and (3) facilitate the development of uniform technical guidelines. An analogous framework for ecological risk assessment could be used to evaluate the consistency and adequacy of individual assessments, to compare assessments for related environmental problems, to identify explicitly the connections between risk assessment and risk management, and to identify environmental research topics and data needs common to many ecological risk assessment problems. The USEPA (1992) described ecological risk assessment as a three-part process consisting of problem formulation, analysis, and risk characterization. The term "analysis" was further subdivided into characterization of exposure and that of ecological effects.

A more general framework that integrates human health with ecological concerns is depicted in Figure 17.1. The relationship among the four components is a hybrid between the arrangements proposed by the NRC (1983) and USEPA (1992). This framework, and the definitions provided below, have now been formally proposed by the NRC's Committee on Risk Assessment Methodology (NRC, 1993).

Hazard identification may be broadly defined to be the determination of whether a particular hazardous agent is associated with health or ecological effects of sufficient importance to warrant further scientific study or immediate management action. The purpose of hazard identification is to determine whether a "hazard" exists; and, if one is identified, to determine the kinds of additional scientific information required to assess the degree of risk present and evaluate the alternative risk management actions. Typical kinds of information used for hazard identification include quantitative structure–activity relationships (QSARS), short-term toxicity tests, and reviews of existing information about the characteristics of potentially affected ecosystems.

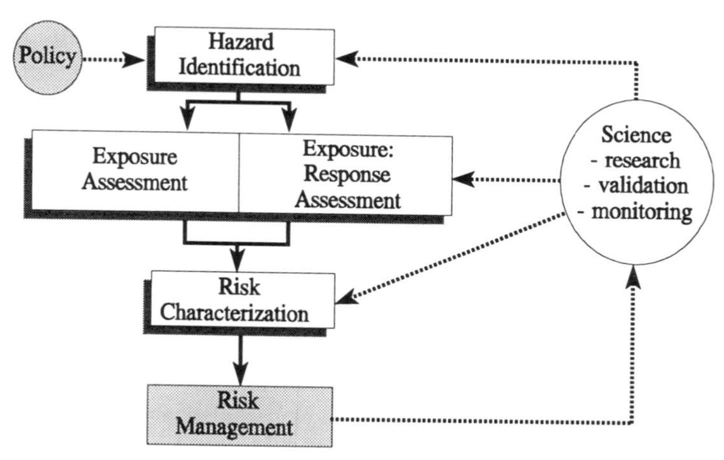

Figure 17.1. A risk assessment framework intended to facilitate integration of human health and ecological risks of toxic chemicals and other environmental stresses (from NRC, 1983, and USEPA, 1992)

Exposure-response assessment may be defined as the determination of the relation between the magnitude of exposure and the probability of occurrence of the effects in question. In the case of toxic chemicals, information included in an exposure-response assessment could include detailed toxicological information (e.g., chronic toxicity, mode of action, sensitivities), mesocosm or field test data, field surveys to compare exposed and unexposed sites, and population or ecosystem modelling. The responses addressed in ecological risk assessments include both direct effects of exposure and the much broader indirect effects, such as secondary poisoning of raptors due to accumulation of pesticide residues.

Exposure assessment may be defined as the determination of the extent of exposure to the hazardous agent in question. As applied to toxic contaminants, this component includes measurement or prediction of the movement, fate, and partitioning of chemicals in the environment. The term can also be legitimately applied to non-chemical stresses, including both physical stresses (such as habitat change and UV radiation) and biological stresses (such as species introductions).

In actual assessments, exposure assessment and exposure-response assessment occur roughly in parallel, and must be closely linked. The arrangement of those components in Figure 17.1, within a single box divided in half by a "permeable membrane," is intended to emphasize the ties between them.

Risk characterization may be defined as the description of the nature and often the magnitude of risk, including attendant uncertainty, expressed in terms comprehensible to decision makers and the public. The purpose of risk characterization is to integrate information obtained from all of the other components and to communicate it to decision-makers in a form comprehensible to non-specialists and relevant to the decision being made. Contents should include (1) a description of the nature and often the magnitude of risk to ecological resources, and (2) qualitative and quantitative characterization of uncertainty. Because the purpose of risk assessment is to support decision-making, communication with decision makers is a critical aspect of risk characterization.

In addition to the four basic components, Figure 17.1 depicts the relationship between risk assessment and environmental management. Policies and regulations determine the scope and content of risk assessments; results of risk assessments provide the scientific foundation for decisions. The risk assessment process should not, however, end when a regulatory decision is made. Follow-up in the form of monitoring (where measurable effects have been predicted), validation studies, and basic research are needed to improve the data and models available to technical risk assessors whenever the same or a similar problem is encountered in the future.

An additional concept relevant to this chapter is the distinction between assessment endpoints and measurement endpoints (Suter and Barnthouse, 1993). Assessment endpoints define the adverse effects to be avoided or the biological resources to be protected. Measurement endpoints are those variables measured in laboratory or field studies. Assessment endpoints are most often defined in terms of population or ecosystem properties such as risks of population extinction, reduction in productivity, or altered species composition. Assessment endpoints should be defined at the beginning of an assessment, as part of hazard identification, in order to guide the acquisition of exposure and effects data. Findings presented to decision makers as part of risk characterization should be expressed in terms of assessment endpoints. Assessment endpoints are rarely directly measurable, because exposure to the hazardous agent has not yet occurred, because the scale and complexity of the system of interest (e.g., a bird population distributed over thousands of km^2) precludes direct measurement, or because the precision of the available measurement techniques is insufficient to detect changes before they become irreversible.

Measurement endpoints may include toxicity tests, field surveys of contaminant distributions, or widely-spaced samples of population or ecosystem characteristics. Most of the technical literature on ecological effects of toxic chemicals deals with measurement rather than assessment endpoints. Over the last decade, great advances have been made in our understanding of modes of action of toxic chemicals, biochemical markers of toxicant exposure, the relative sensitivities of different kinds of organisms, and the changes in populations and ecosystems that can be caused by exposure to toxic chemicals. Such studies, however well performed, are usually insufficient as ecological risk assessments, because the scales of time and space that are amenable to rigorous scientific investigation are usually much shorter and smaller than the scales of interest in environmental management (Suter and Barnthouse, 1993). Additional extrapolation using expert judgement and mathematical models is necessary to relate these measurements to assessment endpoints that can be effectively communicated to decision makers and that can support informed decisions.

17.4 EXAMPLES OF ECOLOGICAL RISK ASSESSMENTS

To illustrate the application of the framework to ecological assessment problems, distinguishing between two major classes of risk assessments is useful: predictive assessments and retrospective assessments. The objective of a predictive assessment is to estimate the potential consequences of an action prior to taking that action. Assessments performed prior to the manufacture of toxic chemicals or the registration of pesticides are typically predictive assessments.

The objective of a retrospective assessment is to determine the causes and consequences of an event that has already occurred. Assessments of the need for environmental restoration of contaminated soil, water, and sediment are among the most common type of ecological risk assessment in the United States. Fisheries management and protection of endangered species also frequently involve retrospective risk assessments.

Predictive ecological risk assessments both in the United States and in the OECD are highly standardized, and are based principally on laboratory-derived toxicity test data and simplified environmental transport models. Hazard identification in these assessments consists of (1) documentation of basic physicochemical characteristics of the material being regulated (e.g., chemical structure, solubility, octanol-water partitioning coefficient), (2) summarization of available information on toxicity, and (3) description of likely environmental pathways, exposed species, and maximal exposures. The objective of the hazard identification step in these assessments is to quickly and efficiently classify chemicals as being clearly harmless, clearly hazardous, or potentially hazardous. Regulatory agencies such as the US Environmental Protection Agency, which must assess several thousand new chemicals every year, rely on standardized hazard identification protocols to facilitate approval of chemicals posing little or no risk (Bascietto *et al.*, 1990). The OECD (1984) has promulgated a similar scheme to test chemicals in the European community. Chemical manufacturers utilize similar protocols (e.g., Kimerle *et al.*, 1983) to identify potentially hazardous chemicals during the development process, before major financial commitments are made.

If a potential hazard is identified, then more rigorous exposure and exposure-response studies may be performed. Aquatic exposure assessments would typically involve modelling of the transport and fate of the chemical in question in generalized streams, ponds, lakes, or estuaries. Many models of the transport and fate of chemicals have been developed and several excellent reviews have been published (Jørgensen, 1984; Cohen, 1986; OECD, 1989; MacKay and Patterson, 1993). Exposure assessment models for terrestrial ecosystems are much less developed. Most often, field tests are used to provide direct measurements of the persistence of chemicals in soil and uptake by vegetation.

Many methods are now available to characterize exposure-response relationships for use in predictive ecological risk assessments. Kendall (1992) recently summarized the approaches used to characterize effects of pesticides on birds. These range from biochemical studies of the relationships between doses and enzyme activities (e.g., cholinesterase) to field tests in which birds are allowed to forage on fields to which pesticides have been applied using realistic application regimes. The USEPA uses a standardized tiered testing scheme in which increasingly realistic tests are performed at different levels (Urban and Cook, 1986).

Most existing predictive ecological risk assessment protocols characterize risks by comparing estimated exposure concentrations to a test-derived effects criterion such as a "no-observed-effects level," or a "lowest-observed-effects level." Depending on the kind of test performed, the criterion may include factors of 10 to 100 uncertainty or safety factors designed to ensure that risks are not underestimated. This procedure is acknowledged as being overly simplistic (Bascietto *et al.*, 1990). Methods to replace the uncertainty factors with empirical estimates of uncertainties inherent in extrapolating test data between species and test types have been developed (Suter *et al.*, 1983; Sloof *et al.*, 1986; Suter *et al.*, 1987; Kooijman, 1987; Volmer *et al.*, 1988; van Straalen and Denneman, 1989; Barnthouse *et al.*, 1990) but not widely implemented. Empirical and theoretical models have also been used to characterize ecological risks in terms of effects on exposed populations and ecosystems (Bartell *et al.*, 1992; Barnthouse *et al.*, 1990; Barnthouse, 1993), but these methods are not now used in regulatory practice.

Retrospective ecological risk assessments are more diverse and more difficult to characterize than are predictive assessments. Suter (1993b) distinguished three types of retrospective assessments: source-driven, exposure-driven, and effects-driven. A source-driven assessment is initiated by the observation of a source of environmental contamination such as an oil spill or a discharge pipe. An exposure-driven assessment is initiated by the observation of contaminated environmental media or biota. An effects-driven assessment is initiated by the observation of adverse changes in ecosystems, populations, or individual organisms. The two common elements in all of these assessments are (1) the possibility (at least in principle) of basing the assessment on actual field data and (2) the diagnosis of the sources of observed contamination or the causes of observed effects usually a major goal.

Assessments of the ecological risks of tributyl tin (Huggett *et al.*, 1992) provide an excellent example of retrospective ecological risk assessment. Tributyltin (TBT) is a chemical biocide often used as an antifouling agent in boat paints (Blunden and Chapman, 1982). Concentrations as low as 1 μg/litre are lethal to larvae of some species, and non-lethal effects have been observed at concentrations as low as 0.002 μg/litre (Huggett *et al.*, 1992). Adverse changes in a variety of marine invertebrate species, including commercially valuable shellfish, were observed in Europe in the early 1980s (Alzieu, 1986; Abel *et al.*, 1986). The changes observed included both declines in abundance of populations and morphological changes in the surviving organisms. Laboratory experiments with the oyster *Crassostrea gigas* showed that concentrations of TBT similar to concentrations found in coastal waters following the introduction of TBT-based antifouling paints caused the same patterns of abnormal shell growth observed in wild oyster populations. Snails in the vicinity of a marina on the York River, Virginia (US), were shown to have an abnormally high incidence of imposex (expression of male characteristics by female organisms), an effect previously observed under laboratory conditions in females of the European oyster *Ostrea edulis* (Huggett *et al.*, 1992). Regulatory agencies in Europe and the US have concluded that the diagnostic link between TBT exposures and adverse effects on shellfish are sufficient to demonstrate a significant ecological risk. Although the US Environmental Protection Agency has not issued regulations for TBT, the US Congress and several states have restricted the types of boats to which TBT-based paints may be applied and the leaching characteristics of the paints that may be used.

TBT is somewhat unusual in that a single chemical was responsible for observed adverse ecological changes and that easily diagnosed morphological changes could be associated with TBT exposures. Suter and Loar (1992) described a much more common situation: assessment of ecological risks associated with a complex mixture of contaminants derived from a large federal facility containing multiple contaminant sources. The objectives of this assessment are to determine whether remedial actions are needed to reduce exposures of organisms to contaminants derived from the US Department of Energy Oak Ridge Reservation, to identify the specific chemicals and sources requiring remediation, and to monitor recovery of on- and off-site ecosystems during the remediation process (expected to last for several decades).

A combination of field and laboratory studies have been employed. The field studies include characterization of contaminant distributions in water, sediment, and biota below known point sources, in the streams draining the reservation, and in the Clinch River adjacent to the Oak Ridge Reservation. Several biological responses are also being measured. Population and community (fish and invertebrates) characteristics are being monitored and correlated with exposure concentrations. Physiological and biochemical responses (i.e., biomarkers) known to be linked to contaminant exposures are also being measured (Shugart *et al.*, 1992).

Ultimately, these diverse types of data will be used to characterize the ecological risks of the current levels of environmental contamination on the Oak Ridge Reservation and to evaluate the relative ecological costs and benefits of potential restoration alternatives.

17.5 ECOLOGICAL RISK ASSESSMENT AND ENVIRONMENTAL SUSTAINABILITY

No discussion of environmental risk assessment written presently can be considered complete without a consideration of how the concept of risk assessment relates to the concept of environmental sustainability. The World Commission on Environment and Development (1987) defined sustainable development as the form of development or progress that "... meets the needs of the present without compromising the ability of future generations to meet their own needs." At the "Earth Summit" in Rio De Janeiro, delegates from 178 nations signed a declaration embracing sustainable development as an environmental management goal. The concept of sustainable development implies that the earth's ecosystems should be managed in such a way that they persist in more or less their present form, and continue to provide material sustenance, services, and aesthetic enjoyment to human societies. This goal has profound implications for environmental management and risk assessment. In sustainability-based environmental management, the focus is on maintaining or improving the quality of the environment, not on restricting discharges or requiring particular waste treatment technologies. Because the environment rather than the chemical or the technology is the focus of management, the management of chemicals must be integrated with management of other stresses on ecosystems. Decisions about remedial actions to restore ecosystems affected by chemical contamination have to account for other existing stresses such as erosion, eutrophication, and direct human exploitation. Most important, sustainable environmental management implies that the welfare of ecosystems is intimately tied to the welfare of human societies. Environmental management must consider both the human and non-human implications of decisions, and more often than not decisions that promote one will promote the other as well.

In many ways the call for sustainable development is reinforcing a trend in environmental management that began well over a decade ago. Most of our major ecosystems have already been substantially altered from their pristine state, sometimes unintentionally, but more often through deliberate actions intended to benefit man. Restoration rather than protection is the appropriate management goal. Attempts to restore these systems must recognize that most have been altered as greatly by erosion, physical alteration, and harvesting as they have by chemical pollution. Water quality management in the United States has been moving away from regulation based on the toxicity of the effluents being discharged to regulation based on measurements of the biotic structures of the ecosystems receiving the effluents (Bascietto *et al.*, 1990).

On a higher level, the USEPA's Science Advisory Board recommended that the Agency adopt an integrated approach to environmental management by using comparative risk assessments to prioritize its research and regulatory programmes. A national Environmental Monitoring and Assessment Program was instituted to measure the state of the major US ecosystems and provide agency decision-makers with information that can be used to allocate programme resources (Hunsaker *et al.*,

1990). Industry is also taking an active role, with many large companies developing their own pollution reduction programmes and attempting to assess the environmental impacts of their products over the full product life cycle, from resource extraction to ultimate end use and disposal (SETAC, 1991).

17.6 IMPLICATIONS FOR CHEMICAL EFFECTS ASSESSMENT

If environmental goals and regulatory approaches change, then assessment and measurement endpoints for ecological risk assessments must also change. From the sustainability perspective, the key management questions relate to the implications of environmental change for human and environmental welfare. Two aspects to the problem of how to formulate ecological risk assessments in terms provide useful answers to these questions: the scale of the technical studies supporting exposure and exposure-response assessments and the units in which risks are characterized and communicated. As noted previously, scientific studies must deal with tractable-sized systems and with durations of no more than a few months for laboratory studies or a few years for field studies. Space and time scales for management are quite different: the unit of management is likely to have an area of hundreds or thousands of km^2, and the planning timeframe is likely to be measured in decades. Extrapolating from scientifically tractable scales to scales of management interest requires the use of statistical and mathematical models (Suter and Barnthouse, 1993). Several different kinds of extrapolations exist. Experimental studies such as toxicity tests performed on single species must be extrapolated to effects on ecosystems containing many interacting species; Bartell *et al.* (1992) and Suter (1993b) have recently summarized methods for making these extrapolations. Methods to quantify characteristics of landscapes that are mosaics of many different ecosystems have now been developed (Turner and Gardner, 1991), and models that link atmospheric transport of pollutants to large-scale ecological effects have been developed and used to support regulation of sulphur and nitrogen emissions (Hordjik, 1991). These techniques cannot substitute for rigorous experimental and observational studies, but they are essential for interpreting the management-scale implications of empirical information.

The second component of the extrapolation problem is the translation of ecological characterizations of exposures and effects into estimates of risks to human-ecological welfare. Conventional descriptors of population and ecosystem status, e.g., numbers, biomass, trophic structure, and productivity, are meaningful to ecologists, but not to decision-makers or to the public. Means must be found to translate these descriptors into measures of gained or lost value to society. Increasingly, the method chosen to make this translation is economic valuation. Economic measures such as marketable discharge permits and pollution taxes are replacing conventional emissions limits and performance standards as the preferred means of regulating industrial pollutant emissions (Schmidheiny, 1992). In the United States, the Comprehensive Environmental Response, Compensation and

Liability Act (CERCLA) permits federal agencies, states, and other institutions responsible for managing natural resources to obtain monetary compensation for damages to those resources caused by oil spills and other forms of chemical contamination (USDI, 1991). These approaches to environmental management require that accurate monetary values be assigned to changes in the quality of the environment. These values must include marketable goods such as timber and fish, indirect services of nature such as pollution abatement and flood control, recreational use, and aesthetic values unrelated to actual use of the resources. A substantial literature on economic valuation of wetland ecosystems already exists (Scodari, 1990); the literature on ecosystem valuation in general is rapidly growing (Orians, 1990; Costanza, 1990). Although generally accepted principles do not yet exist, ecologists can expect eventually to be more frequently asked to express the results of ecological studies in terms that can be used as starting points for economic valuation studies.

A focus on assessing the overall quality of ecosystems, on restoration and maintenance of ecosystem quality, and on management of multiple human influences on ecosystems will probably also require changes in the kinds of data collected in chemical effects studies. The range of activities in chemical effects assessment will have to be broadened to address questions of rehabilitation, of causal relationships, and of impaired human welfare. Key scientific issues include (1) distinguishing the influence of chemicals from the influence of other stressors, (2) predicting and measuring recovery from reduced chemical inputs, and (3) estimating impacts of chemicals on the use of the environment by man, including recreation, flood control, natural biodegradation capacity, and aesthetic enjoyment. The first two of the above issues relate to retrospective risk assessment as defined above. Observed adverse ecological conditions must be causally related to specific sources of stress so that managers can determine which stress to reduce. Koch's postulates (Suter, 1993b) and Hill's criteria (Hill, 1965; USEPA, 1992) have both been suggested as decision rules to determine ecological causal relationships. Both approaches require that empirical data include information on toxicity, mode of action, spatial distribution of contaminants and exposed organisms, temporal sequence (i.e., putative cause must regularly precede observed effect), and exposure-response relationships from both experimental and observational studies.

The expected rate and degree of recovery resulting from different potential restoration activities must be predicted so that the beneficial effects of the alternatives can be evaluated and compared to the costs. This requirement is simply another form of extrapolation, and similarly requires the same kinds of extrapolation techniques described by Bartell *et al.* (1992) and Suter (1993a). If monitoring following institution of restoration actions were routine, our ability to predict restoration following future actions would be greatly enhanced (Yount and Niemi, 1990).

The third issue, that of estimating impacts of chemicals on human use of the environment, implies that ecologists and social scientists must cooperate in designing chemical effects studies. Ecological studies will have to be integrated with socioeconomic studies so that data will be available to interpret the socieconomic implications of the measured or predicted ecological changes.

17.7 CONCLUSION

A framework for ecological risk assessment defines the relationship between ecological science and environmental management. The role of science in risk assessment is to ensure that the actions implemented by environmental managers achieve the goals and objectives defined by society. The science required to achieve the goal of environmental sustainability is substantially broader than the science relevant to past approaches to ensuring chemical safety, which emphasized media-specific concentration limits and technology standards. This changing management perspective implies that eventually the science required to support management of risks related to chemicals in the environment will include (1) more emphasis on diagnostic studies to determine which systems are being affected and by which chemicals, (2) more attention to facilitating recovery of damaged ecosystems subjected to multiple human influences, (3) a greater need for large-scale, long-term studies, and (4) explicit integration of ecological and socieconomic concerns. Increased communication will be required both between ecologists specializing in different kinds of ecosystems and between ecologists and social scientists.

These steps do not require the construction of expensive laboratory facilities or the use of high-performance computers. The process outlined in Figure 17.1 calls only for careful identification of assessment objectives, assembly of multidisciplinary data collection and assessment teams, and rigorous interpretation of results using both quantitative and qualitative methods. These requirements are essentially identical to those for sound environmental impact assessments (EIA), as developed and implemented world-wide since 1970. In fact, the definition of risk assessment proposed here includes EIA as a form of risk assessment concerned with predicting impacts of projects (e.g., power plants or dams) as part of the project planning process. Wherever an infrastructure for EIA exists, that same infrastructure can be used to support ecological risk assessments.

17.8 REFERENCES

Abel, R., *et al.* (1986) In: *Proceedings of IEEE Oceans 1986 Organotin Symposium*, Vol. 4, pp. 1314–1323. Institute of Electrical and Electronics Engineers (IEEE), Washington, D.C.

Alzieu, C. (1986) In: *Proceedings of IEEE Oceans 1986 Organotin Symposium, Vol. 4*, pp. 1130–1134. Institute of Electrical and Electronics Engineers (IEEE), Washington, D.C.

Barnthouse, L.W. (1993) The role of models in ecological risk assessment: a 1990s perspective. *Environ. Toxicol. Chem.* **11**, 1751–1760.

Barnthouse, L.W., Suter II, G.W., and Rosen, A.E. (1990) Risks of toxic contaminants to exploited fish populations: influence of life history, data uncertainty, and exploitation intensity. *Environ. Toxicol. Chem.* **9**, 297–311.

Bartell, S.M., Gardner, R.H., and O'Neill, R.V. (1992) *Ecological Risk Estimation*. Lewis, Boca Raton, Florida, 252 pp.

Bascietto, J., Hinckley, D., Plafkin, J., and Slimak, M. (1990) Ecotoxicity and ecological risk assessment: Regulatory applications at EPA. *Environ. Sci. Technol.* **24**, 10–14.

Blunden, S.J., and Chapman, A. (1982) In: Craig, P.J. (Ed.), *Organometallic Compounds in the Environment*, pp. 110–150. John Wiley & Sons, New York.

Cohen, Y. (1986) *Pollutants on a Multimedia Environment*. Plenum, New York.

Costanza, R. (1990) *Ecological Economics*. Columbia University Press, New York, 525 pp.

Hill, A.B. (1965) The environment and disease: association or causation? *Proc. Royal Soc. Med.* **58**, 295–300.

Hordjik, L. (1991) Use of the RAINS model in acid rain negotiations in Europe. *Environ. Sci. Technol.* **25**, 596–603

Huggett, R.J., Unger, M.A., Seligman, P.F., and Valkirs, A.O. (1992) The marine biocide tributyltin. *Environ. Sci. Technol.* **26**, 232–237.

Hunsaker, C.T., Carpenter, D.L., and Messer, J. (1990) Ecological indicators for regional environmental monitoring. *Bull. Ecol. Soc. Am.* **71**, 165–172.

Jørgensen, S.E. (Ed.) (1984) *Modelling the Fate and Effects of Toxic Substances in the Environment*. Elsevier, Amsterdam.

Kendall, R.J. (1992) Farming with agrochemicals: the response of wildlife. *Environ. Sci. Technol.* **26**, 239–245.

Kimerle, R.A., Werner, A.F., and Adams, W.J. (1983) Aquatic hazard evaluation principles applied to the development of water quality criteria. In: Cardwell, R.D., Purdy, R., and Bahner, R.C. (Eds.) *Aquatic Toxicology and Hazard Assessment, Seventh Symposium*, pp. 538–547. American Society for Testing and Materials, Philadelphia.

Kooijman, S.A.L.M. (1987) A safety factor for LC_{50} values allowing for differences in sensitivity among species. *Water Res.* **21**, 269–276.

MacKay, D., and Patterson, S. (1993) Mathematical models of transport and fate. In: Suter II, G.W. (Ed.) *Ecological Risk Assessment*, pp. 129–152. Lewis, Boca Raton, Florida.

NRC (National Research Council) (1983) *Risk Assessment in the Federal Government: Managing the Process*. National Academy Press, Washington, D.C., 191 pp.

NRC (National Research Council) (1993) *Issues in Risk Assessment*, Vol. 3: Ecological Risk Assessment. National Academy Press, Washington, D.C.

OECD (1984) *OECD Guidelines for Testing of Chemicals. Updated Guidelines.* Organisation for Economic Co-operation and Development, Paris.

OECD (1989) *Compendium of Environmental Exposure Assessment Methods for Chemicals.* Environmental Monographs No. 27. Organisation for Economic Co-operation and Development, Paris.

Orians, G.H. (Ed.) (1990) *The Preservation and Valuation of Biological Resources.* University of Washington Press, Seattle.

Schmidheiny, S. (1992) *Changing Course: A Global Business Perspective on Development and the Environment*. The MIT Press, Cambridge, Massachusetts, 374 pp.

Scodari, P.F. (1990) *Wetlands Protection: The Role of Economics*. Environmental Law Institute, Washington, D.C., 89 pp.

SETAC (Society for Environmental Toxicology and Chemistry) (1991) *A Technical Framework for Life-Cycle Assessments*. Society for Environmental Toxicology and Chemistry, and SETAC Foundation for Environmental Education, Inc., Pensacola, Florida.

Shugart, L.R., McCarthy, J.F., and Holbrook, R.S. (1992) Biological markers of environmental and ecological contamination: An overview. *Risk Anal.* **12**, 353–360.

Sloof, W., van Oers, J.A.M., and De Zwart, D. (1986) Margins of uncertainty in ecotoxicological hazard assessment. *Environ. Toxicol. Chem.* **5**, 841–852.

Suter II, G.W., and Loar, J.M. (1992) Weighing the ecological risk of hazardous waste sites: The Oak Ridge case. *Environ. Sci. Technol.* **26**, 432–437.

Suter II, G.W., Vaughan, D.S., and Gardner, R.H. (1983) Risk assessment by analysis of extrapolation error: A demonstration for effects on fish. *Environ. Toxicol. Chem.* **2**, 369–378

Suter II, G.W., Rosen, A.E., Linder, E., and Parkhurst, D.F. (1987) Endpoints for responses of fish to chronic toxic exposures. *Environ. Toxicol. Chem.* **6**, 793–809.

Suter II, G.W. (1993a) *Ecological Risk Assessment*. Lewis, Boca Raton, Florida, 505 pp.

Suter II, G.W. (1993b) Retrospective risk assessment. In: Suter II, G.W. (Ed.) *Ecological Risk Assessment*, pp. 311–364. Lewis, Boca Raton, Florida, 505 pp.

Suter II, W.G., and Barnthouse, L.W. (1993) Assessment concepts. In: Suter II, G.W. (Ed.) *Ecological Risk Assessment*, pp. 21–47. Lewis, Boca Raton, Florida.

van Straalen, N.M., and Denneman, G.A.J. (1989) Ecological evaluation of soil quality criteria. *Ecotoxicol. Environ.* **18**, 241–245.

Volmer, J., Kordel, W., and Klein, W. (1988) A proposed method for calculating taxonomic-group-specific variances for use in ecological risk assessment. *Chemosphere* **17**, 1493–1500.

Turner, M.G., and Gardner, R.H. (Eds.) (1991) *Quantitative Methods in Landscape Ecology*. Springer-Verlag, New York.

Urban, J.D., and Cook, N.J. (1986) *Ecological Risk Assessment: Standard Evaluation Procedures*. Report No. EPA 540/9-85-001. US Environmental Protection Agency, Hazard Evaluation Division, Washington, D.C. Available from NTIS (National Technical Information Service), Springfield, Virginia 22161, as PB 86-247-657.

USDI (US Department of the Interior) (1991) Natural resource damage assessments; notice of proposed rulemaking. *Fed. Reg.* **56**, 19753-19773 (43 CFR Part 11), 29 April.

USEPA (US Environmental Protection Agency) (1992) *Framework for Ecological Risk Assessment*. Report No. EPA/630/R-92/001. US Environmental Protection Agency, Risk Assessment Forum, Washington, D.C.

World Commission on Environment and Development (1987) *Our Common Future*. Oxford University Press, Oxford, UK, 400 pp.

Yount, J.D., and Niemi, G.J. (1990) Recovery of lotic communities and ecosystems from disturbance—a narrative review of case studies. *Environ. Manag.* **14**, 547–569.

18 Estimation of Damage to Ecosystems

Francois Ramade
University of Paris-Sud, France

18.1 INTRODUCTION

Assessment of the environmental impact of chemicals at the ecosystem level raises theoretical and methodological problems of substantial magnitude.

Historically, the assessment of the potential environmental impact of a chemical has been achieved through laboratory testing—overwhelmingly on single species; recently, multispecies testing has been attempted to increase understanding about the toxicity of chemical (Cairns and Orvos, 1989; Levin *et al.*, 1990).

The prediction of toxicity has also been attempted by using quantitative structure–activity relationships (QSR) (Calamari, 1990), and in some special instances by using dispersion models (MacKay and Paterson, 1981) to compare the potential impact on sensitive species, serving as bioindicators, with the expected lethal concentration for a specified environment. The predictive value of microcosm toxicity tests to predict hazards has been hotly debated, but the testing of new chemicals in mesocosms, especially pesticides, has been more recently developed in several countries, and even regulated by the USEPA in the late 1980s (Vosheel, 1989).

For studies in natural habitats, biomarkers have been developed to improve the estimation of sublethal exposure on populations of critical species in the wild (Peakall, 1992). They provide increased accuracy in the estimation of impacts from chronic exposures to defined chemicals in an environment. The occurrence in nature of tolerant strains of exposed species has also been proposed as an indication that some threshold of toxicity for a substance has been reached (Blanck *et al.*, 1988).

Whatever their usefulness, these methods are deficient in assessment—at the ecosystem level—of the impacts of chemical mixtures such as may be found in industrial effluents (Cairns and Orvos, 1989).

Several environmental toxicologists believe that the properties of an ecosystem could be equated to the sum of the properties of its individual components—a widespread and misleading concept in ecotoxicology. Were this concept valid, the study of bioindicator organisms from among those estimated to be the most sensitive would be adequate for an overall assessment of potential damage. This hypothesis is ecologically unsound for several reasons. First, many detrimental

Methods to Assess the Effects of Chemicals on Ecosystems
Edited by R. A. Linthurst, P. Bourdeau, and R. G. Tardiff

effects (particularly impairment of reproductive performance, reductions in development or growth) may occur at concentrations well below those causing lethality. Second, even if perfectly understood, the toxicity of a chemical on a specific population is of little value in characterizing the toxicity that may be manifest in the whole ecosystem.

Therefore, the current approach of ecological risk assessment must be replaced with a paradigm that assures that the toxicity of chemicals are studied throughout entire ecosystems. Therefore, an urgent need exists to develop and validate methods to more accurately assess impacts of chemicals at the ecosystem level. Such methods will require a strong emphasis on basic ecological research. Some ecological catastrophes (like that of the oil spill from the foundering of the *Exxon Valdez*) have emphasized this need at a practical level, namely, the appraisal of ecological damage to determine fair levels of compensation to those who suffered economic loss.

18.2 ASSESSMENT OF EFFECTS ON COMMUNITY STRUCTURES AND DYNAMICS

Natural communities consist of complex assemblies of hundreds to hundreds of thousands of species that are in dynamic equilibrium and that interact with the complex physico–chemical components of their ecosystems.

The biota specific to an ecosystem play a major role in fundamental ecological processes, and modify its physical and chemical environment. Conversely, the ecosystem influences the composition and diversity of the community. Thus, it is most important for the ecological risk assessment to appraise the effects of chemical stresses on the community structure and dynamics. Such studies require specific methodologies to estimate the potential effects of substances on living resources.

18.3 PREDICTING RISKS IN POPULATIONS

18.3.1 REDUCTIONS IN POPULATION SIZE AND DENSITY

The most obvious and impressive effect of a chemical spill is the elimination of most organisms poisoned in the impacted area. Several such dramatic consequences have been observed recently. For example, on 30 October 1986, in Switzerland, the Sandoz accident killed most fish over several hundred kilometres of the Rhine river downstream from Basel. Even when contamination is less dramatic, ample evidence indicates that the introduction of toxic chemicals in terrestrial or aquatic habitats results in reduction of density, and even complete extinction, of populations of the most sensitive species.

By contrast, chronic exposure to chemicals occurring in terrestrial or aquatic habitats may lead to extensive decline of the exposed populations of animal or plant species. The author's investigations have demonstrated that the permanent exposure of a macroinvertebrate community to low concentrations of organochlorine insecticides occurring in ponds located among field areas intensively sprayed resulted in a sharp decrease in density and in the disappearance of several families of benthic invertebrates, especially those most pollutant-sensitive (Ramade *et al.*, 1983, 1985; Ramade, 1987a).

Mortality from either acute or chronic exposures is the most obvious toxic manifestation and, thereby, the most frequently used in the assessment of impacts on populations. However, a chemical can cause non-lethal damage, such as a decline in fecundity and reproductive success, the impairment of mating in vertebrates, the slowing of growth, a delay in, or an impairment of, metamorphosis in invertebrates, inefficient pollination or gamete proliferation in algae, and partial inhibition of photosynthesis in plants. Any of these changes can have a deleterious effect on populations.

18.3.2 PREDICTING EFFECTS ON POPULATIONS

To minimize the impact of chemical pollution on wild populations and on the whole biota, safety standards for environmental protection have been promulgated. The most widely recognized environmental safety standard for any potentially dangerous substance is the maximum acceptable toxicant concentration (MATC).

The MATC is a measure of the lethal dose of a substance in a test of a single species with a vast array of organisms. An estimate of the concentration that will protect 95 percent of the taxa occurring in a given community exposed to a test chemical is calculated; therefore, only the most sensitive 5 percent are affected. The MATC is the geometric mean of either the no-observed-effect concentration (NOEC) or of the lowest-observed-effect concentration (LOEC) in chronic exposures.

Other similar methodologies have been devised by Van Leeuwen (1990). In the Netherlands, the assessment of risks to ecosystems from toxic chemicals relies on the HC*p* method of Van Straalen and Denneman (1989), where HC*p* is the hazardous concentration for a percentage *p* of the exposed species:

$$HCp = \frac{e^{Xm}}{T} \quad (1)$$

$$T = \exp \left[\frac{3d_m S_m}{\pi^2 \log \frac{(1-\delta_1)}{\delta_2}} \right] \quad (2)$$

where *Xm* is the sample mean of log NOEC for *m* test species, S_m is the sample standard deviation of log NOEC values for *m* test species, δ_1 is the fraction of the ecosystem that is not protected (recommended value $\delta_1 = 0.05$), δ_2 is the probability of overestimating the HC*p* (recommended value 0.05), d_m is the value such that the probability of $(S_m > d_m) = \delta_2$, and *T* is the application factor between *HCp* and e^{Xm}.

Several limitations accompany these methods, including the accuracy which decreases substantially from the population to the community level and even further from the community to the whole ecosystem. A major limitation is the lack of concordance between findings from laboratory tests and those from field investigations. As Cairns and Orvos (1989) pointed out, "this is probably responsible for the rather curious evolutionary process of laboratory tests becoming more and more sophisticated and the methodology more complex and, most important, more isolated from mainstream ecology."

Apart from the admonition that the distribution of sensitivities of the species studied is log-normal, a major criticism of these methods is that they rely on the assumption that a community is protected if the MATC is not exceeded, implying that the NOEC would be exceeded for only 5 percent of the species. This assumption is invalid if any critical species are among the 5 percent affected. Moreover, these predictive tests almost exclusively measure lethality, ignoring other toxic reactions. In many countries, this shortcoming stems from the fact that the so-called predictive tests "have been driven more by regulatory convenience than by sound ecological practices" (Cairns and Orvos, 1989).

The current methods to predict effects at the population level do not identify discrete changes in population size of any critical species; thus, a dominant or key species can trigger major changes in the overall structure of an entire community. The effects of chemicals on key species has been poorly investigated, particularly in some temperate ecosystems (Levin *et al.*, 1990). For example, many plant pollinators, although discrete and minute, may prove to be species essential to a community, so that an undetected decline in their densities could generate detrimental changes jeopardizing the future of the whole ecosystem (Sheehan *et al.*, 1984).

A non-toxicological factor is food shortages that change susceptibility to a toxicant in a prey species, which is thought highly important ecologically, but which is also ignored in the protection of biota against chemical pollution. The decline of the population upon which a predator preys may have considerable impact on the predator population despite the absence of toxicity of the chemical for this species. Several examples demonstrate this phenomenon: (1) in Britain, the decline of the grey partridge due to the rarefaction of non-target insects in cereal fields sprayed by pesticides (Rands, 1985) and (2) the breeding success of the blue tit *(Parus caeruleus)* in forests sprayed by cypermethrin, due to the reduction of caterpillars, that are the basic diet of nestlings (Pascual and Peres, 1992). However, the development of multispecies predictive tests, even when accepted by regulatory authorities, still has not increased efficiency in conserving the exposed community. Accordingly, only the monitoring of polluted ecosystem and site-specific validation of the effects of a given chemical can provide the additional ecological data requested.

18.3.2.1 Reduction in diversity and species richness

The reduction in density and diversity of species in chronically polluted habitats is the primary contributor to alterations in community structure. Indicator species are also relevant to assess effects on communities. Despite several criticisms regarding their use, they provide valuable information on the early effects of chemicals on biota.

The properties required from an effective bioindicator are the ability of its population to respond to discrete changes in the environment induced by chemicals, hypersensitivity to stimuli, reliability and specificity of its responses, rapidity of response, and ease of monitoring. If these prerequisites are met, discrete changes in carefully selected indicators would measure both the rate and magnitude of change induced by a chemical in a community.

18.3.2.2 Species diversity for use as an ecological index

Conceptually, ecological diversity integrates both the number of species (richness) and their relative frequency in a given biota. Decreased diversity has been used to assess gross environmental degradation in ecosystems for several decades. Among those indices that are the most frequently applied to chemical pollution, that of Margaleff are presented:

$$H = \left(\frac{1}{N}\right) \log_2 N! - \sum_{i=1}^{S} \log_2 n_1! \tag{3}$$

where N = total number of individuals, S = total number of species, n_i = number of individuals of the ith species.

The major limitation of this index is its need to count all individuals from the stressed community. Thereafter, for practical applications, ecotoxicologists selected other indices that could be applied to samples, because communities cannot usually be entirely numbered. The index of Lyod, Zad, and Karr was routinely applied to the study of water pollution of the Seine river:

$$H' = \frac{C}{N} (\log_{10} N - \sum_{i=1}^{s} n_i \log_{10} n_i) \qquad (4)$$

where C is the number of class of frequency expressed in bits (for 10 class, C = 3.3219).

Shannon's index has been by far the most widely used in ecotoxicology:

$$H' = - \sum_{i=1}^{s} \frac{n_i}{N} \log_2 \frac{n_i}{N} \qquad (5)$$

Several limitations hamper the effectiveness of the diversity index to assess the effects of chemicals on community structure. Shannon's index, for example, gives the same weight to systematic units of the same abundance, whatever their taxonomic level and *a fortiori* affinities. To avoid these limitations, Osborne, Davis, and Linton have proposed the use of a hierarchical diversity index (HDI) devised by a formula expanded from Pielou to include three taxonomic levels (familial, generic, and specific):

$$HDI = H'(F) + H'(G) + H'(S) \qquad (6)$$

where $H'(F)$ is the familial component of the total diversity, $H'(G)$ is the generic component, and $H'(S)$ is the specific component of the total diversity.

The most universal criticism of the application of the Shannon diversity measure in ecotoxicology is the misleading interpretation of data from communities influenced by a large evenness component (Godfrey, 1978).

Another problem in using the diversity index is the absence of linearity of a community response to a given gradient of increasing pollution, which is never univocal. During an initial stage, at sublethal exposure, some dominant species among the most pollutant sensitive will decrease their abundance, increasing the evenness and, therefore, the diversity index value. Only the onset of lethal exposures triggers the disappearance of species less pollutant-sensitive, thereby lowering the index (Figure 18.1).

Boyle *et al.* (1990) have conducted a theoretical study to assess the validity of

16 indices of water quality. They started with three communities differing in species richness but presenting the same abundance–value distribution curves. Their overall conclusion was that while the Shannon index affords a good representation of changes in species richness, it generally does not provide much representativeness of changes occurring in polluted communities and can even produce misleading conclusions.

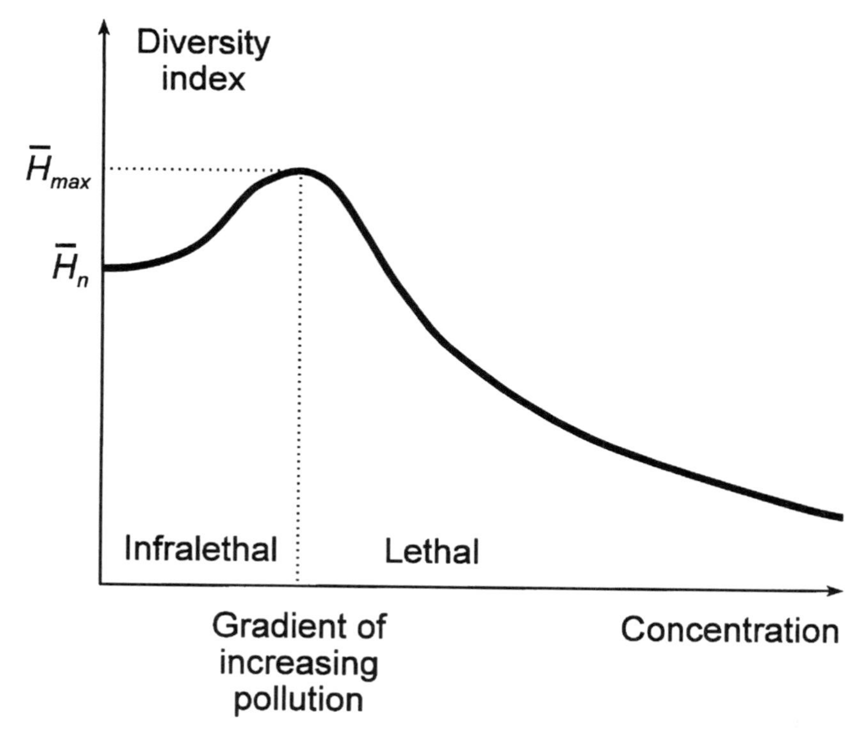

Figure 18.1. Theoretical variation of the diversity index inside a gradient of increasing concentration of a given chemical (Ramade, 1987b).

In conclusion, the diversity index reflects changes in ecosystems structure only during periods of severe stress. In moderately polluted ecosystems, changes in dominance strongly affect equatability, thereby hampering the effectiveness of diversity indices in distinguishing degraded communities from unstressed ones. The index can be improved in estimating changes in abundance in a whole community, because the changes due to chemical pollution on a taxonomic group has been identified previously, and has been carefully studied for its value as indicator. The lichens are a sound illustration of such a use to assess the level of air pollution by SO_2 and predict the ecological effects associated with the observed levels. For example, an index of atmospheric purity (IPA) was derived from biocoenotic surveys revealing the species diversity of lichen communities in relationship with the level of SO_2 pollution):

$$IPA = \frac{S}{100[\Sigma(Q \times f)]} \tag{7}$$

where S = number of lichen species in the area sampled, f = frequency of each species, and Q = index of toxiphobia of each species.

18.3.2.3 Effects on frequency distribution of species.

The means by which species populations are controlled is one of the most important aspects of the ecological niche. The presence of a pollutant in a habitat will either impact the area occupied by each species or the resources used by each species, depending upon the level of tolerance or sensitivity of the species. Consequently, the balance between the various parts of a community will be disturbed as pollutants force modifications in competition leading to the elimination of the most sensitive populations. Therefore, the frequency distribution of species will be distorted to varying degrees by the chronic pollution of an ecosystem. The use of importance value distribution curves can be used methodologically to assess the response of a community to a chemical contaminant.

Ramade *et al.* (1985) compared the importance value distribution curves computed from experimental data from two lentic communities of benthic invertebrates: one from ponds contaminated by insecticides and the other from meadow ponds used as a control (Figure 18.2). The latter fits well on a Preston distribution, whereas the community from fields ponds fits on an intermediary position between Preston's model and a log-linear one. Generally, communities strongly disturbed by any pollutant are prone to fit a log-linear model, which is applicable to communities living in constraining environments.

18.4 PRINCIPLES TO ASSESS THE EFFECTS OF CHEMICALS ON ECOSYSTEMS

An ecosystem is a structural and functional complex array of associations within a living community and with its physico–chemical environment. Consequently, a chemical may be assumed to induce changes in the community structure and function, notwithstanding possible corresponding changes in the physicochemical structure of the contaminated habitats due to changes in activity of its whole biota, particularly its decomposer biomass. Dissecting extensive studies on the impact of acid rain on lakes, the action of chronic pollution by a chemical on photosynthesis ranks among the major deleterious effects of chemicals on fundamental ecological processes.

Discrete actions of a chemical can affect the primary productivity of terrestrial or aquatic ecosystems. For example, it was demonstrated that SO_2 at only 10 ppb in the atmosphere may affect the primary productivity of the Scot pine (*Pinus sylvestris*) forest; no morphological or anatomical damage can be detected at this concentration (Grodzinsky *et al.*, 1984). Phytoplanctonic algae may experience an inhibition of their photosynthetic activity at dosages lower than 1 ppb for some pesticides. Research has shown a substantial decrease in phytoplankton and filamentous algae from ponds contaminated by various herbicides (especially Chlortoluron, Triazine, and Neburon) coming from adjacent fields (Goacoulou and Echaubard, 1987).

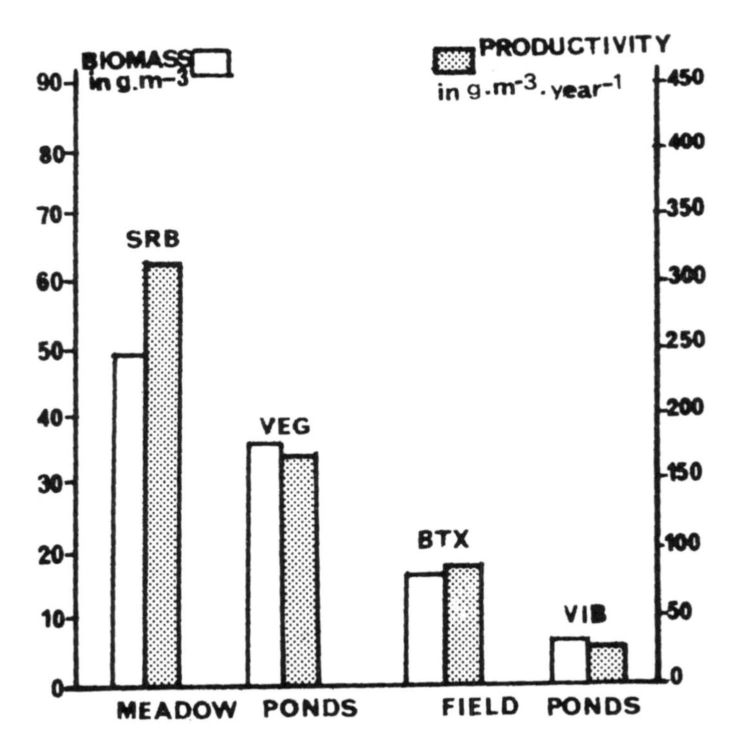

Figure 18.2. Comparison of importance–value curves computed for a community of benthic macroinvertebrates from field ponds chronically polluted by pesticides and from control ponds (Ramade *et al.*, 1985)

Effects on secondary productivity, though less explored, stand as a major parameter by which to assess effects at the ecosystem level. Among the most investigated is the study of the action of acid rain and pesticides on invertebrates and on freshwater fisheries exposed to forest spraying or acidification (Ide, 1967).

For pesticides, a substantial decrease has been demonstrated in benthic macroinvertebrates productivity resulting from contamination of freshwater ponds by organochlorine insecticides run off from surrounding fields, despite the fact that concentrations in water and sediments were well below acutely toxic concentrations. In acidified lakes, a sharp decrease in abundance and biomass of zooplankton has been observed (Almer *et al.*, 1974; Stenson and Oscarson, 1985). A sharp decline has been also observed in commercial salmon fishing as a result of organochlorine insecticide pollution as early as the 1960s. For example, the New Brunswick salmon streams were badly affected by DDT spraying intended to control the spruce budworm. The decrease in the secondary productivity was demonstrated to be not so much a consequence of the insecticide toxicity in fish, but of the collapse of the aquatic insect populations upon which young salmons normally feed (Ide, 1967). Effects of river acidification on the productivity of the salmon fishing industry has been also well documented, and careful studies have shown a strong decline directly related to the level of acidification experienced (Leivestad and Muniz, 1976; Watt *et al.*, 1983).

Some attempts have been made to model the effects of a chemical on some major components of ecosystem structure or function to expand the risk analysis methodology for ecosystems. O'Neill *et al.* (1982) devised a method to extrapolate laboratory toxicity data to aquatic ecosystem effects such as decreased productivity or reduction in fish biomass. Called the standard water column model or SWACOM, this method requires translating laboratory data into changes in the parameters of an entire ecosystem. The extrapolation is accomplished with knowledge of toxicological modes of action, and by simulation of the effects of a toxic substance across different trophic levels—accordingly on the relationship between nutrients, phytoplankton, zooplankton, and fish. Various scenarios are developed, and each models population interactions that alter both the level and the nature of the risk to ecosystem processes. Moreover, the method describes uncertainties associated with laboratory measurements and the extrapolations, and describes risks as probabilities.

Relying on this SWACOM method, O'Neill *et al.* (1982) simulated well the effects of phenol and quinoline in an aquatic ecosystem. However, Cairns *et al.* (1988) stressed that experimental simulation results must be validated to assure a high accuracy of the estimates; many ecosystem simulation units have relatively low environmental realism, making estimates highly imprecise. Consequently, field studies on natural ecosystems are essential to achieve this validation.

Since research on the effects of chemicals carried out in full-scale ecosystems may prove lengthy, complex, expensive, and inaccurate due to the vast number of variables, the mesocosm approach was developed in the 1970s to estimate the effects of chemicals on ecosystems (Mauck *et al.*, 1976; Vosheel, 1989). The problem that still impedes assessment of the effects of chemicals by the mesocosm method stems from the expense that limits its use in ecotoxicology. Research is progressing (Caquet *et al.*, 1992) across Europe to develop mesocosms that would be less expensive while still providing adequate representation of natural ecosystems.

18.4.1 EFFECTS ON DECOMPOSERS AND NUTRIENTS CYCLING

Another major parameter of ecosystem functioning is related to nutrient cycling. Impairment of the activity of decomposers by a chemical may lead to major alterations and even the collapse of the whole ecosystem. For example, air pollutants disrupt the nutrients cycle of saprophagous insects that feed on the dead litter in forest ecosystems, and the consumption of dead organic matter by the soil Arthropods is slowed considerably. Other studies have shown effects of the same magnitude on leaf decomposition in the waters of acidified lakes. Therefore, possible effects of pollutants must be assessed on microorganisms that decompose dead organic matter in assorted ecosystems as part of a sound ecological risk assessment.

The appraisal of possible effects on the biogeochemical cycles of major nutrients is also a major issue in risk assessments at the ecosystem level. Attempts have been made to judge impairment of the nitrogen cycle at ecosystem level: Mathes and Shultz-Berendt (1988) reported how Aldicarb alters nitrification at the agroecosystem level. From previous studies, Mathes and Weidemann (1990) proposed an integrated approach to assess changes caused by chemicals on terrestrial ecosystems, including comparisons between two parts of an ecosystem, one with the test substance and the other without to serve as a control.

This approach has several prerequisites, especially a suitable indicator system. The spatio–temporal scale of exposure and temporal variability must be considered. The selection of reference organisms should incorporate both short-lived and long-lived species.

18.5 CONCLUSIONS

Major conclusions from this review include the following:

1. The structural and functional complexity of ecosystems needs further research to identify key species and to improve the knowledge of bioindicators as a tool to identify the effects of chemicals on communities.
2. The need exists to accurately assess toxic effects and requires additional study to ascertain the impact of chemicals on fundamental ecological processes such as primary and secondary productivity at the decomposer level, on nutrient cycles, and on biogeochemical cycles.
3. Since the detrimental effects of chemicals are not usually restricted to an ecosystem but impact areas that include both terrestrial and aquatic habitats, landscape ecology must be included in any assessment of risk.

18.6 REFERENCES

Almer, B., Dickson, W., and Eckstrom, C. (1974) Effects of acidification on Swedish lakes. *Ambio* **3**, 30–36.

Blanck, H., Wanberg, S.T, and Molander, S. (1988) *Pollution-Induced Community Tolerance, A New Ecotoxicological Tool*, pp. 219–230. Special Technical Publication No. 988. American Society for Testing and Materials, Philadelphia.

Boyle, T.C., Smillie, G.M., Anderson, J.C., and Beeson, J.R. (1990) A sensitivity analysis of nine diversity and seven similarity indices. *J. Water Pollut. Control Fed.* **62**, 749–762.

Cairns, J., and Orvos, D. (1989) Ecological consequence assessment: predicting effects of hazardous substances upon aquatic ecosystems using ecotoxicological engineering. In: *Ecological Engineering: An Application to Ecotechnology*. John Wiley & Sons, New York, 411 pp.

Cairns, J., Smith, E.P., and Orvos, D. (1988) The problem of validating simulation of hazardous exposure in natural systems. In: Barnett, C.C., and Holmes, W.M. (Eds.) *Proceedings of a Summer Conference of the Society Comp. Simul. Int., San Diego, California*, pp. 448–454. San Diego.

Calamari, D. (Ed.) (1990) *Chemical Exposure Prediction. Abstracts of a Special Workshop.* European Science Foundation, Strasbourg, 224 pp.

Caquet, T., Thybaud, E., LeBras, S., Jonot, O., and Ramade, F. (1992) Fate and biological effect of lindane and deltamethrin in freshwater mesocosms. *Aquat. Toxicol.* **23**, 261–278.

Goacoulou, J., and Echaubard, M. (1987) Influence des traitements phytosanitaires Sur les biocoenoses limniques: le phytoplancton. *Hydrobiologia* **148**, 269–280.

Godfrey, P.J. (1978) Diversity as a measure of benthic macroinvertebrates community response to water pollution. *Hydrobiologia* **57**, 111–122.

Grodzinsky, W., Weiner, J., Maycock, P.F. (1984) *Forest Ecosystems in Industrial Regions.* Springer-Verlag, Berlin, 223 pp.

Harris, J., Sager, P., Regier, H.A., and Francis, G.C. (1990) Ecotoxicology and ecosystem integrity: the Great Lakes examined. *Environ. Sci. Technol.* **24**(5), 598–603.

Ide, F.P. (1967) Effects of forest spraying with DDT on aquatic insects of salmon streams in New Brunswick. *J. Fish. Res. Board. Can.*, **24**(9), 769–805.

Jackson, D.R., and Watson, A.P. (1977) Disruption of nutrients pools and transport of heavy metals in a forested watershed near a lead smelter. *J. Environ. Qual.* **6**, 331–338.

Keenleyside, M.H.A. (1967) Effects of forest spraying with DDT in New Brunswick on food of young Atlantic salmon. *J. Fish. Res. Board Can.* **24**(4) 807–822.

Leivestad, H., and Muniz, I.P. (1976) Fish kill at low pH in a Norwegian river. *Nature* **259**(5542), 391–392.

Levin, S.A., Harwell, M.A., Kelly J.R., and Kimball, K.D. (Eds.) (1990) *Ecotoxicology: Problems and Approches*. Springer Verlag, New York, 547 pp.

MacKay, D., and Paterson, S.A. (1981) Calculating fugacity. *Environ. Sci. Technol.* **15**, 1006–10021.

Mauck, W.L., Mayer, F.L., and Holz, D.D. (1976) Simazine residues dynamics in small ponds. *Bull. Environ. Contam. Toxicol.* **16**, 1–8.

Mathes, K., and Schultz-Berendt, V.M. (1988) Ecotoxicological risk assessment of chemicals by measurements of nitrification combined with a computer simulation model of N-cycle. *Tox. Assess. Int. J.* **3**, 271–286.

Mathes, K., and Weidemann, G. (1990) A baseline-ecosystem approach to the analysis of ecotoxicological effects. *Environ. Safety* **20**, 197–202.

Niederlehner, B.R., Pratt, J.R., Buikema, A.L., and Cairns, J. (1986) Comparison of estimates of hazards derived at three levels of complexity in community toxicity testing. STP 920, pp. 30–48. American Society for Testing and Materials, Philadelphia.

O'Neill, R.V., Gardner, R.H., Barnthouse, G.W., Suter, S., Hildebrand, S.G., and Gehrs, C.W. (1982) Ecosystem risks analysis: a new methodology. *Environ. Toxicol. Chem.* **1**, 167–177.

Osborne, L.L., Davies, R.W., and Linton, K.J. (1980) Use of hierarchical diversity indices in lotic community analysis. *J. Appl. Ecol.* **17**, 567–580.

Pascual, J.A., and Peres, J. (1992) Effects of forest spraying with two application rates of Cypermethrin on food supply and on breeding success of the blue tit (*Parus coeruleus*). *Environ. Toxicol. Chem.* **11**, 1271–1280.

Peakall, D. (1992) *Animal Biomarkers as Pollution Indicators.* Chapman and Hall, London, 291 pp.

Pimentel, D., and Edwards, C.A. (1982) Pesticides and ecosystems. *Bioscience* **32**, 595–600.

Ramade, F. (1977) *Ecotoxicologie.* Masson, Paris, 214 pp.

Ramade, F. (1987a) Proposal of ecotoxicological criteria for the assessment of the impact of pollution on environmental quality. *Toxicol. Environ. Chem.* **13**, 189–203.

Ramade, F. (1987b) *Ecotoxicology.* John Wiley & Sons, Chichester, UK 262 pp.

Ramade, F. (1992) *Precis d'Ecotoxicologie.* Masson, Paris, 97 pp.

Ramade, F., Echaubard, M., LeBras, S., and Moreteau, J.C. (1983) Influence des traitements phytosanitaires sur les Biocoenoses limniques. *Acta Oecol. Oecol. Appl.* **4**, 3.

Ramade, F, Moreteau, J.C, LeBras, S., and Echaubard, M. (1985) Influence des limniques traitements phytosanitaires sur les. Biocoenoses II, comparaison de la structure des peuplements propres aux biotopes etudies. *Acta Oecol. Oecol. Appl.* **6**, 227.

Rands, M.R.W. (1985) Pesticides use in cereal and the survival of grey partridge chicks: a field experiment. *J. Appl. Ecol.* **22**, 49–54.

Sheehan, P.J. (1984) Effects on community and ecosystems structure and dynamic. In: *Effect of Pollutants at Ecosystem Level*, pp. 510–599. Scope 22. John Wiley & Sons, Chichester, UK.

Sheehan, P.J., Miller, D.R., Butler, G.C., and Bourdeau, P. (Eds.) (1984) *Effect of Pollutants at Ecosystem Level*, p. 443. Scope 22. John Wiley & Sons, Chichester, UK.

Stenson, J.E., and Oscarson, N.G. (1985) Crustacean zooplankton in an acidified lake. *Ecol. Bull.* **37**, 224–238.

Van Leeuwen (1990) Ecotoxicological effects assessment in the Netherlands: recent developments. *Environ. Manag.* **14**(6), 779–792.

Van Straalen, N.M., and Denneman, C.A.J. (1989) Ecotoxicological evaluation of soil quality criteria. *Ecotoxicol. Environ. Safety* **18**, 241–251.

Verneaux, J. (1976) Fondements biologiques et ecologiques de l'etude de la qualite des eaux continentales. In: Pesson, P. (Ed.) *La Pollution des Eaux Continentales.* Gauthier-Villars, Paris, 263 pp.

Vosheel, J.R. (1989) *Using Mesocosms to Assess the Aquatic Ecotoxicological Risk of Pesticides: Theory and Practice.* Misc. Pubs. No. 75. Entomological Society of America, Lanham, Maryland, 88 pp.

Watt, W.D., Scott, C.D., and White, W.J. (1983) Evidence of acidification of Nova Scotia rivers and its impact on Atlantic salmon *Salmo salar. Can. J. Fish. Aquat. Sci.* **40**, 462–473.

19 Methods for Economic and Sociological Considerations in Ecological Risk Assessment

Robert Costanza
University of Maryland, USA

Peter P. Principe
US Environmental Protection Agency, USA

19.1 INTRODUCTION

Ecological risk assessment remains a jigsaw puzzle for which the picture being created and all the right pieces are present remain uncertain. This chapter addresses three significant issues related to ecological risk assessment: (1) the definition of ecosystem health, (2) the scale at which such assessments should be conducted, and (3) the use of ecological benefits as an analytical paradigm.

The ecosystem health metric proposed is a comprehensive, multiscale, dynamic, hierarchical measure of system resilience, organization, and vigour that closely tracks the concept of sustainability. Assessment scale is an important issue, because it tends to define the scope of the policy options considered for mitigation. Currently, an overemphasis exists on population and process-level analyses at the expense of the ecosystem and ecoregion levels. As with the health metric proposed, assessment of ecosystems at multiple levels is important to insure that the cure is no worse than the disease. Finally, a somewhat different perspective to assess ecological systems is discussed. By considering changes in an ecosystem's delivery of ecological benefits (goods and services), the assessment may be able to answer more directly the question of significance.

19.2 A DEFINITION OF ECOSYSTEM HEALTH

The term "health" is commonly used in reference to ecosystems by both scientific and policy documents, but a satisfactory definition of ecosystem health remains to be developed. While the framework of human health may provide a starting point,

Methods to Assess the Effects of Chemicals on Ecosystems
Edited by R. A. Linthurst, P. Bourdeau, and R. G. Tardiff

severe limitations are imposed on the parallel between human health and ecological health (Norton, 1991b; USEPA, 1992a). In addition to its use in ecological assessments, a definition of ecosystem health also provides a means to aid the integration of the analyses of ecological and economic systems (Haskell *et al.*, 1992).

As a starting point, this analysis begins with five axioms of ecological management (Norton, 1991a):

1. The Axiom of Dynamism. Nature is more profoundly a set of processes than a collection of objects; all is in flux. Ecosystems develop and age over time.
2. The Axiom of Relatedness. All processes are related to all other processes.
3. The Axiom of Hierarchy. Processes are not related equally, but unfold in systems within systems, which mainly differ regarding the temporal and spatial scale on which they are organized.
4. The Axiom of Autopoiesis. The autonomous processes of nature are creative, and represent the basis for all biologically-based productivity. The vehicle of that creativity is energy flowing through systems that in turn find stable contexts in larger systems, that provide sufficient stability to allow self-organization within them through repetition and duplication.
5. The Axiom of Differential Fragility. Ecological systems, which form the context of all human activities, vary in the extent to which they can absorb and equilibrate human-caused disruptions in their autonomous processes.

These axioms regularly recur, even if implicitly, in the following discussion, and they are essential elements of the definition of ecosystem health proposed below.

Ecosystem health is often framed in terms of human health (Rapport, 1989, 1992). While both are complex systems, medical science has a large body of knowledge and expert systems (in the form of doctors) available to advance diagnosis (Schaeffer *et al.*, 1988; Schaeffer and Cox, 1992). Such analytical tools are absent for ecosystems. However, ecosystems have been studied extensively with respect to their stability and resilience (Pimm, 1984; Holling, 1986).

Six major concepts are most often used to describe ecosystem health (Costanza 1992):

1. homeostasis,
2. the absence of disease,
3. diversity or complexity,
4. stability or resilience,
5. vigour or scope for growth, and
6. balance between system components.

Each concept represents a piece of the puzzle, but none is sufficiently comprehensive, especially in terms of being able to deal with many different levels of ecological systems.

Homeostasis is the simplest and most popular definition of system health: any and all changes in the system represent a decrease in health. The greatest difficulty with this approach is in differentiating between naturally occurring stresses and external (including anthropogenic) stresses. This definition is best used for warm-blooded vertebrates, since they are homeostatic and since normal ranges can be more easily determined from large populations. However, for ecological systems, all changes (or even any given change) cannot be assumed to be bad. The best example of this is succession—for the initial state, succession is an irreversible change, and one that might be necessary for the system to be sustained. Given that ecosystems are constantly changing, this definition does not deal with a fundamental characteristic of ecosystems.

The definition of ecosystem health as the absence of disease has several failings. First, while various (including anthropogenic) ecosystem stresses can be described, their mere existence does not indicate that they are adverse stresses. A separate, independent definition of ecosystem health would be required. Second, this definition yields only a dichotomous result that is inadequate to characterize complex systems.

The notion of basing a definition of ecosystem health on a system's diversity or complexity rests on the assumption that these characteristics are predictors of stability or resilience and that these are indicators of ecosystem health. Presently, the analytical basis is insufficient to use this concept, but network analysis may yield a more sophisticated framework for incorporating system diversity into a definition of ecosystem health (Wulff *et al.*, 1989; Ulanowicz, 1992).

Stability and resilience are key measures of ecosystem health, since healthy organisms and systems have the ability to recover from stresses or to use the stress in some creative manner to improve their status. A failing is that these measures do not characterize the level of system organization or the level at which the system is functioning.

Odum (1971) has suggested that the level of a system's metabolism (energy flow) is an indicator of its ability to deal with stresses. The concept of ecosystem balance is based in Eastern traditional medicine, and the notion that a healthy system is one that maintains a proper balance between its parts. However, the proper balance can only be determined by some independent measure of ecosystem health.

Based on these framing concepts, a practical definition of ecosystem health must have four essential characteristics. First, it must integrate the definitions described above into one that combines system resilience, balance, organization (diversity), and vigour (metabolism). Second, it must represent a comprehensive description of the ecosystem. Third, it must use weighting factors to compare and aggregate different components of the system. Finally, it must be temporally and spatially hierarchical (Costanza, 1992).

Such a definition would be: "An ecological system is healthy and free from distress syndrome, if it is stable and sustainable—that is, if it is active, and maintains its organization and autonomy over time, and is resilient to stress"

(Haskell *et al.*, 1992). Accordingly, a diseased system is one that is not sustainable, and will eventually cease to exist—clearly illustrating the importance of the temporal and spatial aspects of the definition. Distress syndrome refers to the irreversible processes of system breakdown leading to death.

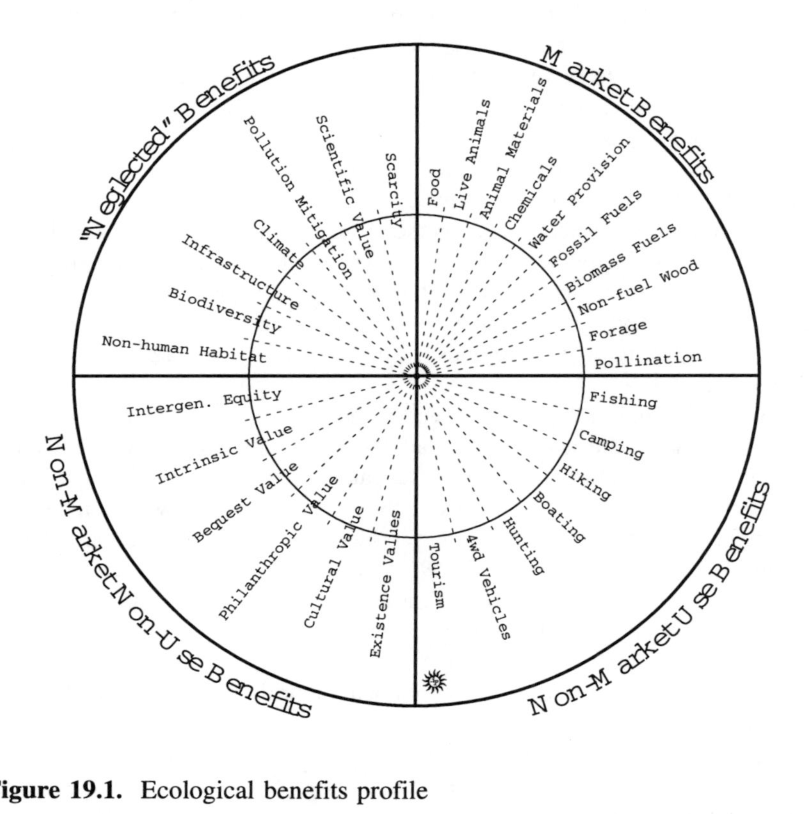

Figure 19.1. Ecological benefits profile

Two very important tools for making operational this definition are network analysis and simulation modelling. Network analysis, in this context, refers to all variations of the analysis of ecological and economic networks. It has the potential for yielding an integrated, quantitative, hierarchical treatment of all complex systems, including ecosystems and combined ecological–economic systems. An important area of network analysis is the development of common pricing mechanisms for ecological and economic systems. In complex systems with many interdependencies, a problem with mixed units is often present. Ecological analyses have ignored this problem by choosing a single commodity as an index; yet this ignores interactions between commodities, and is consequently unrealistic and quite limiting.

Evaluating the health of complex systems demands a pluralistic approach (Norgaard, 1989; Rapport, 1989) and an ability to integrate and synthesize the many diverse perspectives that may be present. An integrated, multiscale, transdisciplinary, and pluralistic approach is required to quantitatively model systems (including organisms, ecosystems, and ecological–economic systems). Achieving such a capability requires the ability to predict the dynamics of ecosystems under stress (Costanza *et al.*, 1990) as well as advances in high-performance computing.

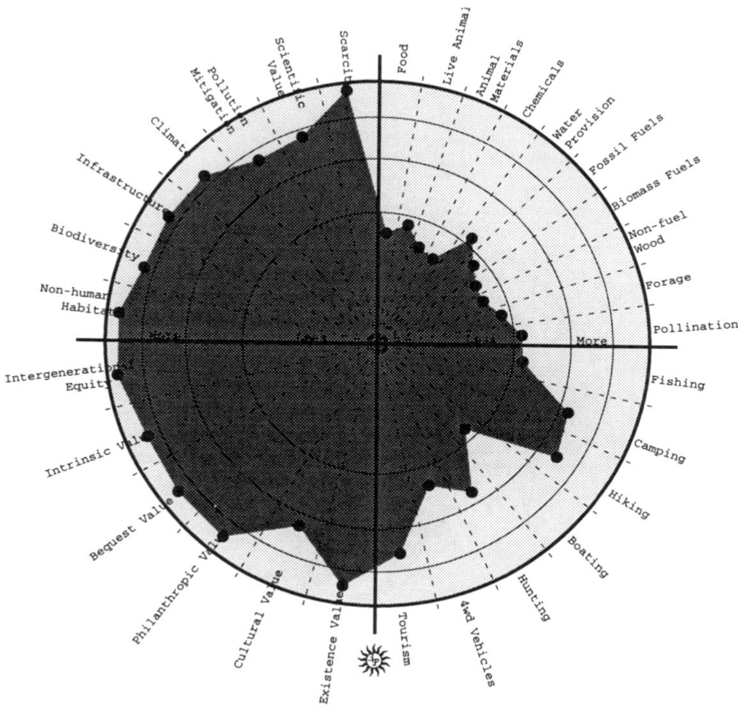

Figure 19.2. Old forest growth with forest intact (points are not data)

19.3 REGIONAL SCALE ECOLOGICAL ASSESSMENT

At several points during the previous discussion, the issue of scale has been raised. Assessment of scale is important, because scale tends to define the scope of the policy options considered for mitigation. Currently, an overemphasis exists on population and process-level analyses at the expense of the ecosystem and ecoregion levels. As with the health metric proposed above, assessment of ecosystems will be important at multiple levels to insure that policy decisions do not result in undesired ecological consequences.

Also, guarding against haphazard aggregation of measures across ecological levels of organization will be important. Different scales of ecological systems may be driven by very different dynamics (Norton *et al.*, 1991) so the best indicators or metrics for one level may be inadequate or misleading at another scale.

While the concept of multiscale analyses is logical and desirable, it poses significant difficulties since most ecological research deals with very small geographic areas (usually 1 m^2). Recognizing the need for ecological assessments to deal with much larger landscapes, some ecologists began in the 1980s to argue that regional-level ecology was important to understand the smaller scale (Allen *et al.*, 1984), and that ecological assessments must be capable of assessing at the regional level (Hunsaker *et al.*, 1990). This view was strongly endorsed by EPA's ecological risk assessment peer-review panel, when it noted:

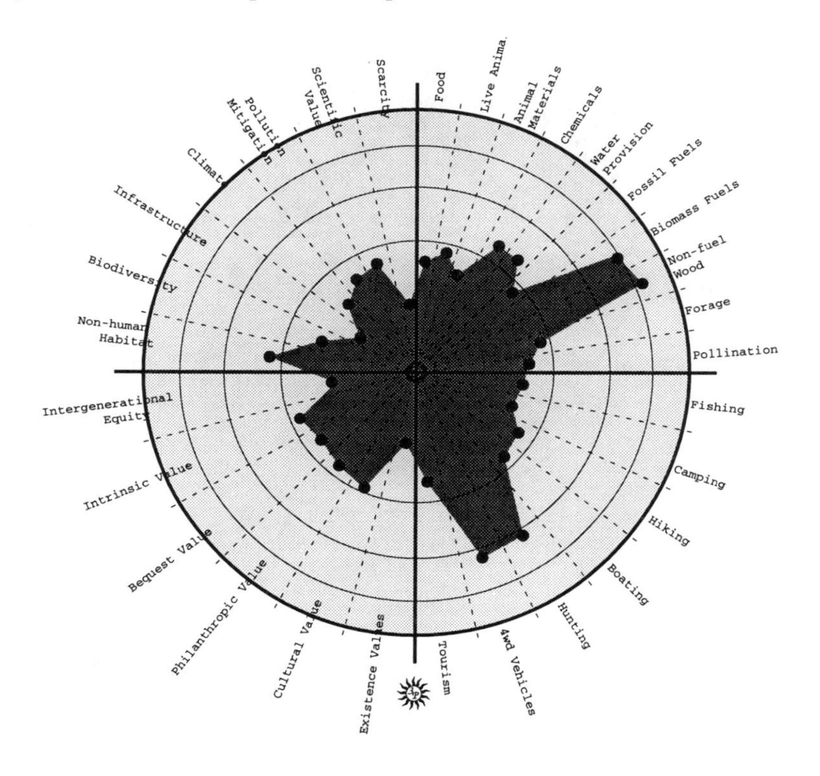

Figure 19.3. Old forest growth with forest clear-cut (points are not data)

"Ecological risk assessment, unlike human health risk assessment, must address a diverse set of ecological systems, from tropical to Arctic environments, deserts to lakes, and estuaries to alpine systems... . Ecological risk assessment may occur over much wider temporal and spatial scales than those for human health risk assessment" (USEPA, 1992).

Meeting this need for multiscale analysis will require the same type of research required to provide the foundation for the definition of ecosystem health, namely, network analysis and simulation modelling. Analyses such as these will be much more feasible because of recent advances in high-powered computing and visualization techniques.

An important element of multiscale analysis of ecosystems, and even single scale analyses, is the proper characterization of uncertainty. Even with the significant advances in modelling, large amounts of uncertainty will exist with respect to anthropocentric effects on ecosystems (Funtowicz and Ravitz, 1991). Developing the means and the methods to characterize this uncertainty in a meaningful manner for policy makers should be a major research area for ecological assessment (Costanza, 1987, 1991; Perrings, 1987, 1989, 1991; Costanza and Perrings, 1990).

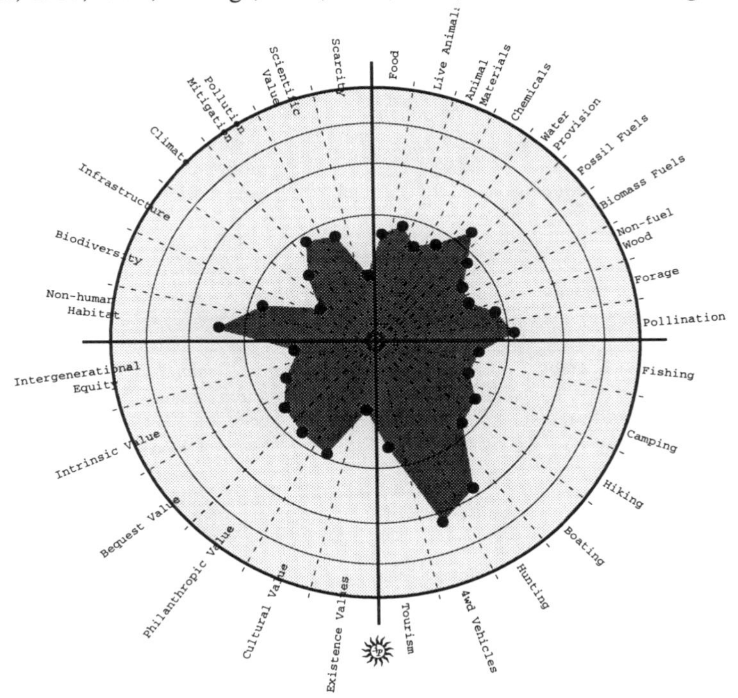

Figure 19.4. Old forest growth growth when recovery begins

19.4 THE ECOLOGICAL BENEFITS PARADIGM

Even when unable to assess ecological health across scales, space, and time, some often-asked issues remain, such as the significance of the findings. Even if public policy advances were at the point where the maintenance of ecological health is considered an important goal, trade-offs and choices will be needed by policy makers. This situation strongly suggests that simply characterizing the risk of potential outcomes will be an inadequate response. By developing the ability to characterize ecological benefits more completely and by characterizing the impact from the loss of ecosystem health on the delivery of those benefits, ecological assessments will make great strides towards resolving the significance matter.

Table 19.1. Ecological benefits

Market Benefits (first wave)

1. Food
2. Live animals(non-food)
3. Animal materials
   a. Hides
   b. Feathers
   c. Pearls
4. Non-animal commercial inputs
   a. Chemicals
   b. Fertilizers
   c. Peat
   d. Metals
   e. Minerals
5. General water provision
6. Fossil fuels
7. Other fuels (biomass)
8. Wood materials(other than fuel)
9. Livestock forage
10. Pollination

Non-market use benefits (second wave)

1. Recreational uses
   a. Fishing
   b. Camping
   c. Hiking
   d. Boating
   e. Hunting
   f. Four-wheel drive vehicles
2. Tourism

Non-market, non-use benefits (third wave)

1. Existence values
2. Historical,heritage,cultural, and spiritual values
3. Philanthropic values
4. Bequest values
5. Intrinsic values
6. Intergenerational equity

Other benefits (fourth wave)

1. Habitat benefits to non-humans
   a. General habitat
   b. Endangered species habitat
   c. Provision of migratory corridor
   d. Competition testing and design (evolution)
2. Preservation of genetic diversity
   a. Medicinal applications
   b. Vetrinary applications
   c. Crop disease/pest protection
   d. Crop improvement
   e. Biotechnology applications
   f. Other potential applications
3. Infrastructure maintenance
   a. Ground water recharge/discharge
   b. Mineral cycling
   c. Nutrient cycling/nutrient uptake
   d. Energy fixation/ photosynthesis
   e. Carbon sequestration
   f. Organic production export
   g. Erosion control
   h. Sediment trapping
   i. Soil generation
   j. Flood control
   k. Wave buffering
   l. Other physical services (windbreak, shade)
4. Climatic effects
   a. Regional effects (temperature, humidity, rainfall, storm buffering)
   b. Micro-climate (local) effects (temperature, etc.)
   c. Global effects
5. Contaminant/pollutant effects
   a. Decomposition
      i. Organic waste breakdown
      ii. Pollutant detoxification
   b. Contaminant transport/ dilution
   c. Contaminent storage
6. General scientific and research value
7. Scarcity/uniqueness

Current ecological assessment documents and frameworks clearly state that policy makers must be consulted to develop the ecological risk assessment endpoints and that the assessment must ascertain the significance of observed changes.

The selection of assessment endpoints relates in part to policy interests (e.g., to specified regulatory endpoints or to public concerns); thus, changes in assessment endpoints must be related ultimately to changes in parameters of the ecosystem that humans care about (anticipating the significance issue) (EPA, 1992a).

The products that will result from the process clearly will not be couched in terms with which policy makers are most comfortable nor in the metrics that they will understand and be able to communicate to their various constituencies. This tendency to describe scientific findings in terms that are, in the view of the policy maker, either arcane or in multiple metrics has been referred to as multidimensionality. The result is that the findings are described in a manner so detailed and fragmented that no one can grasp the overall implications (McKelvey and Henderson, 1991).

Benefits are ecological goods and services, and have been compared to ecological structure and function (Westman, 1977). Ecological goods and services can be described as those benefits that humans derive from ecological systems. For example, cut trees provide lumber (an economic and ecological good), while uncut trees take up air pollutants (an economic and ecological service). The uptake benefit will be lost when the trees are cut for lumber, and vice versa.

To economists, the term "benefits" often denotes a monetized valuation of an economic good or service. However, in this context, the term "benefit" is used to refer to all ecological goods and services whether or not their value has been monetized. Since the monetization step is often controversial, leaving it aside permits these efforts to focus on the scientific questions surrounding the identification and quantification of benefits. However, an analytical loss will exist in the absence of monetizing (a common metric with which to measure and express the magnitude of the benefits).

Ecological benefits occur in four groups: (1) market benefits (first wave), such as lumber, for which economic markets exist; (2) non-market use benefits (second wave), such as recreational benefits, for which no direct markets exist; (3) non-market, non-use benefits (third wave), such as the existence value and bequest value; and (4) fourth wave benefits are those that would fit into any of the three previous categories, but which have not routinely included in previous benefits analyses, such as pollution uptake, climate modification, habitat, and biodiversity (Principe, 1992). A more complete listing of benefits that would be included in each of these four categories is shown in Table 19.1.

The benefits shown in this table are very highly aggregated. However, in some instances, a less-aggregated list might be more appropriate for each type of ecosystem or region to be evaluated. These benefits are the sort of metrics with which most policy makers are conversant (except for those benefits they had never really considered). If the assessment can describe the extent to which stresses diminish or increase the delivery of these benefits, a policy maker will be in a much better position to understand the consequences of choosing among the available options. Thus, the matter of significance is addressed in the most direct possible way.

While a variety of graphical methods can depict the status and change in magnitude of these benefits, a polar-type chart may be best to demonstrate the technique. The polar chart has some appeal, because of its division into four quadrants. By placing first wave benefits in quadrant one, second wave benefits in quadrant two, etc., the status of each wave's benefits can be clearly shown (Figure 19.1).

This graphical analysis might illuminate a policy maker's choices; for instance, consider the choice to harvest the trees in a hypothetical old-growth forest. At the outset, the benefits profiles shown in the following figures are not based on data—they only illustrate the use of the graphical technique.

A virgin old-growth forest might have a benefits profile similar to that shown in Figure 19.2. Most of the benefits are realized as non-market benefits (i.e., in quadrants 2, 3, and 4). The benefits profile shown in Figure 19.3 might reflect the change in benefits resulting from a decision to harvest the old-growth trees. Most of the non-market benefits have been lost, but the market benefits from timber sales have increased the first quadrant benefits. Continuing the analysis for another year (Figure 19.4) provides an important insight: after the trees are harvested, not only are the non-market benefits lost, but so are the market benefits. The resulting benefits profile in Figure 19.4 is much smaller than the benefits profile in Figure 19.1. If the benefits profiles were based on actual data, a policy maker would gain several important insights into the consequences of the impending decision that otherwise might have been obscured or lost by other forms of analysis.

While these graphs are clearly oversimplified for the purposes of illustration, the technique holds considerable promise to making ecological risk-benefit decisions better informed. Further, a significant opportunity exists not only to characterize many of the ecological properties that scientists fear are never considered by policy makers, but also to have scientists provide these policy makers with a product that will answer the very questions the policy maker must answer in terms they alone can understand. Yet another likely benefit is the identification of research areas anticipated to have the greatest effect on reducing uncertainty. Finally, this technique readily adapts to a variety of different scales, thereby providing an important degree of flexibility for both scientists and policy makers.

19.5 REFERENCES

Allen, T.F.H., O'Neill, R.V., and Hoekstra, T.W. (1984) *Interlevel Relations in Ecological Research and Management: Some Working Principles from Hierarchy Theory.* General Technical Report RM-110. US Department of Agriculture, Rocky Mountain Forest and Range Experiment Station, Fort Collins, Colorado.

Costanza, R. (1987) Social traps and environmental policy. *BioScience* **37**, 407–412.

Costanza, R. (1991) Ecological economics: A research agenda. *Struct. Change Econ. Dynam.* **2**(2), 335–357.

Costanza, R. (1992) Toward an operational definition of ecosystem health. In: Costanza, R., Norton, B.G., and Haskell, B.D. (Eds.) *Ecosystem Health: New Goals for Environmental Management*, pp. 236–253. Island Press, Washington, D.C.

Costanza, R., and Perrings, C. (1990) A flexible assurance bonding system for improved environmental management. *Ecol. Econ.* **2**, 57–76.

Costanza, R., Sklar, F.H., and White, M.L. (1990) Modeling coastal landscape dynamics. *BioScience,* **40**, 91–107.

Funtowicz, S.O., and Ravitz, J.R. (1991) A new scientific methodology for global environmental issues. In: Costanza, R. (Ed.) *Ecological Economics: The Science and Management of Sustainability*, pp. 137–152. Columbia University Press, New York.

Haskell, B.D., Norton, B.G., and Costanza, R. (1992) What is ecosystem health and why should we worry about it? In: Costanza, R., Norton, B.G., and Haskell, B.D. (Eds.) *Ecosystem Health: New Goals for Environmental Management*, pp. 1–18. Island Press, Washington, D.C.

Holling, C.S. (1986) The resilience of terrestrial ecosystems: local surprise and global change. In: Clark, W.C., and Munn, R.E. (Eds.) *Sustainable Development of the Biosphere*. Cambridge University Press, Cambridge.

Hunsaker, C.T., Graham, R.L., Suter II, G.W., O'Neill, R.V., Barnthouse, L.W., and Gardner, R.H. (1990) Assessing ecological risk on a regional scale. *Environ. Manag. 14*, 325–332.

McKelvey, R., and Henderson, S. (1991) The science-policy interface. In: Barker, J.R., and Tingey, D.T. (Eds.) *Air Pollution Effects on Biodiversity*, pp. 280–292. Van Nostrand Reinhold, New York.

Norton, B.G. (1991a) *Toward Unity Among Environmentalists*. Oxford University Press, New York.

Norton, B.G. (1991b) Ecological health and sustainable resource management. In: Costanza, R. (Ed.) *Ecological Economics: The Science and Management of Sustainability*. Columbia University Press, New York.

Norton, B.G., Ulanowicz, R.E., and Haskell, B.D. (1991) *Scale and Environmental Policy Goals*. Report to the US Environmental Protection Agency, Office of Policy, Planning, and Evaluation, Washington, D.C.

Norgaard, R.B. (1989) The case for methodological pluralism. *Ecol. Econ.* **1**, 37–57.

Odum, H.T. (1971) *Environment, Power, and Society*. John Wiley, New York.

Perrings, C. (1987) *Economy and Environment: A Theoretical Essay on the Interdependence of Economic and Environmental Systems*. Cambridge University Press, Cambridge.

Perrings, C. (1989) Environmental bonds and the incentive to research in activities involving uncertain future effects. *Ecol. Econ.* **1**, 95–110.

Perrings, C. (1991) Reserved rationality and the precautionary principle: technological change, time and uncertainty in environmental decision-making. In: Costanza, R. (Ed.) *Ecological Economics: The Science and Management of Sustainability*, pp. 176–193. Columbia University Press, New York.

Pimm, S.L. (1984) The complexity and stability of ecosystems. *Nature* **307**, 321–326.

Principe, P.P. (1992) Estimating systemic ecosystem benefits: a new approach. Paper presented to the Second Meeting of the International Society for Ecological Economics, Stockholm, August 1992.

Rapport, D.J. (1989) What constitutes ecosystem health? *Perspect. Biol. Med.* **33**, 120–132.

Rapport, D.J. (1992) What is clinical ecology? In: Costanza, R., Norton, B.G., and Haskell, B.D. (Eds.) *Ecosystem Health: New Goals for Environmental Management*. Island Press, Washington, D.C.

Schaeffer, D.J., and Cox, D.K. (1992) Establishing ecosystem threshold criteria. In: Costanza, R., Norton, B.G., and Haskell, B.D. (Eds.) *Ecosystem Health: New Goals for Environmental Management*. Island Press, Washington, D.C.

Schaeffer, D.J., Herricks, E.E., and Kersier, H.W. (1988) Ecosystem health: 1. Measuring ecosystem health. *Environ. Manag.* **12**, 445–455.

Ulanowicz, R.E. (1992) Ecosystem health and trophic flow networks. In: Costanza, R., Norton, B.G., and Haskell, B.D. (Eds.) *Ecosystem Health: New Goals for Environmental Management*. Island Press, Washington, D.C.

USEPA (1992) *Peer Review Workshop Report on a Framework for Ecological Risk Assessment*, p. 4. Report No. EPA/625/3-91/022. US Environmental Protection Agency, Risk Assessment Forum, Washington, D.C., February.

Westman, W.E. (1977) How much are Nature's services worth? *Science* **197**, 960–964.

Wulff, F., Field, J.G., and Mann, K.H. (1989) *Network Analysis of Marine Ecosystems: Methods and Applications*. Coastal and Estuarine Studies Series. Springer-Verlag, Heidelberg.

Index